The Linux DevOps Handbook

Customize and scale your LInux distributions to accelerate your DevOps workflow

Damian Wojsław

Grzegorz Adamowicz

BIRMINGHAM—MUMBAI

The Linux DevOps Handbook

Copyright © 2023 Packt Publishing

Group Product Manager: Preet Ahuja
Publishing Product Manager: Preet Ahuja
Senior Editor: Runcil Rebello
Technical Editor: Nithik Cheruvakodan
Copy Editor: Safis Editing
Project Coordinator: Ashwin Kharwa
Proofreader: Safis Editing
Indexer: Hemangini Bari
Production Designer: Ponraj Dhandapani
Marketing Coordinator: Rohan Dobhal

First published: November 2023

Production reference: 1171023

Published by Packt Publishing Ltd.
Grosvenor House
11 St Paul's Square
Birmingham
B3 1RB, UK.

ISBN 978-1-80324-566-9

www.packtpub.com

To my wife, my kids, and all the people who have helped me on my journey to become the best version of me.

– Damian Wojsław

To my mum and dad, who supported me with all my interests, including computers. To my wife and daughter, who are my biggest supporters and motivation in all endeavors.

– Grzegorz Adamowicz

Contributors

About the authors

Damian Wojsław has been working in the IT industry since 2001. He specializes in the administration and troubleshooting of Linux servers. Being a system operator and support engineer, he has found the DevOps philosophy to be a natural evolution of the way SysOps work with developers and other members of the software team.

I would like to thank my wife and kids for their patience and support.

Grzegorz Adamowicz has been working in the IT industry since 2006 in a number of positions, including systems administrator, backend developer (PHP and Python), systems architect, and site reliability engineer. Professionally, he is focused on building tools and automations for the projects he is involved in. He also engages with the professional community by organizing events such as conferences and workshops. Grzegorz has worked in many industries, including oil and gas, hotel, Fintech, DeFI, automotive, and space.

I want to thank my wife, Agnieszka, for her never-ending support, and everyone I've had the pleasure of working with throughout my career – most of what I know professionally, I've learned from you.

About the reviewer

Marcin Juszkiewicz is a seasoned software engineer on a journey since 2004. Over the years, he has worked on building and optimizing code in various projects. His career has spanned from embedded Linux systems to ARM servers. In his spare time, Marcin dons the hat of a system admin, a role he embraces with the same enthusiasm as his daily job. Through his extensive journey as a software engineer and part-time system administrator, Marcin has garnered a wealth of knowledge and a reputation for his meticulous approach to coding and system optimization.

Table of Contents

3

Intermediate Linux   45

4

Automating with Shell Scripts   67

Part 2: Your Day-to-Day DevOps Tools

5

Managing Services in Linux   91

9

A Deep Dive into Docker   189

Part 3: DevOps Cloud Toolkit

10

Monitoring, Tracing, and Distributed Logging   225

11

Using Ansible for Configuration as Code   251

Preface

DevOps has become a critical component of modern software development and delivery. It has revolutionized the way we build, test, deploy, and operate software systems. DevOps is not just a set of tools and practices but also a culture and mindset that focuses on collaboration, communication, and automation.

This book is designed to be a comprehensive guide to DevOps, covering everything from choosing the right Linux distribution to avoiding pitfalls in DevOps. Each chapter in this book provides detailed information and practical examples to help you understand the concepts and apply them to real-world scenarios.

Who this book is for

This book is designed for individuals who have already gained some knowledge and experience in the field of software development and IT operations and are now seeking to further expand their knowledge of DevOps and Linux systems.

If you are not well versed in Linux systems, this book will provide you with the necessary guidance and tools to quickly learn and become proficient in managing Linux-based infrastructures. You will gain an understanding of the Linux operating system, its architecture, and its fundamental concepts.

Furthermore, this book emphasizes learning about public cloud technologies with a focus on AWS. If you are interested in learning how to use AWS to build and manage scalable and reliable systems, this book will provide you with the necessary knowledge and tools to get started.

Whether you are new to DevOps or have already gained some experience, this book provides a solid foundation for learning more complex concepts. It covers a wide range of topics, from the basics of Linux systems to more advanced DevOps practices such as configuration and infrastructure as code and CI/CD.

What this book covers

Chapter 1, Choosing the Right Linux Distribution, discusses the GNU/Linux history, and the differences between popular distributions.

Chapter 2, Command-Line Basics, guides you through the usage of a command line and common tools we'll be using throughout the book.

Chapter 3, Intermediate Linux, describes more advanced features of GNU/Linux that you will find useful.

Chapter 4, Automating with Shell Scripts explains how to start writing your own scripts using the Bash shell.

Chapter 5, Managing Services in Linux, discusses different ways of managing services in Linux and shows you how to define your own services using systemd.

Chapter 6, Networking in Linux, describes how networking works, how to control different aspects of network configurations, and how to use command-line tools.

Chapter 7, Git, Your Doorway to DevOps, discusses what Git is and how to use Git's version control system, including less commonly known Git features.

Chapter 8, Docker Basics, explores the containerization of your services or tools, and running and managing containers.

Chapter 9, A Deep Dive into Docker, discusses the more advanced features of Docker, including Docker Compose and Docker Swarm.

Chapter 10, Monitoring, Tracing, and Distributed Logging, discusses how to monitor your services, what tools you can use in the cloud, and how to do a basic setup.

Chapter 11, Using Ansible for Configuration as Code, looks at Configuration as Code with the use of Ansible; it'll guide you through the basic setup and more advanced features of Ansible.

Chapter 12, Leveraging Infrastructure as Code, discusses what **Infrastructure as Code (IaC)** is, what the popular tools are, and how to manage your infrastructure using Terraform.

Chapter 13, CI/CD with Terraform, GitHub, and Atlantis, takes IaC one step further with the use of the **Continuous Integration (CI)** and **Continuous Deployment (CD)** of infrastructure with the use of Terraform and Atlantis.

Chapter 14, Avoiding Pitfalls in DevOps, discusses challenges you may encounter with your work in DevOps.

To get the most out of this book

You will need Debian Linux or Ubuntu Linux installed on a virtual machine or as your main operating system on your computer. Other software we use is either already installed as a default toolset, or we will show you where to get it to install it on your system.

It is assumed that you have some basic knowledge of Linux and its command-line interface. Familiarity with shell scripting and basic programming concepts would also be helpful. Additionally, some understanding of IT infrastructure and how it is managed is recommended, as well as some exposure to software development practices.

This book is aimed at beginners in the DevOps world, and it assumes that you are eager to learn about the tools and concepts that are commonly used in this field. By the end of this book, you will have gained a solid understanding of how to manage infrastructure using IaC tools such as Terraform and Atlantis, as well as how to automate repetitive tasks using Ansible and Bash scripting. You will also learn how to set up logging and monitoring solutions to help you maintain and troubleshoot your infrastructure.

Software/hardware covered in the book	Operating system requirements
Bash	Linux OS has it preinstalled
Ansible	Python 3 or newer
Terraform	Linux OS
AWS CLI	Python 3 or newer
Docker	Linux OS

If you are using the digital version of this book, we advise you to type the code yourself or access the code from the book's GitHub repository (a link is available in the next section). Doing so will help you avoid any potential errors related to the copying and pasting of code.

Download the example code files

You can download the example code files for this book from GitHub at `https://github.com/PacktPublishing/The-Linux-DevOps-Handbook`. If there's an update to the code, it will be updated in the GitHub repository.

We also have other code bundles from our rich catalog of books and videos available at `https://github.com/PacktPublishing/`. Check them out!

Conventions used

There are a number of text conventions used throughout this book.

`Code in text`: Indicates code words in text, database table names, folder names, filenames, file extensions, pathnames, dummy URLs, user input, and Twitter handles. Here is an example: "While logged in as root, your prompt will end with the # sign. When logged in as a normal user, it will present you with $."

A block of code is set as follows:

```
docker build [OPTIONS] PATH | URL | -
```

When we wish to draw your attention to a particular part of a code block, the relevant lines or items are set in bold:

```
docker build [OPTIONS] PATH | URL | -
```

Any command-line input or output is written as follows:

```
chmod ug=rx testfile
```

Bold: Indicates a new term, an important word, or words that you see onscreen. For instance, words in menus or dialog boxes appear in **bold**. Here is an example: "**Ansible Galaxy** is a community-driven platform that hosts an extensive collection of Ansible roles and playbooks."

> **Tips or important notes**
> Appear like this.

Get in touch

Feedback from our readers is always welcome.

General feedback: If you have questions about any aspect of this book, email us at customercare@packtpub.com and mention the book title in the subject of your message.

Errata: Although we have taken every care to ensure the accuracy of our content, mistakes do happen. If you have found a mistake in this book, we would be grateful if you would report this to us. Please visit www.packtpub.com/support/errata and fill in the form.

Piracy: If you come across any illegal copies of our works in any form on the internet, we would be grateful if you would provide us with the location address or website name. Please contact us at copyright@packt.com with a link to the material.

If you are interested in becoming an author: If there is a topic that you have expertise in and you are interested in either writing or contributing to a book, please visit authors.packtpub.com.

Share Your Thoughts

Once you've read *The Linux DevOps Handbook*, we'd love to hear your thoughts! Scan the QR code below to go straight to the Amazon review page for this book and share your feedback.

https://packt.link/r/1803245662

Your review is important to us and the tech community and will help us make sure we're delivering excellent quality content.

Download a free PDF copy of this book

Thanks for purchasing this book!

Do you like to read on the go but are unable to carry your print books everywhere?

Is your eBook purchase not compatible with the device of your choice?

Don't worry, now with every Packt book you get a DRM-free PDF version of that book at no cost.

Read anywhere, any place, on any device. Search, copy, and paste code from your favorite technical books directly into your application.

The perks don't stop there, you can get exclusive access to discounts, newsletters, and great free content in your inbox daily

Follow these simple steps to get the benefits:

1. Scan the QR code or visit the link below

https://packt.link/free-ebook/9781803245669

2. Submit your proof of purchase
3. That's it! We'll send your free PDF and other benefits to your email directly

Part 1: Linux Basics

In this opening part of the book, we will focus on Linux distributions and the basic skills you will need to efficiently use Linux operating systems. You will also learn about the basics of shell scripting to automate day-to-day tasks.

This part has the following chapters:

- *Chapter 1, Choosing the Right Linux Distribution*
- *Chapter 2, Command-Line Basics*
- *Chapter 3, Intermediate Linux*
- *Chapter 4, Automating with Shell Scripts*

1

Choosing the
Right Linux Distribution

In this chapter, we will dive into the Linux world from the very beginning. We will briefly touch on Linux history, explain what a distribution is, and explain what to take into account when choosing one for production use. You are not expected to know anything about Linux, its administration, or the cloud. If you don't understand some words that we use, worry not. There shouldn't be a lot of confusing terminology in this chapter, and if there is, we will explain it in later chapters. When you finish reading this chapter, you should be able to understand why there are so many Linuxes out there, how much you should expect to pay for it, and how to think about choosing the right Linux for yourself.

In this chapter, we will cover the following main topics:

- What is Linux and what is a Linux distribution?
- What can you use to help you make the right decision?
- Several major Linux distributions that are quite popular today

Technical requirements

This chapter doesn't have any technical requirements. We won't run any commands or install any software yet. This will come in the later chapters.

The code we're presenting in the book is available in the public GitHub repository for your consideration at this address: `https://github.com/PacktPublishing/The-Linux-DevOps-Handbook`.

What exactly is a Linux distribution?

Linux is the standard operating system for cloud workloads. However, there is not a single Linux operating system that goes by that name. Its ecosystem is quite complex. This comes from how it came to be originally.

Long before Linux was conceived by its creator, Linus Torvalds, there was *Unix*. Unix source code was – for legal reasons – licensed to anyone who bought it, thus making it very popular among many institutions. This included universities. The code, however, was not entirely free. This didn't sit well with many people, who believed that software should be free – as in speech or beer – including source code. In the 1980s, a completely free and open implementation of Unix was born under the aegis of the *GNU Project*. Its goal was to develop an operating system that gave complete control over a computer to the user. The project was successful in that it was able to produce all the software required to run an operating system, except one thing – the kernel.

A **kernel** of the operating system is, in short, the core that operates hardware and manages the hardware and programs for the user.

In 1991, Finnish student Linus Torvalds famously announced his hobby kernel – Linux. He called it at that time "*just a hobby – won't be big and professional like GNU.*" It wasn't supposed to get big and popular. The rest is history. The Linux kernel became popular among developers and administrators and became the missing piece of the GNU Project. Linux is the kernel, the GNU tools are the so-called userland, and together they make the GNU/Linux operating system.

The preceding short story is important to us for two reasons:

- While the GNU userland and the Linux kernel is the most popular combination, you'll see it is not the only one.

- Linux delivers a kernel and the GNU Project delivers userland tools, but they have to be somehow prepared for installation. Many people and teams had separate ideas on how to do it best. I will expand on this thought next.

The way that a team or a company delivers a GNU/Linux operating system to end users is called a **distribution**. It facilitates operating system installation, the means to manage software later on, and general notions on how an operating system and the running processes have to be managed.

What makes distributions different?

The open nature of Linux and the GNU Project made it possible for almost anyone to create their own distribution. One of the things that made new users dizzy was the sheer amount of **Operating System (OS)** versions they could use. The surefire way to start a holy war between Linux users is by asking which distribution is the best.

One of the ways we can group Linux distributions is the format in which they deliver the software (**packages**) and additional software used to install and remove that software (**package managers**). There are a number of them, but the two most prevalent are **RPM (RPM Package Manager)** and *DEB* packages. Packages are more than just an archive with binaries. They contain scripts that set the software up for use – creating directories, users, permissions, log rules, and a number of other things that we will explain in later chapters.

The RPM family of distributions starts with **Red Hat Enterprise Linux** (**RHEL**), created and maintained by the Red Hat company. Closely related is *Fedora* (a free community distribution sponsored by Red Hat). It also includes *CentOS Linux* (a free version of RHEL) and *Rocky Linux* (another free version of RHEL).

The DEB distributions include *Debian* (where the DEB packages originate from) – a technocracy community project. From the Debian distribution arose a number of distributions based on it, using most of its core components. Most notable is *Ubuntu*, a server and desktop distribution sponsored by Canonical.

There are also distributions that use packages with minimum automation, most notably *Slackware*, one of the oldest existing Linux distributions.

There are distributions that give the user a set of scripts that compile actual software on the hardware it will be used on – most notably, *Gentoo*.

Finally, there is a distribution that is actually a book with a set of instructions that a user can follow to build the whole OS by hand – the *Linux From Scratch* project.

Another way of grouping distributions is by their acceptance of closed software – software that limits the distribution of source code, binaries, or both. This can mean hardware drivers, such as the ones for NVIDIA graphic cards, and user software, such as movie codecs that allow you to play streamed media and DVD and Blu-Ray discs. Some distributions make it easy to install and use them, while some make it more difficult, arguing that we should strive for all software to be open source and free (as in speech and as in beer).

Yet another way to differentiate them is the security framework a given distribution uses. Two of the most notable ones are *AppArmor*, used mainly by Ubuntu, and *SELinux* (from the USA's National Security Agency), used by, among others, Red Hat Enterprise Linux (and its derivatives) and Fedora Linux.

It's also worth noting that while most Linux distributions use the GNU Project as the userland, the popular one in the cloud Alpine Linux distribution uses its own set of software, written especially with minimum size in mind.

Looking at how the distribution is developed, it can be community-driven (without any commercial entity being an owner of any part of the process and software – Debian being one prime example), commercial (wholly owned by a company – RHEL being one example and SuSE another), and all the mixes in between (Ubuntu and Fedora being examples of a commercially owned distribution with a large body of independent contributors).

Finally, a way we can group distributions is by how well they facilitate the cloud workload. Here, we can look at different aspects:

- **The server side**: How well a given distribution works as an underlying OS for our infrastructure, virtual machines, and containers
- **The service side**: How well a given distribution is suited to run our software as a container or a virtual machine

To make things even more confusing and amusing for new adopters, each distribution can have many variants (called **flavors**, **variants**, or **spins**, depending on the distribution lingo) that offer different sets of software or default configurations.

And to finally confuse you, dear reader, for use on a desktop or laptop, Linux offers the best it can give you – *a choice*. The number of graphical interfaces for the Linux OS can spin the head of even the most experienced user – KDE Plasma, GNOME, Cinnamon Desktop, MATE, Unity Desktop (not related to the Unity 3D game engine), and Xfce. The list is non-exhaustive, subjective, and very limited. They all differ in the ease of use, configurability, the amount of memory and CPU they use, and many other aspects.

The number of distributions is staggering – the go-to site that tracks Linux distributions (`https://distrowatch.com/dwres.php?resource=popularity`) lists 265 various Linux distributions on its distribution popularity page at the time of writing. The sheer number of them makes it necessary to limit the book to three of our choosing. For the most part, it doesn't make a difference which one you choose for yourself, except maybe in licensing and subscription if you choose a commercial one. Each time the choice of distribution makes a difference, especially a technical one, we will point it out.

Choosing a distribution is more than just a pragmatic choice. The Linux community is deeply driven by ideals. For some people, they are the most important ideals on which they build their lives. Harsh words have been traded countless times over which text editor is better, based on their user interface, the license they are published with, or the quality of the source code. The same level of emotion is displayed toward the choice of software to run the WWW server or how to accept new contributions. This will inevitably lead to the way the Linux distribution is installed, what tools there are for configuration and maintenance, and how big the selection of software installed on it out of the box is.

Having said that, we have to mention that even though they have strong beliefs, the open-source community, the Linux community included, is a friendly bunch. In most cases, you'll be able to find help or advice on online forums, and the chances are quite high that you will be able to meet them in person.

To choose your distribution, you need to pay attention to several factors:

- Is the software you wish to run supported on the distribution? Some commercial software limits the number of distributions it publishes packages for. It may be possible to run them on unsupported versions of Linux, but this may be tricky and prone to disruptions.

- Which versions of the software you intend to run are available? Sometimes, the distribution of choice doesn't update your desired packages often enough. In the world of the cloud, software a few months old may already be outdated and lack important features or security updates.

- What is the licensing for the distribution? Is it free to use or does it require a subscription plan?

- What are your support options? For community-driven free distributions, your options are limited to friendly Linux gurus online and in the vicinity. For commercial offerings, you can pay for various support offerings. Depending on your needs and budget, you can find a mix of support options that will suit your requirements and financial reserves.

- What is your level of comfort with editing configuration files and running long and complex commands? Some distributions offer tools (both command-line and graphical) that make the configuration tasks easier and less error-prone. However, those tools are mostly distribution-specific, and you won't find them anywhere else.

- How well are cloud-related tools supported on a given distribution? This can be the ease of installation, the recency of the software itself, or the number of steps to configure for use.

- How well is this distribution supported by the cloud of your choosing? This will mean how many cloud operators offer virtual machines with this distribution. How easy is it to obtain a container image with this distribution to run your software in it? How easy do we suspect it to be to build for this distribution and deploy on it?

- How well is it documented on the internet? This will not only include the documentation written by distribution maintainers but also various blog posts and documentation (mainly tutorials and so-called *how-to* documents) written by its users.

So far, you've learned what a Linux distribution is, how distributions differentiate from one another, and what criteria you can use to actually choose one as the core of the system you will manage.

In the next section, we will look deeper into each distribution to get to know the most popular ones better, giving you a first glimpse of how each one works and what to expect.

Introducing the distributions

After that bit of a lengthy but condensed history of the Linux OS, it is time to finally explore the few we have chosen to cover in this book. In this section, we will cover the factors we just listed, as we believe they are important in making a decision. Please remember though that while we strive to present you with objective facts and valuations, we cannot escape our own subjective views. Always evaluate on your own before you choose, as it's highly possible that you will stick with this distribution for many years to come.

A point to note is that we won't be covering distributions comprehensively. We will only try to create a foundation on which you, dear reader, must build through research.

Also, while you are learning, do not be afraid to hop from distribution to distribution. Only through real-life experiences will you fully understand which one covers your needs best.

Debian

Debian (`https://www.debian.org/`) is one of the oldest active Linux distributions. Its development is led by the community-supported Debian Project. It is known for two things – the sheer number of packages that the distribution provides and the slow release of stable versions. The latter has improved in recent years and stable releases are published every two years. Software is delivered in archives called packages. Debian packages' names have a `.deb` file extension and are colloquially called **debs**. They are kept online in repositories and repositories are broken down into pools. Repositories offer almost 60,000 packages with software in the latest stable release.

Debian always has three versions available (so-called **branches**) – stable, testing, and unstable. The releases are named after characters from the *Toy Story* movie franchise. The latest stable release – version 11 – is called Bullseye.

The **unstable branch** is the rolling branch for developers, people who like living on the edge, or those who require the newest software more than they require stability. Software is accepted into the unstable branch with minimal testing.

The **testing branch** is where, as the name implies, the testing happens. A lot of testing happens here, thanks to the end users. Packages come here from the unstable branch. The software here is still newer than in the stable branch but not as fresh as in the unstable branch. A few months before the new stable release, the testing branch is frozen. It means that no new software will be accepted, and new versions of the already accepted packages are allowed only if they fix bugs.

After a few months, testing becomes the **stable branch**. The software is updated only for security fixes.

This distribution is available for many hardware platforms – Intel, ARM, PowerPC, and so on. Along with unofficial ports, there are a multitude of hardware platforms on which you can install it.

Debian is viewed as the most stable distribution there is, and it is used as a platform for various compute clusters, so it is generally installed on bare-metal servers somewhere in a rack in a data center and intended for use consistently over many years.

According to W3Techs (`https://w3techs.com/technologies/details/os-linux`), Debian makes up for 16% of all servers running on the internet. Its derivative, Ubuntu, runs 33% of them. Together, they account for 49% of all servers. This makes administration skills related to Debian highly marketable.

Ubuntu Linux

The Ubuntu Linux distribution (`https://ubuntu.com/`) is widely credited for making Linux popular on personal computers, and rightly so. Sponsored by Canonical, its mission was to make Linux easily usable for most people. It was one of the first, if not the first, Linux versions to distribute non-free and non-open binary drivers and libraries that made desktop use simpler and more comfortable.

Famously, the first bug report opened for Ubuntu distribution by Mark Shuttleworth (Canonical and Ubuntu founder) was, "*Microsoft has majority market share.*"

The distribution itself is based on Debian Linux, and in the beginning, being fully binary-compatible was one of the major objectives. As the development has progressed, this has lost some of its importance.

This distribution is developed by the community and Canonical. The main source of income for the company is premium services related to Ubuntu Linux – support, training, and consultations.

Due to the very close-knit relationship between Debian Linux and Ubuntu Linux, many developers and maintainers for one distribution serve the same roles in the other one. This results in a lot of software being packaged for both distributions in parallel.

Ubuntu has three major flavors – Desktop, Server, and Core (for the internet of things). Desktop and Server may differ slightly in how services are configured out of the box, and Core differs a lot

The software is distributed in `.deb` packages, the same as with Debian, and the sources are actually imported from the Debian unstable branch. However, this doesn't mean you can install Debian packages on Ubuntu or vice versa, as they are not necessarily binary-compatible. It should be possible to rebuild and install your own version.

There are four package repositories per release – the free and non-free software supported officially by Canonical is called *main* and *restricted*, respectively. Free and non-free software delivered and maintained by the community is called *universe* and *multiverse*, respectively.

> **Important note**
>
> A word of advice – a widely accepted practice of system upgrades between major versions is to wait for the first sub-release. So, if the currently installed version of the distribution is 2.5 and the new version 3.0 is released, it is wise to wait until 3.1 or even 3.2 is released and upgrade then. This is applicable to all the distributions we list here.

The **Long-Term Support** (**LTS**) versions are supported for five years. A new LTS version is released every two years. It is possible to negotiate extended support. This gives a very good timeline to plan major upgrades. A new Ubuntu version is released every six months.

Ubuntu Linux is widely adopted in education and government projects. Famously, the city of Munich, between 2004 and 2013, migrated over 14,000 municipal desktop computers to a variant of Ubuntu with the KDE desktop environment. While the migration saw disturbances politically – other operating system vendors lobbied strongly against this migration – it was considered a success technically.

Ubuntu is the Linux of choice for personal computers. Canonical works very closely with hardware vendors, notably Lenovo and Dell, but lately also with HP, to ensure full compatibility between the distribution and the computers. Dell sells its flagship laptops with Ubuntu preinstalled.

Several sources cite Ubuntu Linux as the most installed Linux distribution on servers and personal computers. The actual number can only be estimated, as Ubuntu doesn't require any subscription or registration.

As a byproduct of Ubuntu Linux's popularity, software vendors, more often than not, offer `.deb` packages of their software, if they release a Linux version. This is especially true for desktop software.

The amount of unofficial versions, clones, or modified distributions based on Ubuntu is staggering.

Ubuntu has a very active community, both organized and unorganized. It's quite easy to get a hold of a group of users near your city. This also directly translates to the amount of tutorials and documentation on the internet.

Ubuntu Linux, especially under a support plan, is installed as a foundation for many cloud computing infrastructure deployments. Many telecoms, banking, and insurance companies have chosen Ubuntu Server as their foundation.

Red Hat Enterprise Linux (RHEL)

RHEL (`https://www.redhat.com/en/technologies/linux-platforms/enterprise-linux`) is a spiritual successor of Red Hat Linux and is developed and maintained by Red Hat Inc. (`https://www.redhat.com/`). Its main target is the commercial entities market. It is possible to use RHEL for free for development or in production with up to 16 servers (at the time of writing). However, the main advantage of this distribution is the enormous pool of articles that help solve issues and the assistance of support engineers, although the latter can only be acquired through a paid support plan.

RHEL is considered a very stable and solid distribution. It is one of the main choices for banks, insurance companies, and financial markets. It lacks many popular desktop software packages, but on the server side of things, especially as an OS to run other commercial applications, it is a first-class citizen.

The software is distributed in online repository packages that end with `.rpm`, hence the name **RPMs**. The main tool to administer the packages is RPM, with more sophisticated tools – **yum**, and lately its successor, **dnf** – also available.

In the true spirit of an open source-based company, Red Hat makes sources for its distribution available. This has led to the creation of a famous free and open clone of RHEL – *CentOS*. Until fairly recently, it had been quite a popular choice for people who wanted to use RHEL but didn't want to, or couldn't, pay a subscription. In 2014, CentOS joined the Red Hat company, and in 2020, Red Hat announced that the versioned releases of CentOS would no longer be available; there would only be the so-called rolling release, which constantly updates packages and does not mirror the RHEL releases. This resulted in a very heated reaction from CentOS users. The original CentOS founder, Gregory Kurtzer, started another clone of RHEL called *Rocky Linux*. Its main objective is the same as the original CentOS – to deliver a free, open, and community-driven distribution, fully binary-compatible with RHEL.

The RHEL distribution delivers stable versions every few years and supports them for 10 years, starting from release 5. The full support, however, is offered only for a few years. For the rest of the time, Red Hat provides only security fixes and critical updates for your systems, with no new package versions being introduced. Still, this life cycle is what users with large installations or mission-critical systems came to like.

As with Ubuntu, it is possible to negotiate extended support time.

The Red Hat company has a turbulent relationship with the open source community. While the company mostly plays fair, there have been some decisions that the community didn't like. Lately, it was Red Hat's decision to change the CentOS release model to a rolling release (`https://lists.centos.org/pipermail/centos-announce/2020-December/048208.html`).

RHEL, like Ubuntu, is the chosen foundation for commercially supported deployments of cloud infrastructure.

Fedora Linux

Fedora (`https://fedoraproject.org/wiki/Fedora_Project_Wiki`) is a distribution associated with the Red Hat company. While more than 50% of its developers and maintainers are community members not affiliated with Red Hat, the company holds full stewardship over the development. It is a RHEL upstream, which means that this is the real development frontend for the actual RHEL. It doesn't mean that everything from Fedora is included in the release of RHEL. However, following Fedora closely will yield insight into the current direction of the RHEL distribution.

Contrary to RHEL, for which Fedora is the foundation, the new releases happen every six months. It uses the same package type as RHEL, RPM.

Fedora is considered a fast-paced distribution. It quickly adopts the newest and bleeding-edge versions of packages.

CentOS

CentOS (`https://centos.org`) used to be the go-to free version of RHEL. The name is an acronym for **Community Enterprise Operating System**. Its main goal was to be fully binary-compatible with RHEL and adhere to the same releases and numbering scheme. In 2014, CentOS joined Red Hat, but it was promised that the distribution would keep its independence from the company while benefiting from development and testing resources. Unfortunately, in 2020, Red Hat announced that CentOS 8 would be the last numbered release, and from then on, CentOS Stream would be the only variant. CentOS Stream is a midstream version. This means it sits in the middle between bleeding-edge and fast-paced Fedora and stable and production-ready RHEL. The difference between CentOS Stream and CentOS is that Stream is a development variant, while CentOS was simply a rebuilt and repackaged mirror of the actual final product, RHEL.

All the knowledge, skills, and experience gained when working with RHEL are 100% applicable to CentOS. Since CentOS is the third most-deployed Linux distribution on servers, according to W3Techs (`https://w3techs.com/technologies/details/os-linux`), the skills are very marketable.

Rocky Linux

As a response to the situation with the CentOS distribution, its founder announced the creation of Rocky Linux (`https://rockylinux.org/`). The goals are the same as the original CentOS. The release scheme and numbering follow RHEL. Shortly after the announcement, the GitHub repository of Rocky Linux became top trending (`https://web.archive.org/web/20201212220049/https://github.com/trending`). Rocky Linux is 100% binary-compatible with CentOS. The project has released a set of tools that easily migrate from CentOS to Rocky Linux without reinstalling the system.

The distribution is quite young, having been founded in 2020, and its popularity is still to be determined. It has made a lot of noise in the community, and it seems that a steady stream of CentOS users have moved to Rocky Linux as their preferred choice.

A very important contribution to the open source world from the Rocky Linux project is the build system. It ensures that even if Rocky Linux shuts down, the community will be able to easily start up a new RHEL clone.

All the skills, knowledge, and articles for RHEL and CentOS are 100% applicable to Rocky Linux. All the software that runs on RHEL and CentOS should run without any modifications on Rocky Linux too.

Alpine

Alpine Linux (`https://alpinelinux.org/`) is an interesting one. The main programming library and most basic command-line tools are not from the GNU Project. Also, the services management system, currently **systemd** in most distributions, is uncommon. This makes some of the administration skills from other major distributions non-applicable. The strength of Alpine lies in its size (which is rather small), its security-first mindset, and one of the fastest boot times among existing Linux distributions. Those characteristics, with the boot time being admittedly more important, make it the most popular choice for containers. If you run containerized software or build your own container images, it is very likely that it is on Alpine Linux.

Alpine has its roots in the **LEAF** (**Linux Embedded Appliance Framework**; see: `https://bering-uclibc.zetam.org/wiki/Main_Page`) project – a Linux distribution that fits on a single floppy disk. LEAF is currently a popular choice for embedded markets, routers, and firewalls. Alpine is a bigger distribution, but that sacrifice had to be made, since developers wanted to include several useful but rather large software packages.

The package manager is called *apk*. The build system is borrowed from another distribution called Gentoo Linux. As Gentoo builds software as it installs it, the portage obviously contains a lot of logic around building software that is used as a part of an OS.

Alpine can run from RAM entirely. There's even a special mechanism that allows you to initially only load a few required packages from the boot device, and it can be achieved using Alpine's **Local Backup Utility** (**LBU**).

As mentioned before, this is a preferred distribution for container images. You won't see it running on a large server installation often, if at all. When we cross over to the cloud world, chances are you'll see a lot of Alpine Linux.

Having said that, every single one of those distributions has a variant for the cloud as a container base image – a way to run your software in the true cloud way.

In this chapter, you learned the basics of popular Linux distributions and how they are different from one another. You should now have some understanding of what you can choose from and what consequences you will need to face – good and bad. To give you an even better idea of how to interact with some cherry-picked Linux distributions, we will look at how to interact with a system using your keyboard in *Chapter 2*.

Summary

The short list in this chapter is just a tiny portion of the Linux distributions available. The list is largely based on the popularity and marketability of skills, as well as our own experience and knowledge that we acquired over the years. They are not, by any means, the best or only choices that you have.

We tried to point out where the main strengths lie and what a user's relationship is with respective distributions.

It's not likely we were able to answer all your questions. Each of the Linux distributions from our list has its own book out there, and there is even more knowledge on blogs, wikis, and YouTube tutorials.

In the next chapter, we will dive into the magical world of the command line.

2
Command-Line Basics

In this chapter, we're going to dive right into the Linux command line. We will explain what makes it so powerful and, by extension, important to every system administrator and DevOps person. But more importantly, we will start teaching you the most useful commands and a way to use them efficiently. Along the way, we will be adding other core Linux concepts, as they will be required to understand the chapter.

In this chapter, we're going to cover the following main topics:

- What a command line is and how it works
- Why it is so important to feel comfortable working with the command line
- Basic commands for Linux system administration

It is not possible to introduce all the commands and tools in a single chapter. What follows is our choice of the most basic tools you will need to know. Managing the Linux system is a separate book topic on its own. It so happens that Packt does have several publications on that.

Technical requirements

It is highly recommended to have a Linux system installed and ready for use. We recommend it to be a virtual machine or a laptop that you can safely reinstall from scratch in case something goes horribly wrong. It will let you follow the examples in the book and perform any kind of exercise that we give you.

We are not going to cover an installation. Every distribution may use its own installer, either graphical or text (depending on the distribution and which variant you've picked). You'll need to note down or remember the name of your user (conveniently called username or login) and password. There are ways to get into the system if you have physical access and you don't know either the login or password, or both, but they are way outside of the scope of this book.

Our main distribution in the book is Debian. However, you should be alright with any of the major ones we covered in the previous chapter, *Choosing the Right Linux Distribution*, as long as it isn't Alpine.

The Linux command line – shell

The natural environment for a Linux system administrator is the command line. You'll never hear anyone call it that, however. The correct name is the **shell**, and from now on, that's how we're going to address it in the book.

The **shell** is a program that accepts input from a user (mostly keyboard strokes, but there are other ways, and you can even use a mouse pointer), interprets it, and, if it's a valid command, executes it, providing the user with the result or with error information if they've made a mistake or if the commands couldn't complete their execution properly.

There are a few ways to access the shell:

- Log in to the Terminal (screenshot in *Figure 2.1*)

> **Note**
>
> You'll also see the term **console**. There is a difference between the Terminal and a console. A **console** is a physical device that lets users interact with the computer. It is the physical input (nowadays, mostly keyboard) and output (nowadays, a monitor, but in the beginning, the output was printed out). **Terminal** is a console emulator, a program that lets users perform the same tasks.

- Open a Terminal window in a graphical interface, if you have one
- Log in remotely over a secure connection from another device (phone, tablet, or your computer)

The shell is a very powerful environment. It may seem cumbersome at first to do everything by typing commands, but soon you'll learn that the more complex the task, the more you can do with the shell, more easily than with graphical interfaces. Every Linux system administrator worth their salt will know how to do tasks in the shell and how to manage the system through it, and I wouldn't risk much by betting that they will prefer the shell to any GUI.

Getting to know your shell

The shell is a program, and as such, there is not a single shell. Instead, there are a number of more or less popular shells that bring forth their own view on how things should be done.

By far the most popular and default in most Linux distributions is **Bash** (**Bourne again shell**). There are other shells you may want to be aware of:

- **sh**: The original Steve Bourne shell. It is *the* shell, the very first one ever written that we know of. While it lacks many interactive features that users came to appreciate from other, more modern shells, *sh* is known for its speed of script execution and small size.
- **ksh**: Developed as an evolution of the Bourne shell, it is a superset of its predecessor. Thus, all scripts written for *sh* will run in *ksh*.

- **csh**: The C shell. The name comes from its syntax, which closely follows the C programming language.
- **zsh**: The Z shell. It should be well known to macOS users, as it is a default on this operating system. It is a completely modern shell, providing a lot of features that you'd expect from it: command history, arithmetic operations, command completion, and so on.

We won't trouble you much with shell variants and history. If you are interested in how Bash came to be, refer to this Wikipedia article: `https://en.wikipedia.org/wiki/Bash_(Unix_shell)`.

In our book, we are working with Bash. As mentioned earlier, it is the default shell for most Linux distributions, it offers *sh* compatibility mode, it has all the features you'd expect from a modern shell, and the amount of books, articles, how-tos, and other material for extending your knowledge about it is staggering.

The first task that we will be performing is logging in to your shell. Depending on your chosen installation method, you may need to power up the local virtual machine, your physical machine, or a cloud-based **Virtual Private Server** (**VPS**). If you have picked a server installation without a graphical user interface, you should see something similar to the following screenshot:

Figure 2.1 – Login screen

You are presented with a login screen, where you can provide your username and password. Once successfully logged in, you are presented with a command prompt, which confirms that you've just successfully started your shell.

The way the prompt looks is configurable and may be different depending on your chosen distribution. There is, however, one thing that will stay the same, and we advise you to never change it. In the world of Linux, there are two types of users: *normal users* and the *superuser*. While the login for normal users can be anything as long as it adheres to the Linux user naming conventions, the superuser is called **root**. To log in to the root account, you'd type in `root` as the username/login and its password next.

The superuser account is named that way for a reason. In most Linux distributions, it's the omnipotent user. Once logged in as `root`, you can do anything you wish, even deleting all files, including the operating system itself.

To help you distinguish between being a normal user and the `root` one, the prompt will present you with a clue. While logged in as `root`, your prompt will end with the # sign. When logged in as a normal user, it will present you with $.

While we're at it, the # (hash) sign is also a so-called comment sign. If you happen to paste a command or type it from the internet, if it starts with # or $, it is your clue as to which type of user should run this command. You should omit this sign, especially as # in front will prevent the command from running.

In addition to the sign that ends the prompt, many distributions will prepend the username to it, making sure you know which user you are. Given a user admin on a Linux system called `myhome`, the default prompt for Debian 11 will look like this:

```
$admin@myhome:~$
```

For the `root` user, it would look like this:

```
root@myhome:~#
```

For the record, there are more ways to check your identity, but we will leave that for *Chapter 4*.

I call to thee

So far, so good. You have logged in, you know who you are, you can type, and you can read. But how do you actually run programs?

In the Linux lingo, running a program is executing it or calling it. Actually, **calling** is mostly used when referring to system commands or shell built-in commands, and **executing** is used when talking about something that is not a part of the distribution—so-called third-party programs or binary.

Before I can tell you how to execute a program or call a command, I'll have to explain a little bit about filesystem structure and a very important system variable called PATH.

The filesystem structure

Since it may be your first time, we are going to step back a little bit and explain how the filesystem is structured (in other words, how directories are organized in typical Linux).

Linux follows the Unix philosophy that states that everything is a file. (There are exceptions, but not many.) The consequence is that almost every aspect of the operating system is reflected either as a file or a directory. Memory states, processes' (running programs) states, logs, binaries, and device drivers all live within this structure. This also means that almost every aspect of your Linux system can be edited or inspected using just normal text editing tools.

In the directory tree, its structure always starts with a / folder, called the **root directory**. Every drive, network share, and system directory lives in a hierarchy that starts from the root.

The process of making a network share or local drive available to the system or user is called **mounting** and the resource that is made available is **mounted**. Remember that word, as we're going to be using it from now on. Different from Microsoft Windows, where your drives appear as a separate letter with each being its own **root directory**, in Linux, your hard drive will be mounted somewhere within the directories hierarchy. As an example, the user home directory can be stored on a separate hard drive, mounted under the `/home/` directory. You would never be able to tell it when browsing the filesystem structure. The only way to tell would be to inspect mounted drives and partitions using the following commands:

```
$ mount
```

or

```
$ df
```

We are going to elaborate on them in *Chapter 3*, so for now, just know they exist.

The name of the uppermost directory is /. We already covered that. The separator between folders nested in another folder is also /. So `/usr/bin` means a `bin` directory that exists in the `usr` directory, and the `usr` directory exists in the / directory. Simple.

There is a very nice command that lets us inspect the directory structure, conveniently called `tree`. It may not be present in your system. If so, don't be worried; it's not so important that you run it rather than go through our explanation. In *Chapter 3*, when we introduce the installation of packages, you can revisit and play around. By default, the `tree` command will flood your screen with the full hierarchy, making it difficult to read and follow. There is an option, however, that lets us limit the depth we will be inspecting:

```
admin@myhome:~$ tree -L 1 /
/
|-- bin -> usr/bin
|-- boot
|-- dev
|-- etc
|-- home
|-- lib -> usr/lib
|-- lib32 -> usr/lib32
|-- lib64 -> usr/lib64
|-- libx32 -> usr/libx32
|-- lost+found
|-- media
|-- mnt
|-- opt
|-- proc
|-- root
```

```
|-- run
|-- sbin -> usr/sbin
|-- srv
|-- sys
|-- tmp
|-- usr
`-- var


22 directories, 0 files
```

There are several important concepts to cover here; we won't be explaining all the directories at this time, however. The first time a given folder becomes important, in *Chapter 3* onward, we will briefly touch upon it.

First, the calling of `tree`. You saw my prompt, which tells me I am running as user admin, on a system named myhome and that I am not a `root` user (*the dollar sign at the end*). If you want to run the `tree` command, you will skip the prompt. Next is the actual call: `tree` with the `-L` option and number `1`; this instructs the program to only print one level of depth. In other words, it will not go any deeper into the directories. Finally, the `/` symbol tells the program to start printing at the very beginning of the filesystem—the `root` folder.

Next, you'll notice that some rows have this mysterious arrow pointing from one name to another. This denotes a shortcut. There are two types of shortcuts, hard and symbolic. Directories can only have a symbolic link. In the preceding output, the `/bin` directory is a link to a `/usr/bin` directory. For all intents and purposes, they can be treated as one. There are technical and historical reasons for the existence of this link. In days past, tools living in `/bin` and `/sbin` were used to mount a `/usr` partition and then allow access to `/usr/bin` and `/usr/sbin`. Nowadays, this task is handled earlier in the boot process by other tools and the requirement is no longer necessary. The structure is kept for backward compatibility with tools that may require the existence of both `/bin` and `/sbin` directories. More details can be found at `https://refspecs.linuxfoundation.org/FHS_3.0/fhs/index.html` or `https://www.linuxfoundation.org/blog/blog/classic-sysadmin-the-linux-filesystem-explained`.

Since we already touched on the `/bin` and `/sbin` directory, let's explain the difference. The `/usr/bin` directory contains **binaries**—in other words, commands that can be of interest to every user in the system. The `/usr/sbin` directory contains so-called **system binaries**. Those are commands that should concern only the `root` user. It will also contain binaries for system processes (called **daemons**)—programs that run in the background and do important work for the running system.

The `/root` directory is the home of the superuser. This is where all its configuration files lay.

The interesting one is the `/home` directory. This is where all the user home directories exist. When I created my admin user for my home machine, it was placed in the `/home/admin` folder.

Of importance to us at this moment will also be the `/etc/` directory. It contains all the configuration files for the whole system: the source for online package repositories, default shell configuration, system name, processes that start at the boot, system users and their passwords, and time-based commands. On a freshly installed Debian 11 system, the `/etc/` directory contains about 150 subdirectories and files, and each subdirectory may contain more folders inside.

The `/tmp` folder contains temporary files. They only live when the system is booted and will be deleted the moment it is shut down or restarted. The nature of those files is often very volatile; they can hop into existence and disappear very fast or can be modified very often. It is not uncommon for this directory to only exist in the computer's RAM. It's the fastest storage device your system has and will automatically purge on restart or power off.

As mentioned earlier, additional drives are mounted under this structure. We could have a separate hard drive for home directories. The whole `/home` folder may live on it or even on a networked hard disk array. As mentioned, the `/tmp` directory is often mounted in the RAM. Sometimes, the `/var` directory (which contains things that can change often in the system but are not supposed to be purged, such as logs) is mounted on a separate drive. One of the reasons is that the contents of `/var`, and especially of `/var/log` (where the system logs live), can grow very fast and take all the available space, making accessing the system impossible or very tricky.

Finally, there are two important and special directories that exist everywhere you go:

- `.`: A single dot means the **current directory**. If you choose to go to a `.` folder, you'll end up where you are. It's useful, however, as you'll see in *Chapter 3*.

- `..`: Two dots mean the directory above us—the **parent directory**. If you choose to go to a `..` folder, you'll end up one level above where you started. Note that for the `/` directory, both `.` and `..` mean the same: `/`. You can't go any higher than the root.

Running a program

Now that we know some things about the folder hierarchy, let's briefly touch on executing programs.

There are three basic ways to execute a program in a shell:

- **Call it by its name**: This requires that it is placed in a directory that is in a `PATH` variable (explained in *Chapter 3*).

- **Call it by an absolute path**: The absolute path starts with `/`. When using the absolute path, you have to list all the directories that lead to the program, including the leading `/`. An example execution may look like this:

```
/opt/somecompany/importantdir/bin/program
```

- **Call it by a relative path**: This will not start with a dot or two dots. This way of executing a program requires that you know where in the filesystem you are and where in the filesystem this program is. It's useful if you are in the same directory as the binary or very close to it:

  - To call a program in the same directory as your user, you precede it with a dot and a slash: `./`. This shortcut means the **current directory**. A sample call could look like `./myprogram` or `./bin/myprogram`. The latter would mean: let's start a program called `myprogram` that is in a `bin` directory that is in the current directory.

  - To call a program in a directory somewhere else in the system using a relative path, we will have to use the two dots, meaning the parent folder. Let's say you are logged in to your home directory, `/home/admin`, and want to execute a program in `/opt/some/program/bin`; you'd call `../../opt/some/program/bin/myprogram`. The two dots and slash mean moving up.

If this looks arcane, it is because it is a little bit. Fortunately, everything will start coming together as the book progresses.

The command to teach you all commands

You should get into the habit of searching the internet whenever you have a question or problem. Most of the issues you are going to run into are already fixed or explained out there. However, there's one command that can save your life—or at least lots of time. You should make another habit of using it often – even if you are sure you know the correct syntax, you might just discover a better way of completing your task. This command is as follows:

```
$ man
```

The man command is a shorthand for *manual* and it is exactly what it says: it is a manual for whatever command you want to learn about. To learn more about the man command, simply call the following:

```
$ man man
```

The output you'll see should be similar to the following:

```
MAN(1)
                                    Manual pager utils
                                                                        MAN(1)


NAME
       man - an interface to the system reference manuals

SYNOPSIS
       man [man options] [[section] page ...] ...
       man -k [apropos options] regexp ...
```

```
man -K [man options] [section] term ...
man -f [whatis options] page ...
man -l [man options] file ...
man -w|-W [man options] page ...
```

DESCRIPTION
        man is the system's manual pager. Each page argument given
to man is normally the name of a program, utility or function. The
manual page associated with each of these arguments is then found and
displayed. A section, if

        provided, will direct man to look only in that section of the
manual. The default action is to search in all of the available
sections following a pre-defined order (see DEFAULTS), and to show
only the first page found,

        even if page exists in several sections.

I have cut it for brevity. A well-written man page will have several sections:

- name: This is where the name of the command is stated. If the command exists under various names, they will all be listed here.

- synopsis: This will list possible ways of calling the command.

- description: This is the purpose of the command.

- examples: This will show several examples of the command call, to make the syntax clearer and give some ideas.

- options: This will show all the available options and their meaning.

- getting help: This is how to get a shortened version of the command synopsis.

- files: If the command has configuration files or consumes files known to exist in the filesystem, they will be listed here (for man, I have listed /etc/manpath.config and /usr/share/man).

- bugs: This is where to look for bugs and report new ones.

- history: This will show all the current and previous authors of the program.

- see also: These are programs that are in some way tied to the command (for man: apropos(1), groff(1), less(1), manpath(1), nroff(1), troff(1), whatis(1), zsoelim(1), manpath(5), man(7), catman(8), and mandb(8)).

Many commands will contain a lot of additional sections, specific to this program.

The man pages will contain a lot of knowledge, and sometimes the names will duplicate. This is where the mysterious numbers in brackets come into play. The man pages are broken down into sections. To quote the man page about man:

1. Executable programs or shell commands

2. System calls (functions provided by the kernel)

3. Library calls (functions within program libraries)

4. Special files (usually found in /dev)

5. File formats and conventions, for example, /etc/passwd

6. Games

7. Miscellaneous (including macro packages and conventions), for example, man(7), groff(7)

8. System administration commands (usually only for root)

9. Kernel routines [non-standard]

Let's take, for example, printf. There are several things that are called printf. One of them is a library function of the C programming language. Its man page will live in section 3.

To read about that library function, you have to tell man to look into section 3:

```
admin@myhome:~$ man 3 printf
PRINTF(3)
                                    Linux Programmer's Manual
                                                            PRINTF(3)


NAME
        printf, fprintf, dprintf, sprintf, snprintf, vprintf, vfprintf,
vdprintf, vsprintf, vsnprintf - formatted output conversion
If you don't, what you'll get is a shell function for printing—totally
useless in C programming:
admin@myhome:~$ man printf
PRINTF(1)
                                    User Commands
                                                            PRINTF(1)


NAME
        printf - format and print data


SYNOPSIS
        printf FORMAT [ARGUMENT]...
        printf OPTION
```

Most commands and shell programs have a shorter synopsis called help. Usually, it can be invoked by running binary with the -h or --help option:

```
admin@myhome:/$ man --help
Usage: man [OPTION...] [SECTION] PAGE...

  -C, --config-file=FILE     use this user configuration file
  -d, --debug                emit debugging messages
  -D, --default              reset all options to their default values
      --warnings[=WARNINGS]  enable warnings from groff
```

I have cut the output for brevity, but you get the drift.

> **Note**
>
> The short options are preceded with one dash while longer ones are preceded with two. --help is not one long dash but two standard ones.

The man and –help commands should become your friends even before searching online. A lot of questions can be quickly answered just by looking at the help output. Even if you are a seasoned administrator, you are allowed to forget command syntax. There is an online source of endless Linux guides called *The Linux Documentation Project*: https://tldp.org. Make it your bookmark.

Know your environment

The way your system behaves is controlled by several factors. One of them is a set of variables called **environment**. They set things such as what language your system is going to use when talking to you, how entries during listing will be sorted, where the shell is going to look for executables, and many more. The exact set of variables depends on the distribution.

The full set of all the environment variables that your shell has set can be printed using the env command:

```
admin@myhome:/$ env
SHELL=/bin/Bash
PWD=/
LOGNAME=admin
XDG_SESSION_TYPE=tty
MOTD_SHOWN=pam
HOME=/home/admin
LANG=C.UTF-8
PATH=/usr/local/bin:/usr/bin:/bin:/usr/local/games:/usr/games
```

If you know the variable you want to inspect, you can use the echo command:

```
admin@myhome:/$ echo $PATH
/usr/local/bin:/usr/bin:/bin:/usr/local/games:/usr/games
```

Notice that when you are using a variable, you have to precede its name with a dollar sign, hence $PATH.

Where in the PATH am I?

Since we mentioned PATH, let's talk briefly about it. **Path** can mean two things:

- A place in the system that leads to something: a binary, a file, or a device
- An environment variable that lists places where the shell is going to look when trying to execute a program

You already know a little bit about the first kind of path. We explained the absolute path and relative path. There is a command that lets you move around, and it's called cd (shortcut for **change directory**). If you call cd without an argument, it will take you to your home directory. If you call it with an argument, it will move you to the specified folder, given that it exists, you specified the path properly, and you have the right to access it. Let's see a few examples:

- Checking which directory we're currently in:

```
admin@myhome:~# pwd
/home/admin
```

- Changing directory to /home/admin/documents/current:

```
admin@myhome:~# cd documents/current
admin@myhome:~/documents/current#
```

- Changing one directory up from a current level:

```
admin@myhome:~/documents/current# cd ..
admin@myhome:~/documents#
```

- Changing the directory to the user home directory:

```
admin@myhome:~/documents# cd
admin@myhome:~# pwd
/home/admin
```

Know your rights

The most basic security mechanism in Linux is based on defining a combination of rights for a set of entities. The rights are as follows:

- `read`
- `write`
- `execute` (read contents when talking about a directory)

And the entities are as follows:

- The owner of the file or directory
- The group that owns the file or directory
- All the other users and groups

This is a crude security system. It's sufficient for small servers and desktop uses but, for more complex setups, it is sometimes too restraining. There are other additional systems, such as **Access Control Lists** (**ACLs**), AppArmor, SELinux, and more. We are not going to cover them in this book.

With the use of the previous systems, we can still achieve quite a lot regarding our system security.

How do those rights and ownership work? We use the `ls` command (list files and directories):

```
admin@myhome:/$ ls -ahl
total 36K
drwxr-xr-x 3 admin admin 4.0K Aug 20 20:21 .
drwxr-xr-x 3 root  root  4.0K Aug 20 14:35 ..
-rw------- 1 admin admin  650 Aug 20 20:21 .Bash_history
-rw-r--r-- 1 admin admin  220 Aug  4  2021 .Bash_logout
-rw-r--r-- 1 admin admin 3.5K Aug 20 14:47 .Bashrc
-rw-r--r-- 1 admin admin    0 Aug 20 14:40 .cloud-locale-test.skip
-rw------- 1 admin admin   28 Aug 20 16:39 .lesshst
-rw-r--r-- 1 admin admin  807 Aug  4  2021 .profile
drwx------ 2 admin admin 4.0K Aug 20 14:35 .ssh
-rw------- 1 admin admin 1.5K Aug 20 14:47 .viminfo
```

The output is presented in nine columns.

The first column presents the type of the entry and the rights in a concise manner, but let's jump to the third and fourth. Those columns inform us about who is the owner of the file and what user group the file belongs to. All files and directories must belong to a user and to a group. In the preceding output, most of the files belong to the user admin and to the group admin. The exception is the `..` directory, which belongs to the `root` user and `root` group. It is common to denote this pair in the form `user:group`.

The next columns describe the size (with a directory, it describes the size of the entry, not how much space the directory contents take), the date of the last change, the time of the last change, and the name of the entry.

Now, let us go back to the first column. It informs us what an owner, a group, and all other users in the system are allowed to do with the given file or directory:

- The letter d means that we are dealing with a directory. A dash (–) means it's a file.

- Next is a group of nine one-letter symbols that denote who can do what with the given entry:

The first three letters denote what the file or directory owner can do with it. r means that they can read it, w means they can write to it, and x means they can execute it as a program. In the case of a text file with x set, the shell will try to run it as a script. The caveat is when we are dealing with a directory. x means that we can change a current working directory to it. It is possible to be able to go into a directory (x set) and not be able to see what is in it (r not set).

The same three letters explain the group rights in the second grouping.

The same set explains all other users' rights in the third grouping.

In the preceding output, .Bash_history is a file (it has a dash in the first field); the owner of the file (user admin) can read from it and write to it. It is possible to be able to write to a file (for example, a log file) but not be able to read from it. The file cannot be executed as a script (a dash). The next six dashes inform us that neither users assigned to the group admin nor any other user or group in the system can do anything with this file.

There is one exception, and it is the root user. Without stepping up with ACLs and tools such as SELinux, you cannot limit root's omnipotence in the system. Even for a file or directory that has no rights assigned (all dashes), root has full access to it.

The administration of the ownership and the rights is done by means of two commands:

- chown: This command allows you to change the ownership of a file or directory. The name is a shortcut for *changing owners*. The syntax is quite simple. Let's take this opportunity to exercise a bit with the notation in Linux help:

```
chown [OPTION]... [OWNER][:[GROUP]] FILE...
```

There is an unwritten convention that most, if not all, commands' help adhere to:

- Text without any brackets must be typed as shown.

- Anything inside [ ] brackets is optional. In the chown command, the user and group are optional, but you must supply at least one.

- Text in < > brackets is mandatory but is a placeholder for whatever you will have to supply.

- Brackets { } denote a set of options and you're supposed to choose one. They can be separated by a vertical line, |.

* Three dots after an element means this element can be supplied multiple times. In the chown case, it's the name of the file or directory.

The following is a set of ownership changes that I apply to a file:

* I change the ownership from admin to testuser, but I leave the group unchanged. Please note that changing actually requires the use of the root account (via the sudo command, explained in *Chapter 3*).

* I change the ownership back to admin, but change the group to testuser:

```
admin@myhome:~$ ls -ahl testfile
-rw-r--r-- 1 admin admin 0 Aug 22 18:32 testfile
admin@myhome:~$ sudo chown testuser testfile
admin@myhome:~$ ls -ahl testfile
-rw-r--r-- 1 testuser admin 0 Aug 22 18:32 testfile
admin@myhome:~$ sudo chown admin:testuser testfile
admin@myhome:~$ ls -ahl testfile
-rw-r--r-- 1 admin testuser 0 Aug 22 18:32 testfile
```

In the preceding output, we can see that a successful call to a command produces no output (chown) unless the output is the purpose of the command (ls). This is one of the basic rules that Linux follows. In the following output, we can see what happens when the command terminates with an error—insufficient rights to change the group:

```
admin@myhome:~$ chown :testuser testfile
chown: changing group of 'testfile': Operation not permitted
```

One other way to run the chown command is to point to a reference file, as in the following example:

```
admin@myhome:~$ sudo chown --reference=.Bash_history testfile
admin@myhome:~$ ls -ahl testfile
-rw-r-r-1 admin admin 0 Aug 22 18:32 testfile
admin@myhome:~$ ls -ahl .Bash_history
-rw------- 1 admin admin 1.1K Aug 22 18:33 .Bash_history
```

Using the --reference option, we point to a file that will be a matrix for our changes. This will become more interesting once we move on to the next chapter.

* chmod: Similar to the chown command, chmod (short for *change mode*) is the one you'll be looking for to change the rights of the assigned user and group:

```
admin@myhome:~$ chmod --help
Usage: chmod [OPTION]... MODE[,MODE]... FILE...
  or:  chmod [OPTION]... OCTAL-MODE FILE...
  or:  chmod [OPTION]... --reference=RFILE FILE...
```

The chmod command will accept options, mandatory mode, optionally more modes, and a list of files that the change is supposed to be applied to. As with the chown command, we can specify a reference file that will have its mode copied over.

In the first form, you will specify the mode for either a user, a group, others, or all of them using the following syntax:

```
chmod [ugoa...] [-+=] [perms...] FILE...
```

Here, the following meanings apply:

- u: User, in other words, the owner of the file
- g: Group that owns the file
- o: Others—all others
- a: All, meaning everyone
- -: Remove the specified right
- +: Add the specified right
- =: Make the rights exactly as specified

Let's see some examples.

This adds read and write permissions to the testfile file for the file owner:

```
chmod u+rw testfile
```

This removes the execute right to the testfile file for everyone who is not the file owner and is not in the group that owns the file:

```
chmod o-x testfile
```

This gives the user and the group read and execute rights to the testfile file:

```
chmod ug=rx testfile
```

In the syntax summary we have pasted, the middle row is interesting. The octal mode means that we can specify the mode by means of numbers. This is especially useful in scripts, as dealing with numbers is easier. Once you memorize the mode numbers, you may find it easier to use the octal chmod.

The formula for the numeric setting of file mode is simple:

- 0: no rights (---)
- 1: execute mode (--x)
- 2: write mode (-w-)
- 4: read mode (r-)

To set the mode on a file or directory, you will use a sum of the rights you want to apply, starting with a leading 0, which will inform chmod that you are setting octal mode. The syntax is as follows:

```
chmod [OPTION]... OCTAL-MODE FILE...
```

There is a very important difference between this form and the letter one—you must specify the mode for all three types of entities: user, group, and others. You cannot omit any of them. This means that you also won't be using the -, +, or = signs. With the octal syntax, the rights will always be exactly as specified.

To combine the modes, you will add their numbers and use the sum as the final specifier. You will find that this is a very elegant solution. There are no two identical sums of those numbers (combination of rights). Try it:

- Execute and read is 1 and 2 = 3. There is no other combination of the modes that can result in 3.

- Read and write is 2 and 4 = 6. Again, no other combination can result in 6.

Now, let's try some examples.

The owner of the file will have read and write access to the file, while the group and others will have read access:

```
chmod 0644 testfile
```

The owner of the file has all the rights (read, write, and execute), while the group and others can read and execute only:

```
chmod 0755 testfile
```

The leading 0 in the mode is not obligatory.

We have covered the filesystem, directory structure, and basic file permissions related to users and groups. In the next section, we will introduce basic Linux commands.

Interacting with the system

Programs and scripts lying on a hard drive are just files. The moment they get mapped to the memory and start performing their function, they become **processes**. At this stage, you can safely assume that anything running in the system is some kind of a process.

Process this

Processes in Linux have several characteristics that you need to be aware of:

- **Process ID (PID)**: A system-wide unique numerical identifier.
- **Parent process ID (PPID)**: Every single process in the Linux system, except process number 1, has a parent. Process number 1 is an *init* process. It is the program responsible for starting all the system services. A program that starts another program is known as a **parent**. A program started by another program is known as a **child**. When you log in to the system, your shell is a process too, and it has its PID. When you start a program in that shell, your command-line PID will become the parent ID of that program. If a process loses its parent (i.e., the parent process terminates without terminating its child), then the child process is assigned a new parent: process number 1.
- **State**: There are five states that a process can be in:
  - **R**: Runnable or running. There is a distinction between running and runnable, but in the statistics, they are grouped together. A **running** process is one that actively takes CPU time and resources. It is being executed as we speak. A **runnable** process is one that has been put aside while another process takes its place. This is how single-core computers create the illusion of multitasking—executing many programs at once; they rotate every started program on the CPU in time slices that are so short that humans can't notice when the program is actually being executed on the CPU and when it is waiting in the queue.
  - **S**: Interruptible sleep. A program may request an additional resource besides RAM and CPU time. If it needs to wait for the resource to become available, then it goes into a separate queue and *sleeps*. This program will *awake* when the required resource is available and can be interacted with using interrupts—special kinds of signals. We will cover signals in *Chapter 3*.
  - **D**: Uninterruptible sleep. As with the preceding S state, the process went to sleep. It will not, however, accept any interrupts and signals. It will only awake when a requested resource becomes available.
  - **T**: We can instruct a program to stop its execution and wait. This is called a **stopped state**. Such a process can be brought back to execution with the use of a special signal.
  - **Z**: Zombie. When a process ends its execution, it informs its parent process about that. The same happens when the process is terminated. The parent is responsible for removing it from the process table. Until it happens, the process remains in a zombie state, also called **defunc**.
- **User**: The owner of the process, or rather, a user with whose system rights the process is being executed. If that user cannot do something, the process cannot do it either.
- **CPU**: Percentage of the CPU time the process is using, presented as a floating-point number between 0.0 and 1.0.
- **MEM**: The memory usage, again between 0.0 and 1.0, where 1.0 is 100% of system memory.

Each process has more characteristics than the ones we've just covered, but those are absolutely essential. Managing processes is a subject for a separate chapter, if not an entire book.

The tool to inspect the processes is called ps. It seems like a pretty simple command at first sight, but in reality, the man page shows an incredible amount of options. It is important to note that ps itself will only print a snapshot of the system. It won't be monitoring and updating its output. You can combine ps with watch or run top or htop commands to have continuous information about the processes and system load.

In its simplest form, ps will print all processes running with the same user ID as the user that calls it:

```
admin@myhome:~$ ps
    PID TTY          TIME CMD
  24133 pts/1    00:00:00 Bash
  25616 pts/1    00:00:00 ps
```

On my Linux machine, there are only two processes running for my user: the Bash shell and the ps program itself.

The ps command has several interesting options that help interrogate the system about the running processes. Now, specifying options to ps is tricky, because it accepts two syntaxes, one with a dash and one without, and some options have different meanings depending on the dash. Let me quote the man page:

```
        1   UNIX options, which may be grouped and must be preceded by
    a dash.
        2   BSD options, which may be grouped and must not be used with
    a dash.
        3   GNU long options, which are preceded by two dashes.
    Note that ps -aux is distinct from ps aux.   The POSIX and UNIX
    standards require that ps -aux print all processes owned by a user
    named x, as well as printing all processes that would be selected by
    the -a option.   If the
        user named x does not exist, this ps may interpret the command
    as ps aux instead and print a warning.   This behavior is intended to
    aid in transitioning old scripts and habits.   It is fragile, subject
    to change, and thus
        should not be relied upon.
```

I have cut some output for brevity. This distinction may become important once you start working with shell scripts, as they may adopt any of the three syntaxes. Whenever in doubt, use the dash notation:

- -f: So-called **full output**. It will print the ID of the user that runs the command, PID, PPID, **CPU utilization** (C), **start time of the process** (STIME), **terminal** (TTY) to which it is attached, **how long it has been running** (TIME), and the command that started the process:

```
admin@myhome:~$ ps -f
UID          PID    PPID  C STIME TTY          TIME CMD
admin      24133   24132  0 16:05 pts/1    00:00:00 -Bash
admin      25628   24133  0 17:35 pts/1    00:00:00 ps -f
```

- -e: All processes of all users:

```
admin@myhome:~$ ps -e
    PID TTY          TIME CMD
      1 ?        00:00:04 systemd
      2 ?        00:00:00 kthreadd
[...]
  25633 ?        00:00:00 kworker/u30:0-events_unbound
  25656 pts/1    00:00:00 ps
```

- -ef: To see all processes in the long output format:

```
admin@myhome:~$ ps -ef
UID          PID    PPID  C STIME TTY          TIME CMD
root           1       0  0 Aug20 ?        00:00:04 /sbin/init
root           2       0  0 Aug20 ?        00:00:00 [kthreadd]
```

- -ejH: A nice process tree. The more indented CMDs (the last column of the output) are children of the less indented ones:

```
admin@myhome:~$ ps -ejH
    PID   PGID    SID TTY          TIME CMD
      2      0      0 ?        00:00:00 kthreadd
      3      0      0 ?        00:00:00   rcu_gp
      4      0      0 ?        00:00:00   rcu_par_gp
      6      0      0 ?        00:00:00   kworker/0:0H-events_
highpri
      9      0      0 ?        00:00:00   mm_percpu_wq
```

There are many more options available, especially to control which fields are of interest. We will come back to them in later chapters.

There is a command that has a name that can be very misleading, the `kill` command. It is used to send so-called signals to the running processes. Signals are a way of telling processes to perform a kind of action. One of them actually kills the program, terminating it at once, but that's just one of many.

To list existing signals, use the `kill -l` command:

```
admin@myhome:~$ admin@myhome:~$ kill -l
 1) SIGHUP 2) SIGINT 3) SIGQUIT 4) SIGILL 5) SIGTRAP
 6) SIGABRT 7) SIGBUS 8) SIGFPE 9) SIGKILL 10) SIGUSR1
11) SIGSEGV 12) SIGUSR2 13) SIGPIPE 14) SIGALRM 15) SIGTERM
16) SIGSTKFLT 17) SIGCHLD 18) SIGCONT 19) SIGSTOP 20) SIGTSTP
21) SIGTTIN 22) SIGTTOU 23) SIGURG 24) SIGXCPU 25) SIGXFSZ
26) SIGVTALRM 27) SIGPROF 28) SIGWINCH 29) SIGIO 30) SIGPWR
31) SIGSYS 34) SIGRTMIN 35) SIGRTMIN+1 36) SIGRTMIN+2 37) SIGRTMIN+3
38) SIGRTMIN+4 39) SIGRTMIN+5 40) SIGRTMIN+6 41) SIGRTMIN+7 42)
SIGRTMIN+8
43) SIGRTMIN+9 44) SIGRTMIN+10 45) SIGRTMIN+11 46) SIGRTMIN+12 47)
SIGRTMIN+13
48) SIGRTMIN+14 49) SIGRTMIN+15 50) SIGRTMAX-14 51) SIGRTMAX-13 52)
SIGRTMAX-12
53) SIGRTMAX-11 54) SIGRTMAX-10 55) SIGRTMAX-9 56) SIGRTMAX-8 57)
SIGRTMAX-7
58) SIGRTMAX-6 59) SIGRTMAX-5 60) SIGRTMAX-4 61) SIGRTMAX-3 62)
SIGRTMAX-2
63) SIGRTMAX-1 64) SIGRTMAX
```

Okay, this list is cute, but it tells us nothing. How can we tell what each of those signals does? Here's a bit of detective work. First, since we see those in the `kill -l` output, let's invoke the `man kill` command and see whether there's anything that can explain them to us:

```
EXAMPLES
       kill -9 -1
              Kill all processes you can kill.

       kill -l 11
              Translate number 11 into a signal name.

       kill -L
              List the available signal choices in a nice table.

       kill 123 543 2341 3453
              Send the default signal, SIGTERM, to all those
processes.

SEE ALSO
```

```
       kill(2), killall(1), nice(1), pkill(1), renice(1), signal(7),
sigqueue(3), skill(1)
```

There is an examples section that shows and describes one signal, but in the SEE ALSO section, we can see a reference to a man page signal in section 7. Let us check it out:

```
admin@myhome:~$ man 7 signal
SIGNAL(7)

                                Linux Programmer's Manual
                                                              SIGNAL(7)


NAME
        signal - overview of signals
```

Now, there's a nice page with a table listing all signals available to you in Linux.

So, how do you use this kill command? Do you have to learn about all these signals? The answer is no. There are a handful of signals that you'll be using. And if you forget any, don't hesitate to man them or search the web:

- kill -9 PID: The infamous SIGKILL. This terminates the process whose PID we have specified abruptly, forcing it to omit any cleanup it may do normally. If it has opened file handles, it won't free them; if it has to write any information to a file or synchronize with another program, it won't do it. This should be used sparingly, only when we are sure that we really have to stop a program from running.

- kill PID: If no signal is specified, then by default, SIGTERM is sent. This tells the program to stop running but gracefully—perform all the exit routines and cleanup it needs. If you have to use the kill command, this is the preferred use.

- kill -1: The so-called SIGHUP signal. It was originally used for detecting loss of user connection—hangup of the phone line. Currently, it is often used to tell the process to reread its configuration.

Here is an example of calling kill to terminate a process. I have started a shell script that does nothing except wait for a keyboard input. I called it sleep.sh:

```
admin@myhome:~$ ps aux | grep sleep
admin       24184  0.0  0.2   5648   2768 pts/1     S+    16:09    0:00 /
bin/Bash ./sleep.sh

admin@myhome:~$ pgrep sleep.sh
24184
admin@myhome:~$ kill -9 24184
admin@myhome:~$ pgrep sleep.sh
admin@myhome:~$ ps aux | grep sleep
```

```
admin        24189  0.0  0.0    4732    732 pts/0    S+    16:09    0:00
grep sleep
```

First, I used `ps aux` and searched in the output for the `sleep.sh` process, just to show you that it's there. Next, I used `pgrep` to find the PID of my running script quickly. I have supplied this PID to the `kill -9` command. In effect, `sleep.sh` has been killed. This can be confirmed on another Terminal, where I was running `sleep.sh`:

```
admin@myhome:~$ ./sleep.sh
Killed
```

Had I simply used `kill`, the output would be different:

```
admin@myhome:~$ ./sleep.sh
Terminated
```

There is another way of delivering two of all signals to running programs, but only if currently executing in the shell we're logged into in the foreground; this means it has the control of the screen and keyboard:

- Pressing the *Ctrl + C* keys will send `SIGINT` to the program. `SIGINT` tells the program that the user has pressed the key combination and it should stop. It's still up to the program how the termination should occur.

- Pressing the *Ctrl + D* keys will send `SIGQUIT`—it is like `SIGINT` but also produces a so-called core dump, or a file that can be used in a debugging process.

The common way of denoting those combinations in text is `^c` and `^d` (`^` for the *Ctrl* key), `ctrl+c` and `ctrl+d` (`ctrl` being a shortcut for the *Ctrl* key), and `C-c` and `C-d` (`C` denoting the *Ctrl* key, again).

Looking for something?

There are times when you need to look for a directory or a file on the filesystem. Linux has a set of commands that let you perform this operation. Of them all, `find` is the most powerful. To cover all of its abilities would take a lot more space than we have. You can look for a file or directory of both that have exactly the name you specified, that have part of the name that you specified, that have been modified at a defined point in time, that are owned by a user or a group, and many more scenarios. Additionally, on each file that has been found, an action can be performed, such as rename, compress, or search for a word.

In the following example, we're looking for a file, `signals.h`, in the `/usr` directory:

```
admin@myhome:/$ find / -name os-release
find: '/lost+found': Permission denied
find: '/etc/sudoers.d': Permission denied
```

```
/etc/os-release
find: '/etc/ssl/private': Permission denied
/usr/lib/os-release
[...]
```

First, we call find itself, then we tell it to start looking at the beginning of the filesystem (/), and then we tell it to look for a file named os-release (-name os-release).

You will notice that in the output, which I have cut for brevity, there are errors about files to which find had no proper rights.

If you are unsure about the case of the name, that is, whether it contains lowercase or uppercase letters (remember that Linux is case -sensitive and *OS-Release* is not the same file as *os-release*), you can use the -iname option instead.

If you are sure that what you are looking for is a file, then you can use the -type f option:

```
admin@myhome:/$ find / -type f -name os-release
```

For a directory option, use -type d.

To look for files that match a pattern, say filenames that end with .sh, you'd use a pattern:

```
admin@myhome:/$ find / -type f -name "*.sh"
```

The asterisk means any number of any characters. We enclosed it in the quotes to avoid problems with the shell interpreting the asterisk before find got a chance. We will explain all the special signs (called **globs**) and regular expressions in the next chapter, *Intermediate Linux*.

To delete all found files, you can use the -delete option:

```
admin@myhome:~$ find . -type f -name test -delete
```

To perform an action on the found file, you can use the -exec option:

```
admin@myhome:/$ find / -type f -name  os-release -exec grep -i debian
{} \;
PRETTY_NAME="Debian GNU/Linux 11 (bullseye)"
NAME="Debian GNU/Linux"
ID=debian
HOME_URL="https://www.debian.org/"
SUPPORT_URL="https://www.debian.org/support"
BUG_REPORT_URL="https://bugs.debian.org/"
```

In the preceding example, we have used the command called grep to look for any line that contains the word debian, no matter whether uppercase or lowercase.

This leads us to the grep command.

The grep command is used to look for an occurrence of a pattern inside a file. Upon finding this pattern-matching line, it prints it. The grep command is similar to the find command, except its purpose is to search inside the files, while find only cares for the characteristics of files and directories themselves.

Assume a text file called red_fox.txt with the following content:

```
admin@myhome:~$ cat red_fox.txt
The red fox
jumps over
the lazy
brown dog.
```

As a side note, cat is a command that prints the contents of a specified file to the Terminal.

Let's say we want to find all lines that contain the word the:

```
admin@myhome:~$ grep -i The red_fox.txt
The red fox
the lazy
```

You have probably guessed that the -i option means that we don't care about the case.

But wait. We can search for alternatives. Let's say we care for lines that contain either fox or dog. You would use the -e option for each word you search for or the -E option with all the words in single quotes and separated by the | character:

```
admin@myhome:~$ grep -e fox -e dog red_fox.txt
The red fox
brown dog.
admin@myhome:~$ grep -E 'fox|dog' red_fox.txt
The red fox
brown dog.
```

By adding the -n option, you'll get information on which line the match was found in:

```
admin@myhome:~$ grep -n -E 'fox|dog' red_fox.txt
1:The red fox
4:brown dog.
```

You can look for lines that start with a given word or have the specified pattern at the end of the line.

You can even grep all files in a directory hierarchy. The syntax is a bit different then: the pattern comes first, and the directory comes at the end. Also, you get the name of the file where the match was found:

```
admin@myhome:~$ grep -r "fox" .
./red_fox.txt:The red fox
```

The most power for both `find` and `grep` comes from a concept called **regular expressions** (**regex** or **regexp** for short), which have a book of their own and can be confusing for new users. We are going to explain them in *Chapters 3* and *4*. We will, however, be only introducing the most everyday uses of them.

If you are looking for a program and want to know its full path, there is a command for that, and it is called `whereis`. Here is an example:

```
admin@myhome:~$ whereis ping
ping: /usr/bin/ping /usr/share/man/man8/ping.8.gz
```

The `whereis` command will not only print the whole path to the binary but also a corresponding man page, if it has one installed.

Let's be manipulative

There are four basic operations that can be performed on files and directories:

- Create
- Rename or move
- Remove
- Copy

Each of these operations has a special tool:

- `mkdir`: This command has very simple and limited syntax. Basically, you are telling it to create a directory with a given name. If you are creating nested directories, that is, one folder contains another, all the directories in the path must exist. If they don't, you can create them using the special `-p` option:

```
admin@myhome:~$ mkdir test
admin@myhome:~$ ls -l
total 4
drwxr-xr-x 2 admin admin 4096 Aug 24 15:17 test
admin@myhome:~$ mkdir something/new
mkdir: cannot create directory 'something/new': No such file or
directory
admin@myhome:~$ mkdir -p something/new
admin@myhome:~$ ls -l
total 8
drwxr-xr-x 3 admin admin 4096 Aug 24 15:18 something
drwxr-xr-x 2 admin admin 4096 Aug 24 15:17 test
```

In the preceding example, you can see that I create a directory test directly in my home folder. Next, I try to create a folder, new, inside the something folder. However, the latter doesn't exist and mkdir tells me so and refuses to create the new directory. I use the special -p option to create a whole path to the new directory.

- mv: This is a command that is used to move and rename files and directories. Again, the syntax is pretty simple, although this command offers a little more functionality, such as creating a backup of moved files.

To rename a file or directory, we move it from the current name to the new one:

```
admin@myhome:~$ mv test no-test
admin@myhome:~$ ls -l
total 8
drwxr-xr-x 2 admin admin 4096 Aug 24 15:17 no-test
```

Check out its man page or help message to find out more.

- rm: This command is more interesting, mainly because it offers safety features. With the special -i option, you can tell it to always ask you before removing the file or directory. Normally, rm bails at directories, as in the following example:

```
admin@myhome:~$ admin@myhome:~$ ls -l no-test/
total 0
-rw-r--r-- 1 admin admin 0 Aug 24 15:26 test
admin@myhome:~$ rm no-test/
rm: cannot remove 'no-test/': Is a directory
admin@myhome:~$ rm -d no-test/
rm: cannot remove 'no-test/': Directory not empty
admin@myhome:~$ rm no-test/test
admin@myhome:~$ rm -d no-test/
```

I created a file test inside the no-test directory. rm refuses to remove the folder. I used the -d option, which instructs the command to remove empty directories. However, it still contains a file. Next, I removed the file and then rm -d cleanly deleted the no-test folder. I could have used the -r option, which makes the command remove all directories, even if they are not empty.

- cp: This command is used for copying files and directories. Note that, as with rm, cp will refuse to copy directories unless instructed with the -r option. cp arguably is the most complex and feature-rich command of all, including the ability to back up files, create links (shortcuts) instead of real copies, and more. Check out its man page. In the following example, I copy a something directory to a new directory. Obviously, I have to use the -r option. Next, I create an empty file called file and copy it to newfile. For those, I don't require any options:

```
admin@myhome:~$ ls -l
total 4
```

```
drwxr-xr-x 3 admin admin 4096 Aug 24 15:18 something
admin@myhome:~$ cp something/ new
cp: -r not specified; omitting directory 'something/'
admin@myhome:~$ cp -r something new
admin@myhome:~$ ls -l
total 8
drwxr-xr-x 3 admin admin 4096 Aug 24 15:33 new
drwxr-xr-x 3 admin admin 4096 Aug 24 15:18 something
admin@myhome:~$ touch file
admin@myhome:~$ cp file newfile
admin@myhome:~$ ls -l
total 8
-rw-r--r-- 1 admin admin    0 Aug 24 15:33 file
drwxr-xr-x 3 admin admin 4096 Aug 24 15:33 new
-rw-r--r-- 1 admin admin    0 Aug 24 15:34 newfile
drwxr-xr-x 3 admin admin 4096 Aug 24 15:18 something
```

You should now understand and be able to use basic command-line commands in Linux or a similar system, such as create, copy, and delete files; you can also find content inside text files or find files or directories by name. You also have an understanding of processes in the system you work on. In the next chapter, we will deepen this knowledge.

Summary

We have presented but a fraction of the hundreds of commands that Linux admins may use in their work. As mentioned at the beginning of this chapter, the complete reference is beyond the scope of this book. What we have learned, however, is enough to do basic system use and it builds the base for our next chapter: the more advanced Linux administration topics.

Exercises

- Find out how to apply chown recursively—it means that our chown call should step inside a directory and apply the change to all the items within.

- Find out what the watch command does. Use it with a ps command.

- Find out how to remove a directory.

- For all the commands you have learned here, read their -help output. Open a man page and look through it, especially the examples section.

Resources

- You can read more about SELinux here: `https://www.packtpub.com/product/selinux-system-administration-second-edition/9781787126954`

- This is a very good publication on Linux systems administration: `https://www.packtpub.com/product/mastering-linux-administration/9781789954272`

3

Intermediate Linux

In this chapter, we're going to continue with the introduction to the Linux shell. The topic itself is vast and warrants a book of its own. We will be coming back to the topics from the previous chapter and introducing new ones.

In this chapter, we will cover the following topics:

- Globs

- Automating repetitive tasks

- Software installation

- Managing users

- **Secure Shell** (**SSH**) protocol

Technical requirements

It is highly recommended that you have a Linux system installed and ready for use. We recommend it be a virtual machine or a laptop that you can safely reinstall from scratch in case something goes horribly wrong. This will let you follow the examples in this book and perform any kind of exercise that we give you.

We are not going to cover an installation. Every distribution may use its own installer, be it graphical or text (depending on the distribution and which variant you've picked). You'll need to note down or remember the name of your user (conveniently called **username** or **login**) and password. There are ways to get into the system if you have physical access and you don't know either the login or password or both, but they are way outside the scope of this book.

Our main distribution in this book is **Debian**. However, you should be okay with any of the major ones we covered in the previous chapter, so long as it isn't Alpine.

Globs

There is a lot that a shell can do for you to make your life easier. One of them is allowing for a level of *uncertainty* when typing in arguments on the shell. To that end, the shell defines several special characters that are treated like symbols for something, not like literal input. These are called **global patterns**, or **globs**. The characters that are used in globs are sometimes referred to as **wildcards**.

Do not confuse globs with **regular expressions** (**regexps**). While globs are quite a powerful tool on their own, they are no match for regexps. On the other hand, regexps are not evaluated by bash when it performs pattern matching.

The following table describes shell globs and their meaning. We're going to explain their exact meaning through several examples:

Glob	Meaning
*	Matches any number of any characters (also zero)
?	Matches exactly one character
[...]	Matches any one character from a set inside the brackets

Table 3.1 – Shell globs

The preceding table may be unclear to you, so let's cover some examples:

Example	Meaning
*	This will match any string of any length.
test	This will match anything that has the word test inside it: test.txt, good_test.txt, test_run, and simply test (remember, it matches nothing too).
test*txt	This will match anything that has a name starting with test and ending in txt, so test.txt, testtxt, test_file.txt, and so on.
test?	This will match any occurrence of test plus one character: test1, test2, testa, test, and so on.
test.[ch]	This will match one of two things: test.c or test.h and nothing else.

| `*.[ab]` | This will match any string that ends in a dot and either a or b. |
| `?[tf]` | This will match exactly one character of any kind followed by either t or f. |

Table 3.2 – Shell globs – examples

The true power of globs emerges once you start writing some more complicated strings of commands (so-called one-liners) or scripts.

Some examples of simple commands that get on an entirely new level when combined with globs are `find`, `grep`, and `rm`. In the following example, I am using globs to remove all files that start with anything, then have test followed by a dot, and then log followed by anything. So, the `weirdtest.log`, `anothertest.log1`, and `test.log.3` files will be matched, but not `testlog` and `important_test.out`. First, let's list all the files that contain the word *test* in their name:

```
admin@myhome:~$ ls -ahl *test*
-rw-r--r-- 1 admin admin 0 Sep 17 20:36 importat_test.out
-rw-r--r-- 1 admin admin 0 Sep 17 20:36 runner_test.lo
-rw-r--r-- 1 admin admin 0 Sep 17 20:36 runner_test.log
-rw-r--r-- 1 admin admin 0 Sep 17 20:36 test.log
-rw-r--r-- 1 admin admin 0 Sep 17 20:35 test.log.1
-rw-r--r-- 1 admin admin 0 Sep 17 20:35 test.log.2
-rw-r--r-- 1 admin admin 0 Sep 17 20:35 test.log.3
-rw-r--r-- 1 admin admin 0 Sep 17 20:35 test.log.4
-rw-r--r-- 1 admin admin 0 Sep 17 20:35 test.log.5
-rw-r--r-- 1 admin admin 0 Sep 17 20:35 test.log.6
-rw-r--r-- 1 admin admin 0 Sep 17 20:35 test.log.7
```

You will notice that I have used a wildcard (`*`) to achieve my goal. Now, it's time for the actual removal:

```
admin@myhome:~$ rm *test.log*
admin@myhome:~$ ls -ahl *test*
-rw-r--r-- 1 admin admin 0 Sep 17 20:36 importat_test.out
-rw-r--r-- 1 admin admin 0 Sep 17 20:36 runner_test.lo
```

As demonstrated here, it worked. You will also notice that a properly executed command doesn't print any messages.

In this section, we explained how to use globs – special characters that allow us to match names in the system with some level of uncertainty. In the next section, we are going to introduce mechanisms to automate repetitive tasks.

Automating repetitive tasks

There are times when you'll want to make some tasks repetitive. You may write a script that will create a backup of a database, check users' home directory permissions, or dump current operating system preformance metrics into a file. Modern Linux distributions provide you with two ways of setting these up. There is a third method that allows you to run a task once, at a delayed time (the at command), but here, we're interested in repetitive tasks.

Cron jobs

Cron is a traditional way of running tasks that need to be executed regularly at specified intervals. Usually, they should be obsolete by **systemd timers**, but a lot of software provides repeatability through the use of cron jobs and Alpine Linux won't have this in the name of the minimal-sized distribution.

Cron jobs are essentially commands that are run at predefined intervals. The command and their trigger timers are defined in configuration files that live in the /etc/ directory. The exact number of files and directories differ by distribution. All of them will have a /etc/crontab file. This file usually contains an explanation of the fields within it and several actual commands that you can use as templates. In the following code block, I have pasted explanations from the default /etc/crontab file:

```
# Example of job definition:
# .---------------- minute (0 - 59)
# |  .------------- hour (0 - 23)
# |  |  .---------- day of month (1 - 31)
# |  |  |  .------- month (1 - 12) OR jan,feb,mar,apr ...
# |  |  |  |  .---- day of week (0 - 6) (Sunday=0 or 7) OR
sun,mon,tue,wed,thu,fri,sat
# |  |  |  |  |
# *  *  *  *  * user-name command to be executed
  17 *    * * *   root    cd / && run-parts --report /etc/cron.hourly
```

Normally, there are two ways to set up a cron job. One of them is to put a script that will run your command in one of four directories: /etc/cron.hourly, /etc/cron.daily, /etc/cron.weekly, or /etc/cron.monthly. They should suffice for normal operations. The /etc/crontab file specifies when those will be run.

The second option is to use the crontab command. The crontab command lets a given user create an entry in their crontab file. However, there's a difference between a system-wide crontab file (living in the /etc/ directory) and a per-user one. The user cron file doesn't specify the user field. All entries are run with the user's permission. Let's look at the differences:

- crontab -l lists all cron jobs that the user has defined in their crontab file
- crontab -e lets the user edit the crontab file to add, remove, or modify the jobs

Systemd timer

We are not going to cover **systemd timers** here in detail and will only mention why they may be a better option.

Systemd timer units are newer versions of **cron daemon** jobs. They can do everything that cron can. However, they offer some additional capabilities:

- You can specify the job to run at some time after a system boots.

- You can specify that the job has to run at some interval after some other job has run.

- The dependency on the timer unit can even be a service unit – the normal system service task.

- The granularity is much better. Cron jobs only go down to every minute. Systemd timers can trigger with accuracy defined to a second.

In this section, we have covered cron jobs and mentioned systemd timers as the two most important ways to automate tasks that have to occur regularly. In the next section, we are going to introduce managing software.

Software installation

Depending on which distribution you've chosen and the type of installation you've decided on, your system may lack software that's essential for your everyday work. It may also be that you one day require a piece of software that isn't installed by default.

Linux distributions pioneered something that other operating systems mirrored later on. The common way of installing software on the Linux operating system is by running an appropriate command that will fetch a binary, put it properly on the system, add some configuration if required, and make it available to users. Today, this may not sound revolutionary at all. After all, we're living in a world of Apple App Store, Google Play, and Microsoft Apps. But back in the day when Windows and macOS users had to navigate the internet to find a suitable installer for their software, Linux users could install most of it with a single command.

This is important in the automated environments that DevOps would strive to set up for several reasons:

- The installable software (distributed in packages) is held in repositories maintained by the distribution team. This means that you don't need to know the software's location on the internet; you just need to know its package name and ensure it's in the repository.

- The package standards that we'll cover here (`rpm` and `deb`) *know* dependencies. This means that if the software you're trying to install depends on another not-yet-installed software, it will automatically get pulled and installed.

- The distributions that we'll cover here have security teams. They work with package maintainers to keep them patched against any known vulnerabilities. However, this does not mean that they will actively research vulnerabilities in said packages.

- The repositories are mirrored over the internet. This means that even if one of them fails (goes offline or is DDoS-ed or for any other reason), you can reach its mirror copies from all over the world. This is not necessarily true for commercial repositories.

- You can create a local repository mirror in your LAN if you wish. This will give you the fastest download times at the cost of a lot of hard drive space. Package repositories can be huge.

The amount and versions of the software depend on the distribution in many ways:

- **The policy distribution has to distribute software with different types of licenses**: Some distributions will forbid any software that isn't strictly open and free as defined by the Open Source Initiative. Other distributions will give the user a choice of adding repositories that may contain software that is more restrictive in its licensing.

- **The number of maintainers and maintainership models**: It is quite obvious that distributions can only do as much work as they have man hours to spare. The smaller the team, the less software they can package and maintain. Part of the work is automated, but a lot will always need to be manual. With Debian being a non-commercial distribution, it relies solely on the work of volunteers. Ubuntu and Fedora do have commercial backing and part of the team is even employed by one of the companies: Canonical and Red Hat. **Red Hat Enterprise Linux (RHEL)** is entirely built and maintained by Red Hat employees.

- **The type of repositories you decide to use**: Some software makers distribute their packages in separate repositories, after which you can add them to your configuration and use them as if they were regular distribution repositories. However, if you do so, there are some things to keep in mind:

  - **The software in the third-party repository is not part of the quality effort made by the distribution**: This is solely at the discretion of the repository maintainer – in this case, the software vendor. This will include security fixes.

  - **The software in the third-party repository may not be updated at the same time as the core distribution repositories**: This means that sometimes, there will be conflicts in package versions required by the software and delivered by the distribution. Moreover, the probability of conflict rises with the number of third-party repositories you add to your server.

Debian and Ubuntu

The Debian distribution and its derivative, Ubuntu, use the DEB package format. It was created solely for Debian. We won't be going into its history here, and we will only touch on technical details as required.

The command that works directly with package files is dpkg. It is used to install, remove, configure, and, importantly, build .deb packages. It can only install packages that exist on the filesystem and doesn't understand remote repositories. Let's look at some of the possible actions for dpkg:

- dpkg -i package_file.deb: Installs the package file. This command will go through several stages, after which it will install the software.

- dpkg -unpack package_file.deb: Unpacking means it puts all important files in their respective places but does not configure the package.

- dpkg -configure package: Note that this requires the package name, not the package filename. If, for some reason, packages have been unpacked but not configured, you can use the -a or -pending flag to work on them.

- dpkg -r package: This action removes the software but does not delete configuration files and eventual data it will contain. This can be useful if you plan to reinstall the software in the future.

- dpkg -p package: This action purges the package and removes everything: software, data, configuration files, and caches. Everything.

In the following example, we're installing a nano-editor from a package that was physically downloaded to the system, possibly by clicking a download button on a web page. Note that this is not a very usual way of doing things:

```
root@myhome:~# dpkg -i nano_5.4-2+deb11u1_amd64.deb
(Reading database ... 35904 files and directories currently
installed.)
Preparing to unpack nano_5.4-2+deb11u1_amd64.deb ...
Unpacking nano (5.4-2+deb11u1) over (5.4-2+deb11u1) ...
Setting up nano (5.4-2+deb11u1) ...
Processing triggers for man-db (2.9.4-2) ...
```

More often than not, you'll need to install and remove software using the apt suite of tools:

- apt-cache search NAME will search for a package that contains the given string. In the following example, I'm looking for a package that contains the vim string (vim is one of several popular command-line text editors). The output has been shortened by me for brevity:

```
root@myhome:~# apt-cache search vim
acr - autoconf like tool
alot - Text mode MUA using notmuch mail
[...]
vim - Vi IMproved - enhanced vi editor
[...]
```

- `apt-get install NAME` will install the package whose name you specify. You can install several packages in one line. In the following example, I am installing a C compiler, a C++ compiler, and a Go language suite. Note that the output also contains a list of packages that are required for my desired software to work and that they will be installed to provide that functionality:

```
root@myhome:~# apt-get install gcc g++ golang
Reading package lists... Done
Building dependency tree... Done
Reading state information... Done
gcc is already the newest version (4:10.2.1-1).
The following additional packages will be installed:
  bzip2 g++-10 golang-1.15 golang-1.15-doc golang-1.15-go golang-1.15-
src golang-doc golang-go golang-src libdpkg-perl libfile-fcntllock-
perl libgdbm-compat4 liblocale-gettext-perl libperl5.32 libstdc++-10-
dev perl perl-modules-5.32
  pkg-config
Suggested packages:
  bzip2-doc g++-multilib g++-10-multilib gcc-10-doc bzr | brz git
mercurial subversion debian-keyring gnupg patch bzr libstdc++-10-
doc perl-doc libterm-readline-gnu-perl | libterm-readline-perl-perl
libtap-harness-archive-perl dpkg-dev
The following NEW packages will be installed:
  bzip2 g++ g++-10 golang golang-1.15 golang-1.15-doc golang-1.15-go
golang-1.15-src golang-doc golang-go golang-src libdpkg-perl libfile-
fcntllock-perl libgdbm-compat4 liblocale-gettext-perl libperl5.32
libstdc++-10-dev perl
  perl-modules-5.32 pkg-config
0 upgraded, 20 newly installed, 0 to remove and 13 not upgraded.
Need to get 83.9 MB of archives.
After this operation, 460 MB of additional disk space will be used.
Do you want to continue? [Y/n]
```

The installer stops here and waits for our input. The default action is to accept all the additional packages and follow the installation. By typing in n or N and pressing *Enter*, we can stop the process. The -y switch for the install action will skip that question and automatically proceed to the next step:

- `apt-get update` will refresh the package database with new available packages and new versions.

- `apt-get upgrade` will upgrade all installed packages to the newest versions listed in the database.

- `apt-get remove NAME` will remove the package of the given name. In the following example, we are uninstalling the C++ compiler:

```
root@myhome:~# apt-get remove g++
Reading package lists... Done
Building dependency tree... Done
Reading state information... Done
The following packages were automatically installed and are no longer
required:
  g++-10 libstdc++-10-dev
Use 'apt autoremove' to remove them.
The following packages will be REMOVED:
  g++
0 upgraded, 0 newly installed, 1 to remove and 13 not upgraded.
After this operation, 15.4 kB disk space will be freed.
Do you want to continue? [Y/n]
(Reading database ... 50861 files and directories currently
installed.)
Removing g++ (4:10.2.1-1) ...
```

CentOS, RHEL, and Fedora

The other popular group of distributions uses the RPM package format. The basic tool to interact with packages is `rpm`. The main distribution that uses this format is RHEL, which is made by the Red Hat company. The packages always have the `.rpm` file extension.

They use the `dnf` command to manage packages. There is also the `yum` command, which is the original package manager for the RHEL distribution (and, by extension, the Fedora and CentOS distributions), but it has been removed. `dnf` is a next-generation rewrite of `yum`, with many improvements underneath to make it more robust and modern:

- `dnf install package_name` will install a package of a given name, along with its dependencies.

- `dnf remove package_name` will remove the package.

- `dnf update` will update all packages to the latest versions in the package database. You can specify a package name, after which `yum` will update that package.

- `dnf search NAME` will search for package names containing the `NAME` string.

- `dnf check-update` will refresh the package database.

With that, let's take a look at another Linux distribution that is widely used, especially as a base for Docker images – Alpine Linux.

Alpine Linux

Alpine Linux is loved especially by engineers working mainly with Docker and Kubernetes. As the main web page of the distribution claims, **Small. Simple. Secure.** There are no bells and whistles you can find in Debian-based or Red Hat-based distributions, but the output Docker image is really small and due to their focus on security, you can presume there are no major security vulnerabilities if you've updated all packages to the recent version.

Alpine Linux's main downside (and an upside, depending on your viewpoint) is that it's compiled with the use of the `musl` libraries instead of the widely spread `libc` libraries, although it does use the **GNU C Compiler (GCC)**, just like most other Linux distributions. Most of the time, it's not going to bother you, except when you're installing applications written in Python. That's because some Python packages use extensions written in C, and naturally, pre-built packages are compiled with `libc`, so you will need to carry out extra steps to ensure you have your compile-time dependencies installed before installing any Python dependencies.

The command to interact with packages is **Alpine Package Keeper** (**apk**). Alpine has two types of releases: stable (with a proper version; for example `3.16`) and edge, which is a rolling release (it always has the latest available version of packages).

Additionally, there are three repositories you can use to install packages: `main`, `community`, and `testing`.

You will find officially supported packages in the main repository; all tested packages are placed inside the community repository, and testing is used, well, for testing, which means that there can be some broken or outdated packages or some with security flaws.

Searching for packages

Before searching for or installing any package, it's advisable to download the latest package cache. You can do this by invoking the `apk update` command:

```
root@myhome:~# apk update
fetch https://dl-cdn.alpinelinux.org/alpine/v3.16/main/x86_64/
APKINDEX.tar.gz
fetch https://dl-cdn.alpinelinux.org/alpine/v3.16/community/x86_64/
APKINDEX.tar.gz
v3.16.2-376-g3ff8974e73 [https://dl-cdn.alpinelinux.org/alpine/v3.16/
main]
v3.16.2-379-g3c25b38306 [https://dl-cdn.alpinelinux.org/alpine/v3.16/
community]
OK: 17037 distinct packages available
```

If you're building a Docker image for later use, it's a good idea to remove this cache on the final step of the build process using the `apk cache clean` command.

Sometimes, we don't know the exact name of the package while we're working on creating a new Docker image. The easiest way to find whatever you're searching for in this case is just using the web interface: https://pkgs.alpinelinux.org/packages.

Using the CLI, you will be able to search for partial library names and binary names, though you can specify that what you're searching for is a library using the so: prefix. Other useful prefixes are cmd: for commands and pc: for pkg-config files:

```
root@myhome:~# apk search libproc
libproc-3.3.17-r1
libksysguard-5.24.5-r0
process-cpp-3.0.1-r3
samba-dc-libs-4.15.7-r0
procps-dev-3.3.17-r1
root@myhome:~# apk search vim
neovim-doc-0.7.0-r0
gvim-8.2.5000-r0
vim-tutor-8.2.5000-r0
faenza-icon-theme-vim-1.3.1-r6
notmuch-vim-0.36-r0
kmymoney-5.1.2-r3
faenza-icon-theme-gvim-1.3.1-r6
meson-vim-0.62.1-r0
runvimtests-1.30-r1
graphviz-3.0.0-r0
neovim-0.7.0-r0
py3-pynvim-0.4.3-r3
nftables-vim-0_git20200629-r1
vim-doc-8.2.5000-r0
vim-editorconfig-0.8.0-r0
apparmor-vim-3.0.4-r0
geany-plugins-vimode-1.38-r1
vimdiff-8.2.5000-r0
vimb-3.6.0-r0
neovim-lang-0.7.0-r0
u-boot-tools-2022.04-r1
fzf-neovim-0.30.0-r7
nginx-vim-1.22.1-r0
msmtp-vim-1.8.20-r0
protobuf-vim-3.18.1-r3
vimb-doc-3.6.0-r0
icinga2-vim-2.13.3-r1
fzf-vim-0.30.0-r7
vim-sleuth-1.2-r0
```

```
gst-plugins-base-1.20.3-r0
mercurial-vim-6.1.1-r0
skim-vim-plugin-0.9.4-r5
root@myhome:~# apk search -e vim
gvim-8.2.5000-r0
root@myhome:~# apk search -e so:libproc*
libproc-3.3.17-r1
libksysguard-5.24.5-r0
process-cpp-3.0.1-r3
samba-dc-libs-4.15.7-r0
```

Installing, upgrading, and uninstalling packages

You can perform basic operations on packages by using the add (installing), del (uninstalling), and upgrade commands. While installing, you can also use special prefixes that are available during search operations, but it's recommended to use the exact name of the package instead. Please also note that when adding a new package to the system, apk will choose the latest version of the package:

```
root@myhome:~# apk search -e postgresql14
postgresql14-14.5-r0
root@myhome:~# apk add postgresql14
(1/17) Installing postgresql-common (1.1-r2)
Executing postgresql-common-1.1-r2.pre-install
(2/17) Installing libpq (14.5-r0)
(3/17) Installing ncurses-terminfo-base (6.3_p20220521-r0)
(4/17) Installing ncurses-libs (6.3_p20220521-r0)
(5/17) Installing readline (8.1.2-r0)
(6/17) Installing postgresql14-client (14.5-r0)
(7/17) Installing tzdata (2022c-r0)
(8/17) Installing icu-data-en (71.1-r2)
Executing icu-data-en-71.1-r2.post-install
*
* If you need ICU with non-English locales and legacy charset support,
install
* package icu-data-full.
*
(9/17) Installing libgcc (11.2.1_git20220219-r2)
(10/17) Installing libstdc++ (11.2.1_git20220219-r2)
(11/17) Installing icu-libs (71.1-r2)
(12/17) Installing gdbm (1.23-r0)
(13/17) Installing libsasl (2.1.28-r0)
(14/17) Installing libldap (2.6.3-r1)
(15/17) Installing xz-libs (5.2.5-r1)
(16/17) Installing libxml2 (2.9.14-r2)
```

```
(17/17) Installing postgresql14 (14.5-r0)
Executing postgresql14-14.5-r0.post-install
*
* If you want to use JIT in PostgreSQL, install postgresql14-jit or
* postgresql-jit (if you didn't install specific major version of
postgresql).
*
Executing busybox-1.35.0-r17.trigger
Executing postgresql-common-1.1-r2.trigger
* Setting postgresql14 as the default version
OK: 38 MiB in 31 packages
```

You can also choose to install a specific version of a package instead of the latest version. Unfortunately, it's impossible to install older versions of the package from the same repository because when a new version is deployed, the old version is removed. You can, however, install older versions of packages from other repositories:

```
root@myhome:~# apk add bash=5.1.16-r0 --repository=http://dl-cdn.
alpinelinux.org/alpine/v3.15/main
 (1/4) Installing ncurses-terminfo-base (6.3_p20220521-r0)
 (2/4) Installing ncurses-libs (6.3_p20220521-r0)
 (3/4) Installing readline (8.1.2-r0)
 (4/4) Installing bash (5.1.16-r0)
Executing bash-5.1.16-r0.post-install
Executing busybox-1.35.0-r17.trigger
OK: 8 MiB in 18 packages
```

You can also install a custom package you've prepared beforehand using the following command:

```
root@myhome:~# apk add --allow-untrusted your-package.apk
```

To upgrade all available packages in your system, you can simply invoke the apk upgrade command. However, if you want to upgrade only a specific package, you will need to add its name after the upgrade option. Remember to refresh the package cache beforehand:

```
root@myhome:~# apk update
fetch https://dl-cdn.alpinelinux.org/alpine/v3.16/main/x86_64/
APKINDEX.tar.gz
fetch https://dl-cdn.alpinelinux.org/alpine/v3.16/community/x86_64/
APKINDEX.tar.gz
v3.16.2-376-g3ff8974e73 [https://dl-cdn.alpinelinux.org/alpine/v3.16/
main]
v3.16.2-383-gcca4d0a396 [https://dl-cdn.alpinelinux.org/alpine/v3.16/
community]
OK: 17037 distinct packages available
root@myhome:~# apk upgrade
```

```
(1/2) Upgrading alpine-baselayout-data (3.2.0-r22 -> 3.2.0-r23)
(2/2) Upgrading alpine-baselayout (3.2.0-r22 -> 3.2.0-r23)
Executing alpine-baselayout-3.2.0-r23.pre-upgrade
Executing alpine-baselayout-3.2.0-r23.post-upgrade
Executing busybox-1.35.0-r17.trigger
OK: 8 MiB in 18 packages
```

You can find all other possible operations by invoking apk without any other options. One of the most useful operations is the apk info command. It will print out information about a package or repository (the following output has been abbreviated):

```
root@myhome:~# apk info --all bash
bash-5.1.16-r2 description:
The GNU Bourne Again shell

bash-5.1.16-r2 webpage:
https://www.gnu.org/software/bash/bash.html

bash-5.1.16-r2 installed size:
1308 KiB

bash-5.1.16-r2 depends on:
/bin/sh
so:libc.musl-x86_64.so.1
so:libreadline.so.8

bash-5.1.16-r2 provides:
cmd:bash=5.1.16-r2.
```

In this section, we introduced package managers – the standard way to manage software in Linux distributions. In the next section, we are going to cover managing user accounts.

Managing users

The user in a Linux system is defined by a set of three files:

- /etc/passwd: This file contains information about the user – that is, the user's name, **unique numerical ID (UID)** in the system, the primary group the user belongs to GID, the path to the home directory, and the shell that is loaded when the user logs in. A typical entry looks like this:

    ```
    admin:x:1000:1000:Debian:/home/admin:/bin/bash
    ```

Each line describes exactly one user. Fields are separated by a colon. The second field will only contain anything other than x in very exotic cases. Here, x means that the password is stored separately in the /etc/shadow file. The reason is that permissions for the /etc/passwd file have to be a bit more relaxed so that the login process can work. /etc/shadow can only be read by root and root group and written to only by root:

```
root@myhome:~# ls -ahl /etc/passwd
-rw-r--r-- 1 root root 1.6K Aug 22 18:38 /etc/passwd
root@myhome:~# ls -ahl /etc/shadow
-rw-r----- 1 root shadow 839 Aug 22 18:38 /etc/shadow
```

- /etc/shadow -: This file contains an encrypted password. As mentioned in the preceding bullet, for security reasons, this file can only be read and written to by the root user.

- /etc/group -: This file contains information about user groups that the user belongs to. Groups are exactly that: accounts that have been grouped together so that their permissions can be managed.

You should never have a reason to modify these files by hand and that goes especially for the /etc/shadow file. The only way to properly change its contents is by using the passwd command. We encourage you to read the man page for more information.

Three commands take part in user modification and there are three for managing groups: useradd, userdel, usermod, groupadd, groupdel, and groupmod.

Adding users

useradd adds a user account to the system. Various switches modify the behavior of this command. One of the most common versions of calling the useradd command will add a user, create its home directory, and specify the default shell:

```
root@myhome:~# useradd -md /home/auser -s /bin/bash auser
```

-m tells the command to create a home directory, -d (here, it's been passed with one minus sign alongside the m) tells the command what the home directory should be (notice the absolute path), and -s specifies the default shell. More parameters can be specified and, again, we encourage you to read the man page for more details.

Modifying users

usermod modifies existing user accounts. It can be used to change the group membership of the user, home directory, lock the account, and more. One interesting option here is the -p flag, which lets you non-interactively apply a new password. This is useful in automation, when we may want to update a user password from a script or a tool, not from the command line. However, there is a security risk associated with this: during the command's execution, anyone in the system can list running processes with their parameters and see the password entry. This does not automatically compromise the password as it must be provided as encrypted through the use of the crypt (3) function. However, if an attacker has the encrypted version of the password, they can run a password-cracking program against it and, finally, brute force its clear text version:

```
root@myhome:~# usermod -a -G admins auser
```

The preceding command will add the auser user to a group called admins. The -a option means that the user will be added to the supplementary group (it won't be removed from other groups it's a member of).

Removing users

The userdel command is used to remove a user from a system. It can only remove user entries from system files and leave the home directory intact or delete users with home directories:

```
root@myhome:~# userdel -r auser
```

The preceding command will remove the user, along with its home directory and all files. Note that if the user is still logged in, it will not be removed.

Managing groups

Similarly to managing users, you can add, remove, and modify groups within your Linux system. There are equivalent commands to accomplish this:

- groupadd: Creates a group in the system. Groups can later be used to group users together and specify their execution, directory, or file access rights.
- groupdel: Removes the group from the system.
- groupmod: Changes the group definition.

You can also check which users are currently logged into the system by using the who command:

```
admin@myhome:~$ who
ubuntu     tty1          2023-09-21 11:58
ubuntu     pts/0         2023-09-22 07:54 (1.2.3.4)
ubuntu     pts/1         2023-09-22 09:25 (5.6.7.8)
trochej    pts/2         2023-09-22 11:21 (10.11.12.13)
```

You can also find the UID and GID of the user you're currently logged into using the `id` command:

```
admin@myhome:~$ id
uid=1000(ubuntu) gid=1000(ubuntu)
groups=1000(ubuntu),4(adm),20(dialout), 24(cdrom),25(floppy),
27(sudo),29(audio), 30(dip),44(video),46(plugdev),119(netdev),
120(lxd),123(docker)
```

By executing this command without any options, it will show your user ID and all the corresponding groups the user is in. Alternatively, you can provide a name you'd like to view the UID or GID of:

```
admin@myhome:~$ id dnsmasq
uid=116(dnsmasq) gid=65534(nogroup) groups=65534(nogroup)
```

To view the primary group's ID and UID, you can use the `-u` and `-g` options, respectively:

```
admin@myhome:~$ id -u
1000
admin@myhome:~$ id -g
1000
```

In this section, we introduced commands that are used to manage user accounts and groups in Linux systems. The next section will explain securely connecting to remote systems using SSH.

Secure Shell (SSH) protocol

In the DevOps world, almost nothing runs locally on your laptop or PC. There is one golden standard among ways to reach remote systems and it's the SSH protocol. SSH was developed in 1995 as a secure, encrypted remote shell access tool that would replace plaintext utilities such as **telnet** or **rsh**. The main reason for this is that in distributed networks, it is too easy to eavesdrop on communication and anything that is being transmitted in open text can easily be intercepted. This includes important data such as login details.

The most commonly used SSH server (and the client) in the Linux world is **OpenSSH** (https://www.openssh.com/). Other open source servers that are still maintained at the time of writing are **lsh** (http://www.lysator.liu.se/~nisse/lsh/), **wolfSSH** (https://www.wolfssl.com/products/wolfssh/), and **Dropbear** (https://matt.ucc.asn.au/dropbear/dropbear.html).

SSH is mainly used to log into a remote machine to execute commands. But it's also capable of transferring files (**Secure File Transfer Protocol** (**SFTP**)), forwarding ports, and creating **X Window System** (**X11**), a graphical user interface) connections. It's also capable of acting as a **socket secure** (**SOCKS**) proxy. Typically, an SSH server listens for connections on TCP port 22.

Configuring OpenSSH

After installing the OpenSSH server, your distribution will place a basic configuration inside the `/etc/ssh/sshd_config` file. The most basic configuration looks like this:

```
AuthorizedKeysFile .ssh/authorized_keys
AllowTcpForwarding no
GatewayPorts no
X11Forwarding no
Subsystem sftp /usr/lib/ssh/sftp-server
```

Let's investigate each of them before moving option:

- `AuthorizedKeysFIle` tells our server where to look inside the user directory for the file where you will store all public keys that can be used to connect to this machine as a specified user. So, if you put your public key inside the `AlphaOne` home directory, `/home/AlphaOne/.ssh/authorized_keys`, you will be able to connect as this user using a counterpart private key (more about keys will be covered in the *Creating and managing SSH keys* subsection).

- `AllowTCPForwarding` will enable or disable the ability for all users to forward TCP ports. **Port forwarding** is used to access remote machines that are not available directly on the internet, but you have access to another machine that can connect. This means you're using an SSH box as a so-called jump host to connect to a private network, similar to using a VPN.

- `GatewayPorts` is another option that's connected directly to the port forwarding feature. By allowing `GatewayPorts`, you can expose forwarded ports not only to your machine but also to other hosts within the network you are connected to. Setting this option to `yes` is not recommended for security reasons; you can accidentally expose a private network to a network you happen to be connected to, for example, in a coffee shop.

- `X11Forwarding` has a very specific use case. Usually, you don't want to have a full-fledged GUI on your servers, but if you have that need, by enabling this option, you will be able to log into the remote machine and start a remote graphical application, which will appear to run on your local host.

- `Subsystem` enables you to extend the OpenSSH server with additional features, such as SFTP in this case.

A very important option that is not specified in the preceding command block is `PermitRootLogin`, which, by default, is set to `prohibit-password`. This means that you will be able to log in as a root user, but only if you need to authenticate using a public and private pair of keys. We recommend setting this option to `no` and allowing access to the root user only via the `sudo` command.

That's all. You can, of course, add more advanced configurations, such as using **Pluggable Authentication Modules** (**PAMs**), only allowing logging into specific users or groups, or closing users in a specific directory after logging in (the **ChrootDirectory** option). You can find all available options for the OpenSSH server by invoking the man sshd_config command.

In the same way, you can find out how to configure your SSH client – that is, by running man ssh_config.

Some very useful options for the client are shown here:

```
# Show keys ascii graphics
VisualHostKey yes

# Offer only one Identity at a time
Host *
  ForwardAgent yes
  IdentitiesOnly yes
  IdentityFile ~/.ssh/mydefaultkey

# Automatically add all used keys to agent
AddKeysToAgent yes
```

VisualHostKey, when set to yes, will show ASCII graphics for the public key of the server. Every server you will be connected to will have a unique key, so it's going to be unique graphics. It's useful because, as humans, we're very good at spotting patterns, so if you are connecting to the 1.2.35.2 server but you intend to get into a different system, chances are that you will figure out something is not right just by seeing different graphics than what you expected.

Here's an example:

```
root@myhome:~# ssh user@hosts
Host key fingerprint is
SHA256:EppY0d4YBvXQhCk0f98n1tyM7fBoyRMQl2o3ife1pY
+--[ED25519 256]--+
|      .oB++o      |
|      +o*o ..     |
|      +..o*.+ .   |
|      .o +.= . +  |
|      ..o.S.= = . |
|       .oOooE. .  |
|         .o.o     |
|                  |
|                  |
+----[SHA256]-----+
```

The host option allows you to set specific options for one or multiple servers. In this example, we're enabling forwarding SSH agent and disabling password logins for all servers. Also, we're setting a default private key we will be using to connect to any server. This leads us to SSH keys and ciphers.

The final option, `AddKeysToAgent`, means that whenever you use (and unlock) a key, it will also be added to the SSH agent for future use. That way, you won't need to specify a key to use while connecting and you won't have to unlock the key on every connection attempt.

Creating and managing SSH keys

SSH consists of three components: the transport layer (**Transmission Control Protocol (TCP)/ Internet Protocol (IP)**), the user authentication layer, and a connection layer, which can effectively be multiple connections transferring data independently.

In terms of the different authentication options, you have a basic form of password authentication, which has proved to be not enough. There's also public-key authentication, which we'll talk about here. The two remaining ones are `keyboard-interactive` and **Generic Security Service Application Programming Interface (GSSAPI)**.

Public-key authentication requires us to generate a key that will have two counterparts: private and public. You will put the public key on the server inside `authorized_keys`; the private key will be used to authenticate.

At the time of writing this book, RSA keys are standard for use with SSH. It is secure, but it's recommended to use bigger keys that are 4,096 bits long, but 3,072 bits (the default) are considered sufficient. Bigger keys mean slower connections.

Currently, a better choice is using the `ed25519` type of key, which has a fixed length.

Additionally, all keys can be secured with a password.

The following code shows how to generate both key types:

```
root@myhome:~# ssh-keygen -b 4096 -o -a 500 -t rsa
Generating public/private rsa key pair.
Enter file in which to save the key (/root/.ssh/id_rsa):
Created directory '/root/.ssh'.
Enter passphrase (empty for no passphrase):
Enter same passphrase again:
Your identification has been saved in /root/.ssh/id_rsa
Your public key has been saved in /root/.ssh/id_rsa.pub
The key fingerprint is:
SHA256:eMu9AMUjpbQQ7DJI14MB2UnpfiUqwZJi+/e9dsKWXZg root@614bbd02e559
The key's randomart image is:
+---[RSA 4096]----+
|o=+++.. .        |
```

```
|.o++.o =          |
|O+... + +         |
|*o+ o .+ .        |
|+o.+ oo S    o    |
|..o .  + o E .    |
|  ...    = + .    |
|   . .  .B +      |
|     . ..oo=      |
+----[SHA256]-----+
```

```
root@myhome:~# ssh-keygen -o -a 100 -t ed25519
Generating public/private ed25519 key pair.
Enter file in which to save the key (/root/.ssh/id_ed25519):
Enter passphrase (empty for no passphrase):
Enter same passphrase again:
Your identification has been saved in /root/.ssh/id_ed25519
Your public key has been saved in /root/.ssh/id_ed25519.pub
The key fingerprint is:
SHA256:EOCjeyRRIz+hZy6Pbtcnjan2lucJ2Mdih/rFc/ZnthI
root@614bbd02e559
The key's randomart image is:
+--[ED25519 256]--+
|  . =..          |
|   * o .         |
|  o B .          |
|   * o .         |
|   + o   S       |
|    B o +    E   |
|   o +.*=B o  .  |
|  ...ooB*+= .. + |
|  ..oo=o=o   .=..|
+----[SHA256]-----+
```

Now, to put this newly created key on a server, you will need to either copy it manually to the authorized_keys file or use the ssh-copy-id command, which will do this for you if you already have other means of access, such as password authentication:

```
root@myhome:~# ssh-copy-id -i ~/.ssh/id_ed25519 user@remote-host
```

You can only perform the next login to this server by using a key.

At this point, you should have a good understanding of how SSH works and how to use its most used features. You now know how to create your keys and where to save them on the remote system.

Summary

This chapter concludes our introduction to the basic Linux operations you will need in your daily work. It has by no means been comprehensive enough to explain everything that you need to know to manage your Linux system, but it is enough to get you started and should be enough to help you manage your system. In the next chapter, we are going to cover writing shell scripts from the very beginning and guide you through basic and more advanced topics.

Exercises

Try out the following exercises to test your knowledge of this chapter:

1. In Debian/Ubuntu, install the `vim` package.
2. Create a cron job that will create a file called `/tmp/cronfile` every Saturday at 10:00 A.M.
3. Create a group called `admins` and add an existing user to it.

4

Automating with Shell Scripts

In this chapter, we're going to demonstrate system administration task automation with shell scripts. We are going to illustrate several ways of handling scripting using Bash shell. The plan is to create a script that will automate the creation of the database dump. This task, while easy, will demonstrate how things can go sideways and how to handle those situations.

In this chapter, we will cover the following topics:

- Backing up a database
- Understanding scripting
- Understanding Bash built-ins and grammar
- Understanding the backup script – first steps
- Handling errors and debugging

Technical requirements

For this chapter, you will require a Linux system where you can install packages and are not afraid to break things in the process. To that end, a virtual machine would be most preferred, typically running on an old computer that you can reinstall from scratch. We do not expect to break anything, but during learning, this may happen.

Backing up a database

In terms of most common databases, such as MySQL and PostgreSQL, there are at least two different ways to back up a database:

- Take a database dump by extracting all current data, along with the database schema
- Copy replication logs

In cloud environments, you can also take a snapshot of the disk database where the backup is being saved.

A database dump can also be used as a full backup. Replication logs aren't self-sufficient database dumps, so you will need to combine them with a full backup. This is called an incremental backup.

Doing a full backup can take a long time, especially for big databases. While it's running, the database puts a lock on its data files, so it doesn't save new data on the disk; instead, it stores everything in the replication logs until the database lock is released. For large databases, this operation can take hours. Because of that, we will be creating a full backup once a week and copying all replication logs every hour. Additionally, we will be creating a daily disk snapshot of our AWS database instance.

With this knowledge, we can start creating the most basic version of our script.

Understanding scripting

A shell script is a simple text file filled with commands. Unlike compiled programs, shell scripts are not evaluated before execution, but rather while they are being executed. This makes for a very quick development process – there's no compilation at all. But at the same time, the execution is a bit slower. Also, the errors that the compiler would have caught surface during execution and can often lead to script exiting.

On the upside, there's not much to learn when you are writing a script – much less than when you are writing a program in C or Python. Interacting with system commands is as simple as just typing their names.

Bash lacks a lot of sophistication in programming languages: there are not many data types and structures, there's very rudimentary control of scope, and the memory implementation is not meant to be efficient with scale.

There's not one good rule of thumb for choosing when to write a script and when to develop a program. However, there are some points to consider. A good candidate for a shell script is as follows:

- It's not very long. We can't give you a rule, but when you start to go into hundreds of lines, it may be good to consider using Python or splitting it into several scripts.

- It interacts with system commands, sometimes a lot. You can consider it an automated way of running those commands.

- It doesn't do a lot of data handling. There are few arrays, strings, and numbers. That's all.

- It doesn't do any system calls. Shell scripts are not there for system calls. There's no straightforward way to do it.

The basic structure of a shell script is as follows:

```
#!/bin/bash
echo "Hello world"
```

The first line starts with the so-called she-bang. It's being used to tell the system which interpreter (in this case, it's Bash) to use to run this script. In a lot of scripts found online, she-bang looks like this:

```
#!/usr/bin/env bash
```

Using the `env` command has a big advantage and disadvantage at the same time. The advantage of using it is that it will use a Bash executable that is first in the current user PATH environment variable. It all depends on the purpose. The `env` command also won't let you pass any arguments to the interpreter you choose.

The second line of the preceding script simply displays **Hello world**. It uses a built-in command echo that does just that – it displays whatever text you put as input.

Now, to execute this script, we will need to save it to a file. It's good practice to end this file with a `.sh` or `.bash` suffix. There are two ways to execute this new script – by invoking the interpreter with a script name or by directly executing the script by its name:

```
admin@myhome:~$ bash ./myscript.sh
Hello world
admin@myhome:~$
```

To directly execute the script, we will need to change its permissions so that it can be executed; otherwise, the system won't recognize it as something it can execute:

```
admin@myhome:~$ chmod +x myscript.sh
admin@myhome:~$ ./myscript.sh
Hello world
admin@myhome:~$
```

Similarly, we could easily set Python or any other shell as an interpreter of our script.

For now, let's focus on Bash and look at some built-in Bash commands that we're going to use.

Understanding Bash built-ins and grammar

Let's get back to the fundamentals before we start creating a script. First, we will look into the Bash scripting language syntax and its limitations.

Built-in commands are commands that are integral to Bash and are the main scripting syntax we are going to use. Bash will try to execute any other commands from the system it runs on.

Just like any other interpreted language (for example, Python or Ruby), Bash has unique syntax and grammar. Let's look into it.

Bash, similar to other programming languages, interprets files from top to bottom and from left to right. Each line usually contains one or more commands. You can glue several commands together in one line using a pipe (|) or double pipe character (| |), semicolon (;), or double ampersands (&&). It's useful to remember that double pipes have the same function as logical OR and double ampersands have the same function as logical AND. This way, you can run commands in sequence and execute the next command depending on the outcome of the previous command, without using more complex conditional statements. This is called a list or chain of commands:

```
commandA && commandB
```

In the preceding command, you can see an example of using double ampersands. Here, commandB will only be executed if commandA is successfully executed. We can chain more commands like this by adding | | at the end of the chain.

```
commandA || commandB
```

This example, on the other hand, shows how to use a double pipe. Here, commandB will only be executed if commandA fails to execute.

Bash (or any other shell in Linux) determines whether a command failed to execute or exited with success by using return codes. Every command needs to exit with a positive number – zero (0) is a code for success, and any other code is a failure. If you chain multiple commands with AND (&&) or OR (| |), the return status of the whole line will be determined by the previous command that was executed.

What about a single ampersand (&) after a command? It has an entirely different function – it will execute a command it is after in the background and the script will continue running without waiting for the command to complete. It's useful for tasks that aren't required to be completed for other parts of the program, or running multiple instances of the same command at the same time, such as running a full backup of multiple databases to shorten execution time.

Now that we know how to chain commands, we can jump into another core feature of any programming language – variables.

Variables

In Bash, there are two types of variables: global and local variables. *Global variables* are accessible throughout the whole time the script is running unless the variable is unset. *Local variables* are accessible only within a block of the script (for example, a defined function).

Whenever a script gets executed, it gets an existing set of variables from the currently running shell; this is called an *environment*. You can add new variables to the environment using the export command, and remove variables by using the use unset command. You can also add functions to the environment by using the declare -x command.

All parameters, regardless of whether they're local or global, are prefixed with a dollar sign ($). So, if you have a variable named CARS (case-sensitive), to refer to it, you will need to write $CARS inside your script.

For variables, single or double quotes (or no quotes) matter. If you put a variable in a single quote, it won't be expanded, and a variable name inside quotes will be treated as a literal string. Variables inside double quotes will be expanded, and it's considered a safe way to refer to variables (and concatenate them, or glue them together) because if there is a space or other special character in a string, it won't be significant for the script – that is, it won't be executed.

Sometimes, you will need to concatenate multiple variables. You can do this using curly brackets ({ }). For example, "${VAR1}${VAR2}" will expand to whatever you have VAR1 and VAR2 set to. In that context, you can also use curly brackets to cut or replace parts of the string. Here's an example:

```
name="John"
welcome_str="Hello ${name}"
echo "${welcome_str%John}Jack"
```

The preceding code will display **Hello Jack** instead of **Hello John**. The % operator will only remove characters from the end of the string. If you want to cut variables from the beginning of the string, you would use the # operator in the same way.

If you refer to a variable without any quotes, space inside the value of the variable may break the flow of the script and impede debugging, so we strongly suggest using either single or double quotes.

Parameters

There are two kinds of parameters we could use, but both are a special kind of variable, so each is prefixed with a dollar sign ($).

The first kind of parameters you need to be aware of are *positional parameters* – these are the parameters that are passed to your script or a function inside a script. All parameters are indexed starting with 1 until n, where n is the last parameter. You'll refer to each of these parameters with $1 until $n. You're probably wondering what will happen if you use $0.

$0 contains the name of your script, so it's useful for generating documentation within a script, for example.

To refer to all available parameters starting from 1, you can use $@ (dollar sign, at).

Here are some other special parameters that are commonly used:

- #: Number of positional parameters
- ?: Exit code of the most recently executed foreground command
- $: Process ID of the shell

Loops

You are probably familiar with different types of loops as they are in other programming languages. You can find all of those in Bash, but with a slightly different syntax than you might be familiar with.

The most basic, a `for loop`, looks like this:

```
for variable_name in other_variable; do some_commands; done
```

This variant of the `for` loop sets whatever element inside `other_variable` as a value of `variable_name` and executes `some_commands` for every element it finds. After it's completed, it exits the loop with the status of the last executed command. The `in other_variable` part can be omitted – in that case, the `for` loop executes `some_commands` once for each positional parameter. The use of this parameter will look like this:

```
for variable_name; do some_commands; done
```

The preceding `for` loop will run as many times as you added input variables for your function (or script in this case). You can refer to all positional parameters with $@.

The following is a for loop in a form known from C/C++:

```
for (( var1 ; var2 ; var3 )); do some_commands; done
```

Here's an example use of this syntax:

```
for ((i=1; i<=5; i++)); do echo $i; done
```

The first expression sets the i variable to 1, the second expression is a condition to be met for the loop to still run, and the final expression is incrementing the i variable by 1. Each loop run will display the next value set to the i variable.

The output of this command will look like this:

```
1

2

3

4

5
```

Another useful loop is a `while loop`, which runs as many times as needed until the condition is met (the command we pass to it exits with a success – it returns zero). Its counterpart is a loop called `until`, which will keep running so long as the command we pass to it returns a non-zero status:

```
while some_command; do some_command; done
until some_command; do some_command; done
```

You can create an infinite loop by using a command that will always meet the condition, so for the `while` loop, it can be simply `true`.

The most commonly used blocks of commands are conditionals and are used with `if` statements. Let's take a closer look.

Conditional execution – if statement

The if statement has the following syntax:

```
if test_command
then
    executed_if_test_was_true
fi
```

`test_command` can be any command you can think of, but commonly, the test is wrapped between double or single square brackets. The difference between these two is the former is a system command called `test` (you can check its syntax by executing `man test`), while the latter is a Bash built-in and more powerful test.

The rule of thumb for putting variables between square brackets is using double quotes, so if a variable contains spaces, it won't change the intention of our test:

```
if [[ -z "$SOME_VAR" ]]; then
    echo "Variable SOME_VAR is empty"
fi
```

The `-z` test checks whether the `$SOME_VAR` variable is empty. It evaluates to `true` if the variable is empty and `false` if not.

The following are other commonly used tests:

- `-a`: Logical AND
- `-o`: Logical OR
- `-eq`: Is equal to
- `-ne`: Is not equal to
- `-gt or >`: Is greater than

- `-ge` or `>=`: Is greater than or equal to
- `-lt` or `<`: Is less than
- `-le` or `<=`: Is less than or equal to
- `=` or `==`: Is equal to
- `!=`: Is not equal to
- `-z`: String is null (its length is zero characters)
- `-n`: String is not null
- `-e`: File exists (directory, symlink, device file, or any other file in the filesystem)
- `-f`: File is a regular file (not a directory or device file)
- `-d`: File is a directory
- `-h` or `-L`: File is a symbolic link
- `-r`: File has read permission (for the user running the test)
- `-w`: File has write permission (for the user running the test)
- `-x`: File can be executed by the user that executed the script

Note that tests may behave differently when using a system test (single square brackets, `[ . . . ]`) rather than the built-in one (double square brackets, `[ [ . . . ] ]`).

The double equal comparison operator, while comparing strings when using globbing, will match the pattern or literal string, depending on whether you've quoted the pattern or not.

The following is an example of pattern matching if the string starts with w:

```
if [[ $variable == w* ]];
    echo "Starts with w"
fi
```

When using a system test (a single square bracket) instead of a built-in one, the test will try to find whether the $ variable matches any of the filenames in the local directory (including those with spaces). This can lead to some unpredictable outcomes.

The following is an example of pattern matching if the string is w*:

```
if [[ $variable == "w*" ]];
    echo "String is literally 'w*'"
fi
```

Equipped with this knowledge, we're ready to start creating and running scripts. So, let's jump right to it!

Understanding the backup script – first steps

Now that we know what a script can look like, we can start writing one. You can use your favorite console editor or IDE to do this. Let's create an empty file named run_backups.sh and change its permissions so that they're executable:

```
admin@myhome:~$ touch run_backups.sh && chmod +x run_backups.sh
admin@myhome:~$ ls -l run_backups.sh
-rwxr-xr-x  1 admin  admin  0 Dec  1 15:56 run_backups.sh
```

It's an empty file, so we'll need to add a basic database backup command and proceed from there. We won't be covering granting this script access to a database. We will be backing up a PostgreSQL database and using the pg_dump tool for that purpose.

Let's input a shebang line and a pg_dump command call in our base script:

```
#!/usr/bin/env bash


pg_dump mydatabase > mydatabase.sql
```

To execute this script, we'll need to start the following command:

```
admin@myhome:~$ ./run_backups.sh
```

The dot and slash indicate that we want to execute something that is located in a current directory, and its name is run_backups.sh. Without the initial dot-slash pair, the shell we're running on (here, bash) would look into the PATH environment variable and would try to find our script in one of the directories listed there:

```
admin@myhome:~$ echo $PATH
/usr/local/sbin:/usr/local/bin:/usr/sbin:/usr/bin:/sbin:/bin
```

As you can see, it's a list of directories delimited with a colon.

Now, let's see what our Bash script does when executed:

```
admin@myhome:~$ ./run_backups.sh
./run_backups.sh: line 3: pg_dump: command not found
```

Unless you already have pg_dump installed on your system, you will see this error. It means that Bash didn't find the command we intended to run. It also displays the line where the error occurred. Also, an empty mydatabase.sql file was created.

Normally, we would proceed with creating a Docker image with all the tools we need, and a second one with a PostgreSQL database running. But since this will be covered in *Chapter 8*, let's just proceed and install everything we need on a local machine. Assuming you're on an Ubuntu or a Debian Linux machine, you want to run the following commands:

```
admin@myhome:~$ sudo apt-get update
Get:1 http://archive.ubuntu.com/ubuntu jammy InRelease [270 kB]
[Condensed for brevity]
Get:18 http://archive.ubuntu.com/ubuntu jammy-backports/main amd64
Packages [3520 B]
Fetched 24.9 MB in 6s (4016 kB/s)
Reading package lists... Done
admin@myhome:~$ sudo apt-get install postgresql
Reading package lists... Done
Building dependency tree... Done
Reading state information... Done
The following additional packages will be installed:
  cron libbsd0 libcommon-sense-perl libedit2 libgdbm-compat4 libgdbm6
libicu70 libjson-perl libjson-xs-perl libldap-2.5-0 libldap-common
libllvm14 libmd0 libperl5.34 libpopt0 libpq5 libreadline8
[Condensed for brevity]
Suggested packages:
  anacron checksecurity default-mta | mail-transport-agent gdbm-
l10n libsasl2-modules-gssapi-mit | libsasl2-modules-gssapi-heimdal
libsasl2-modules-ldap libsasl2-modules-otp libsasl2-modules-sql
[Condensed for brevity]
0 upgraded, 42 newly installed, 0 to remove and 2 not upgraded.
Need to get 68.8 MB of archives.
After this operation, 274 MB of additional disk space will be used.
Do you want to continue? [Y/n] y
```

After user confirmation, the database will be installed, configured, and started in the background. We've cut the further output for readability.

After installation, you might need an additional configuration change for the database so that you can connect to the database with another tool called psql, which is a console command that's useful for connecting to PostgreSQL. In the /etc/postgresql/14/main/pg_hba.conf file, we have defined trust relationships and who can connect to the database using multiple mechanisms.

Find the following line:

```
local    all  postgres peer
```

Change it to the following:

```
local    all  all  trust
```

After making this modification, you can restart the database with the following command:

```
admin@myhome:~$ sudo systemctl restart postgresql
* Restarting PostgreSQL 14 database server
```

Now, you should be able to log in to the database and list all available databases:

```
admin@myhome:~$ psql -U postgres postgres
psql (14.5 (Ubuntu 14.5-0ubuntu0.22.04.1))
Type "help" for help.

postgres=# \l
                            List of databases
    Name    |   Owner   | Encoding | Collate  |  Ctype   |   Access
privileges
-----------+----------+----------+---------+---------+---------------
--------
  postgres  | postgres  | UTF8     | C.UTF-8  | C.UTF-8  |
  template0 | postgres  | UTF8     | C.UTF-8  | C.UTF-8  | =c/
postgres
            +
            |           |          |          |          | postgres=CTc/
postgres
  template1 | postgres  | UTF8     | C.UTF-8  | C.UTF-8  | =c/
postgres
            +
            |           |          |          |          |
postgres=CTc/postgres
(3 rows)
postgres=# \q
admin@myhome:~$
```

After logging in, we list all available databases with \l (backslash, lowercase L) and quit the psql shell with \q (backslash, lowercase Q). Once this is set, we can get back to our script and try to run it again:

```
admin@myhome:~$ ./run_backups.sh
pg_dump: error: connection to server on socket "/var/run/
postgresql/.s.PGSQL.5432" failed: FATAL:  role "root" does not exist
```

There's no root role in PostgreSQL, which is an expected error at this point. We will need to use a different role to connect to the database. The default is postgres and the option to pass to pg_dump is -U, the same as what we used with psql. After updating it, our script will look like this:

```
#!/usr/bin/env bash
pg_dump -U postgres mydatabase > mydatabase.sql
```

The final step is to create a database and some actual data so that the output `sql` file won't be empty. The following script will create a database named `mydatabase` and create two tables that contain random data:

```
CREATE DATABASE mydatabase;
\c mydatabase
CREATE TABLE random_data AS SELECT data_series, md5(random()::text)
from generate_series(1,100000) data_series;
CREATE TABLE another_random AS SELECT data_series, md5(random()::text)
from generate_series(1,100000) data_series;
```

The `CREATE DATABASE` line is creating a database named `mydatabase`. The second line indicates we're connecting to this new database. Two additional lines starting with `CREATE TABLE` are both creating tables and filling them with data using built-in PostgreSQL functions. Let's break it down into two distinct queries – `SELECT` and `CREATE`:

```
SELECT data_series, md5(random()::text) from generate_series(1,100000)
data_series;
```

There are a few things going on here:

- The `generate_series()` function is creating a series of integers starting from 1 up to 100,000 – this will generate all our records in a table
- The `data_series` keyword, just before a semicolon, names output from the `generate_series()` function, so it's an actual field name in a table we intend to create
- The `random()` function is generating a value between 0 and 1 – that is, greater than or equal to 0 and less than 1
- The `::text` after the `random()` function is converting output from this function into text
- The `md5()` function is taking the output from `random()::text` and hashing it with an `md5` algorithm, ensuring that we have a unique string and run as many times as the `generate_series()` function's output amount will be (here, this is from 1 to 100,000)
- Finally, `SELECT data_series, md5()` is producing a table with two fields (`data_series` and `md5`) with data generated by both functions

Now, getting back to `CREATE TABLE`, there's a part called `another_random AS` – this will get the output from `SELECT` and create a table for us.

With this knowledge, we can create a `sql` script and execute it using `psql`:

```
admin@myhome:~$ psql -U postgres < create_db.sql
CREATE DATABASE
You are now connected to database "mydatabase" as user "postgres".
```

To check whether we've created something and investigate the data we've created, again, we will need to use `psql` and the `SELECT` query on our new database:

```
admin@myhome:~$ psql (14.1)
Type "help" for help.

postgres=# \c mydatabase
You are now connected to database "mydatabase" as user "postgres".
mydatabase=# \dt
            List of relations
 Schema |      Name       | Type  |  Owner
--------+-----------------+-------+----------
 public | another_random  | table | postgres
 public | random_data     | table | postgres
(2 rows)

mydatabase-# select * from random_data ;
 data_series |                md5
-------------+-----------------------------------
           1 | 4c250205e8f6d5396167ec69e3436d21
           2 | a5d562ccd600b3c4c70149361d3ab307
           3 | 7d363fac3c83d35733566672c765317f
           4 | 2fd7d594e6d972698038f88d790e9a35
--More--
```

`--More-` at the end of the preceding output indicates there are more records to be shown. You can see more data by pressing the spacebar or quit by pressing *Q* on your keyboard.

Once you have created a database and filled it with some data, you can try to run our backup script again:

```
admin@myhome:~$ ./run_backup.sh
admin@myhome:~$
```

There are no errors, so we have probably created a full database dump with success:

```
admin@myhome:~$ ls -l mydatabase.sql
-rw-r--r--    1 root      root       39978060 Dec 15 10:30 mydatabase.
sql
```

The output file isn't empty; let's see what's inside:

```
admin@myhome:~$ head mydatabase.sql
--
-- PostgreSQL database dump
--
```

```
-- Dumped from database version 14.1
-- Dumped by pg_dump version 14.1

SET statement_timeout = 0;
SET lock_timeout = 0;
SET idle_in_transaction_session_timeout = 0;
```

After some SET statements, you should also find CREATE TABLE and INSERT statements. I haven't provided full output here as it would take up a lot of space.

In this section, we learned how to set up a testing environment for our script and made it possible for our script to create a database dump. In the next section, we'll focus more on error handling and checking whether the backup was successful.

Handling errors and debugging

While running our backup script, we can encounter several errors: access to the database might be blocked, the pg_dump process might get killed, we may be out of disk space, or any other error preventing us from completing a full database dump.

In any of those cases, we will need to catch the error and handle it gracefully.

Additionally, we might want to refactor the script to make it configurable, make use of functions, and debug the script. Debugging will prove very useful, especially when dealing with larger scripts.

Let's dive right into it and start with adding a function:

```
#!/usr/bin/env bash

function run_dump() {
  database_name=$1

  pg_dump -U postgres $database_name > $database_name.sql
}

run_dump mydatabase
```

We've added a run_dump function that takes one argument and sets a local variable called database_name with the content of this argument. It then uses this local variable to pass options to the pg_dump command.

This will immediately allow us to back up multiple databases by using a for loop, like so:

```
for dbname in mydatabase mydatabase2 mydatabase3; do
    run_dump $dbname
done
```

This loop will create a full dump of the databases: `mydatabase`, `mydatabase2`, and `mydatabase3`. Backups will be done one by one using this function. We can now put a list of the databases in a variable to make it more configurable. The current script will look like this:

```
#!/usr/bin/env bash

databases="mydatabase"

function run_dump() {
  database_name=$1

  pg_dump -U postgres $database_name > $database_name.sql
}

for database in $databases; do
  run_dump "$database"
done
```

Now, this backup script is becoming more complicated. We need to note a few things that will be happening now:

- If any of the backups should fail, the script will continue running
- If the backups fail due to `pg_dump` not having access to the database, we will overwrite our previous database dump
- We will be overwriting the dump file on every run

Several default settings are considered to be good practice to override in a script. One that would mitigate the first problem we've pointed out is aborting running when any command returns a value that's different than zero (or `true`). This means the command finished running with an error. The option's name is `errexit` and can we override it with the `set` command, a Bash built-in command. We could do this in two ways:

```
set -o errexit
set -e
```

Here are some other options we recommend using:

- `set -u`: This treats any unset variable we try to use in a script as an error
- `set -o pipefail`: When using chaining commands with a pipe, the exit status of this pipeline will be the status of the last command that finished with a non-zero status or zero (a success) if all commands exit successfully
- `set -C` or `set -o noclobber`: If set, Bash won't overwrite any existing file with the redirection commands (for example, `>`, which we're using in our script)

An additional option that is extremely useful is `set -x` or `set -o xtrace`, which causes Bash to print every command before executing said command.

Let's see how it works for a simple Bash script:

```
#!/usr/bin/env bash
set -x
echo "Hello!"
```

Here's the output of executing this script:

```
admin@myhome:~$ ./simple_script.sh
+ echo 'Hello!'
Hello!
```

Let's update our backup script with the recommended Bash settings:

```
#!/usr/bin/env bash
set -u
set -o pipefail
set -C

databases="mydatabase"

function run_dump() {
  database_name=$1

  pg_dump -U postgres $database_name > $database_name.sql
}

for database in $databases; do
  run_dump "$database"
done
```

Now, let's get back to the console to test whether it's still working as we expect:

```
admin@myhome:~$ ./run_backups.sh
./run_backups.sh: line 12: mydatabase.sql: cannot overwrite existing
file
```

We've enabled the `noclobber` option and it has prevented us from overwriting previously made backups. We'll need to rename or delete the old file before we can proceed. This time, let's also enable the `xtrace` option to see what command script is being executed:

```
admin@myhome:~$ rm mydatabase.sql
admin@myhome:~$ bash -x ./run_backups.sh
```

```
+ set -u
+ set -o pipefail
+ set -C
+ databases=mydatabase
+ for database in $databases
+ run_dump mydatabase
+ database_name=mydatabase
+ pg_dump -U postgres mydatabase
```

To mitigate the overwrite existing file error, we could do one of three things:

- Delete the previous file before attempting to run the backup, which would destroy the previous backup.

- Rename the previous backup file and add a current date suffix.

- Make sure that every time we run a script, the dump file has a different name, such as a current date. This would ensure we are keeping previous backups in case we need to get back to a version that's earlier than the last full backup.

In this case, the most common solution would be the final one we've proposed – generate a different backup file name every time the backup runs. First, let's try to get a timestamp with the local date and time in the YYYYMMDD_HHMM format, where we have the following options:

- YYYY: The current year in a four-digit format

- MM: The current month in a two-digit format

- DD: The day of the month in a two-digit format

- HH: The current hour

- MM: The current minute

We can achieve this by using the date command. By default, it will return the current date, day of the week, and the time zone:

```
admin@myhome:~$ date
Fri Dec 16 14:51:34 UTC 2022
```

To change the default output of this command, we'll need to pass a date format string using formatting characters.

The most common formatting characters for the date command are as follows:

- %Y: Year (for example, 2022)

- %m: Month (01-12)

- %B: Long month name (for example, January)

- %b: Short month name (for example, Jan)

- %d: Day of month (for example, 01-31, depending on how many days are in a certain month)

- %j: Day of year (for example, 001-366)

- %u: Day of week (1-7)

- %A: Full weekday name (for example, Friday)

- %a: Short weekday name (for example, Fri)

- %H: Hour (00-23)

- %I: Hour (01-12)

- %M: Minute (00-59)

- %S: Second (00-59)

- %D: Display date as mm/dd/yy

To format the date in a format we want, we'll need to use formatting characters, %Y%m%d_%H%M, and pass it to the date command for interpretation:

```
admin@myhome:~$ date +"%Y%m%d_%H%M"
20221216_1504
```

To pass the output string to a variable in our script, we'll need to run the date in a subshell (a Bash process executed by our Bash script):

```
timestamp=$(date +"%Y%m%d_%H%M")
```

Let's put it inside our script and use a timestamp variable to generate an output filename:

```
#!/usr/bin/env bash

set -u
set -o pipefail
set -C

timestamp=$(date +"%Yum'd_%H%M")
databases="mydatabase"

function run_dump() {
  database_name="$1"

  pg_dump -U postgres "$database_name" > "${database_
name}_${timestamp}".sql
```

```
  }

  for database in $databases; do
    run_dump "$database"
  done
```

If you have spotted curly braces between variables in a `pg_dump` line, you're probably wondering why we need them. We use curly braces to make sure that the variable name will be correct when we expand a variable to a string. In our case, we're preventing Bash from trying to search for a variable name, `$database_name_`, that doesn't exist.

Now, each time we run our backup script, it will try to create a new file with the current date and time of the backup start time. If we're running this script every day, the number of files will increase over time and will eventually fill our disk space. So, we'll also need our script to remove old backups – say, 14 days and older.

We can achieve this by using the `find` command. Let's find all files beginning with a database name, followed by an underscore, and ending with `.sql` that are older than 14 days:

```
admin@myhome:~$ find . -name "mydatabase_*.sql" -type f -mtime +14
./mydatabase_20221107.sql
```

The `find` command has a peculiar syntax that is a bit different than other command-line tools have, so let's describe what each option means:

- `.` (a dot): This is a directory where we want to search for our files. Dot means *a current directory*.

- `-name`: This option can take a full string or wildcards such as `*` or `?` and it's looking for filenames in this case. It's case-sensitive. If we're not certain that the file or directory we're looking for is uppercase or lowercase, we could use the `-iname` option instead.

- `-type f`: This indicates we're looking for a regular file. Other options are as follows:
  - d: Directory
  - l: Symbolic link
  - s: A socket file
  - p: FIFO file
  - b: Block device
  - c: A character device

- `-mtime +14`: The modification time of this file should be older than 14 days. This option can also take other units (seconds, `-s`, weeks, `-w`, hours, `-h`, and days, `-d` – it's the default if no unit is given).

To delete found files, we have at least two options: the -delete option or the rm command via an -exec find option. Let's see how this looks in both cases:

```
admin@myhome:~$ find . -name "mydatabase_*.sql" -type f -mtime +14
-delete
admin@myhome:~$ find . -name "mydatabase_*.sql" -type f -mtime +14
-exec rm -- {} \;
```

- In this case, a safer choice would be to use -execdir instead of -exec. The difference is subtle but important: -exec will not be executed in the same directory where the found file is located, but -execdir will, which makes it a safer option in edge cases.

Let's deconstruct what we have after the -exec option:

- rm: This is a CLI tool for removing files or directories.

- -- (double dash): This indicates that it will take arguments from stdin, or the output of the find command.

- {}: This will substitute for a filename we've found.

- \; (backslash, semicolon): This will take multiple commands to be executed by -exec. A backslash is an escape character that prevents this semicolon from being interpreted as a separator for the next command. The find utility uses ; or + to terminate the shell commands, so we could note it as " ; ", \+, or + (without quotes).

- The -delete option is good for removing files, but it always returns true, so it will fail silently if, for example, our script doesn't have permission to remove any files. It's relatively safe to use it in our script, so we'll go ahead with it.

Now, let's embed this into our script and see the final iteration of it:

```
#!/usr/bin/env bash

set -u
set -o pipefail
set -C

timestamp=$(date +"%Y%m%d_%H%M")
databases="mydatabase"

function cleanup_old_backups() {
  database_name="$1"

  find . -type f -name "${database_name}_*.sql" -mtime +14 -delete
}
```

```
function run_dump() {
  database_name="$1"

  pg_dump -U postgres "$database_name" > "${database_
name}_${timestamp}".sql
}

for database in $databases; do
  cleanup_old_backups "$database"
  run_dump "$database"
done
```

Here, we've added a function called `cleanup_old_backups` that will run before creating a new dump to free some space for fresh files. We're calling this function inside a for loop just before `run_dump`. This script can be run automatically by a `cron daemon` or `systemd cron` service; we'll cover this in more detail in *Chapter 5, Managing Services in Linux*.

In this section, we learned about the recommended options to enable in shell scripts and how to enable debugging options. We know now how to create functions and loops. Additionally, we partially touched on PostgreSQL and how to create a testing database.

The next chapter will take us deeper into various Linux services and how to manage them using `init` and `systemd`.

Summary

Shell scripting is a very common way of automating periodically running tasks in a Linux system. Sometimes, it evolves to a bigger system chained together with multiple Bash scripts and Python programs to complete complex tasks using multiple smaller tasks that do one thing at the same time in a very reliable way.

In modern systems, you will probably see as much Bash as Python scripts.

In this chapter, we learned how to create an executable script, as well as how to create a simple backup script that handles errors and generates a new filename on each run. We also added a function that removes old backups so that we can avoid filling the disk space. Additionally, as a side effect, we learned how to create a new PostgreSQL database and allow access to it from a local system.

In the next chapter, we'll learn how to create Linux services and how to manage them.

Exercises

Try out the following exercises to test your knowledge of this chapter:

1. Write a function that will list all databases and feed that list to the for loop we've created.
2. Change a date timestamp into another format of your choosing.
3. Catch any errors the find function could return (for example, it couldn't delete a file).

Part 2: Your Day-to-Day DevOps Tools

In this second part, we will learn about Linux internals. Starting with managing services and networking, we will then move on to look at the most common tools, such as Git and Docker.

This part has the following chapters:

5

Managing Services in Linux

In this chapter, we're going to explain services (programs running in the background as daemons) in more depth. We're going to explain `init` scripts and `systemd` units. We are also going to cover Alpine Linux `rc` commands that manage services.

The chapter covers the following topics:

- Understanding Linux services in detail
- A few words about Upstart, an alternative

Technical requirements

For this chapter, you will need a Linux system at hand where you can execute privileged commands, either using `sudo` or jumping straight to the root account (although we particularly recommend the first one). You'll also need a Linux text editor of your choice that will produce pure text files. If you are going to edit on a Windows system, use a text editor that enables you to save Unix files. We recommend editing in the command line with one of your favorite command-line text editors: `vim`, `emacs`, `joe`, `nano`, or whatever suits you.

Understanding Linux services in detail

Unless you are running some kind of low-level embedded device on your desktop—and we strongly doubt it—your operating system manages a multitude of tasks to create a comfortable and productive environment for you. Be it Mac OS X, Linux, Windows, or FreeBSD, they all run a multitude of background programs that together provide a useful system. The same goes for server flavors of those operating systems. A background program or background process (in Unix and Linux called, fluffily, a daemon) means a program that is not attached to any input (keyboard, mouse, etc.) or output (monitor, terminal, etc.). This way, they can start working even when no one is logged in to the system and keep working when the user logs out. They can also run under the privileges of a user who can never log in to the system, making their execution much safer.

The number of services running on your Linux system will, in large part, depend on the distribution and, in even larger part, on the system's purpose.

The history of Linux service management

As you can probably imagine, the task of managing the system services—the programs required to run so that your computer is usable to you—is not a trivial one. The software that runs them must be stable and robust. This, and the fact that the system is expected to start rarely, especially on servers, made the Linux-adopted solution survive for decades. The increasing need to start services in parallel (through the emergence and widespread use of many threaded CPUs) and more intelligently gave way to several implementations of the new `init` system, which we will cover next.

systemd

`systemd` is a service manager for Linux that is able to manage services that start with the operating system. It replaces traditional `init` scripts. It is responsible for starting and stopping system services, managing system state, and logging system events. It has become the default `init` system for many popular Linux distributions, including CentOS, Fedora Linux, **Red Hat Enterprise Linux** (**RHEL**), and Ubuntu Linux.

This service manager is responsible for controlling the initialization of the system itself (services required for the Linux OS), starting and stopping system services, and managing system resources. It provides another way of managing services and other system components, and it allows system administrators to configure and customize the behavior of their systems in a more standardized way than with **System V** (**SysV**) `init`.

One of the key features of `systemd` is its ability to start services in parallel, which can significantly reduce the boot time of a system. It also includes a number of tools for managing and monitoring system services.

Another thing `systemd` is praised for is the uniformity of service configurations it has finally brought to the Linux world. In every Linux distribution, the `systemd` configuration files are delivered to the same path; they look the same. There are, however, still some differences depending on the path binaries are being installed to. `systemd` is also better at knowing whether a process is running, making it more difficult to end up in a situation where we can't start a process because of stale files.

One of the major advantages of `systemd` is its awareness of dependencies. The service (running program under `systemd` control) configuration contains information about all the other services it depends on and can also point to services that depend on it. What's more, the service can inform `systemd` about which targets it requires to run: if your service needs the network to be up and running, you can put this information into the configuration, and `systemd` will ensure your daemon will be brought up only after the network is properly configured.

The following is a list of some of the tools and utilities that are provided as part of `systemd`:

- `systemd`: This is the main system and service manager. It is the main program that controls the initialization and management of services and other system components.

- `systemctl`: This is the command-line utility for managing system services and other system components. It can be used to start, stop, restart, enable, and disable services, as well as to view the status of services and other system components.

- `journalctl`: This is used for viewing and manipulating the system log, which is managed by `systemd`. It can be used to view log messages, filter log messages based on various criteria, and export log data to a file.

- `coredumpctl`: This is a utility, as the name suggests, that helps retrieve core dumps from `systemd`'s journal.

- `systemd-analyze`: This can be used for analyzing the boot performance of a system. It measures the time it takes for a system to boot, as well as the time it takes to identify potential bottlenecks and performance issues.

- `systemd-cgls`: This is a command-line utility for viewing the control group hierarchy on a system. **Control groups**, or **cgroups**, are used by `systemd` to manage system resources and to isolate processes from one another.

- `systemd-delta`: This is a command-line utility for analyzing the differences between the default configuration files provided by `systemd` and any local modifications made to these files.

- `systemd-detect-virt`: This is a command-line utility for detecting the virtualization environment in which a system is running. It can be used to determine whether a system is running in a **virtual machine** (**VM**), a container, or on bare metal.

- `systemd-inhibit`: This is a command-line utility for preventing certain system actions, such as suspending or shutting down the system, from being performed.

- `systemd-nspawn`: This is a command-line utility for running processes in lightweight containers. It can be used to create and manage containers, as well as to execute processes within containers.

This is just a list of more common tools and utilities that are provided as part of `systemd`. There are many more, but we will not cover them here.

Targets

In `systemd`, a **target** is a specific state that the system can be in, and it is represented by a symbolic name. Targets are used to define the high-level behavior of the system, and they are typically used to group together a set of related services and other system components.

For example, `multi-user.target` is a target that represents a system that is ready to provide multi-user access, with networking and other services enabled; `graphical.target` is a target that represents a system that is ready to display a graphical login screen, with a graphical desktop environment and related services enabled.

Targets are typically defined in unit files, which are configuration files that describe the properties and behavior of system components. When a target is activated, `systemd` will start all of the services and other system components that are associated with that target.

`systemd` includes a number of predefined targets that cover a wide range of common system states, and administrators can also define custom targets to suit the specific needs of their systems. Targets can be activated using the `systemctl` command, or by modifying the default target that is set when the system boots.

Here are some examples of predefined `systemd` targets:

- `poweroff.target`: Represents a system that is shutting down or powered off
- `rescue.target`: Represents a system that is running in rescue mode, with minimal services enabled
- `multi-user.target`: Represents a system that is ready to provide multi-user access, with networking and other services enabled
- `graphical.target`: Represents a system that is ready to display a graphical login screen, with a graphical desktop environment and related services enabled
- `reboot.target`: Represents a system that is rebooting
- `emergency.target`: Represents a system that is running in emergency mode, with only the most essential services enabled

Here is an example of a `systemd` unit file that defines a custom target:

```
[Unit]
Description=Unit File With Custom Target

[Install]
WantedBy=multi-user.target
```

This unit file defines a target named `Custom Target` that is meant to be activated as part of `multi-user.target`. The WantedBy directive specifies that the target should be activated when `multi-user.target` is activated.

Here is another example of a `systemd` unit file named `custom.target` that defines a custom target:

```
[Unit]
Description=My simple service
```

```
[Install]
WantedBy=multi-user.target
```

The following is a unit file that is using our `custom.target` target:

```
[Service]
ExecStart=/usr/local/bin/my-simple-service
Type=simple

[Install]
WantedBy=custom.target
```

This unit file defines a target named `Unit File With Custom Target` and a service named `My simple service`. The `ExecStart` directive specifies the command that should be used to start the service, and the `Type` directive specifies the type of service. The `WantedBy` directive in the `[Install]` section of the service unit specifies that the service should be activated when `custom.target` is activated.

Now, as we have touched upon unit files a bit, let's dig into them deeper and see what is possible with them.

Unit files

Unit files are usually stored in the `/lib/systemd/` system directory of Linux OS filesystems. No files in this directory should be altered in any way since they will be replaced by files contained in packages when a service is upgraded using a packet manager.

Instead, to modify a unit file of a specific service, create your customized unit file inside the `/etc/systemd/system` directory. Files in this `etc` directory will take precedence over the default location.

`systemd` is able to provide unit activation using the following:

- **Dependency**: It will start a unit simply by using a pre-built dependency tree handled by `systemd`. At its simplest, you will need to add a dependency with `multi-user.target` or with `network.target` if your service is using networking.

- **Drop-in unit files**: You can easily extend the default unit file by providing a snippet that will override a part of the default unit file (for example, change daemon execution options). Those are the files you'd place inside the `/etc/systemd/system` directory.

- **Templates**: You can also define template unit files. Those special units can be used to create multiple instances of the same general unit.

- **Security features**: You can specify read-only access or deny access to a part of the filesystem, assign `private/tmp` or network access, and limit kernel capabilities.

- **Path**: You can start a unit based on activity on or the availability of a file or directory in the filesystem.

- **Socket**: This a special type of file in the Linux operating system that enables communication between two processes. Using this feature, you can delay starting a service until the associated socket is accessed. You can also create a unit file that simply creates a socket early in a boot process and a separate unit file that uses this socket.

- **Bus**: You can activate a unit by using the bus interface provided by D-Bus. D-Bus is simply a message bus used for **inter-process communication** (**IPC**), most commonly used in GUIs such as GNOME or KDE.

- **Device**: You can also start a unit at the availability of hardware (the `dev` file, located in the `/dev` directory). This will leverage a mechanism known as `udev`, which is a Linux subsystem that supplies device events.

Once you've started a service, you will probably want to check whether it's up and healthy by taking a look at the log files. This job is taken care of by `journald`.

Logging

Every service managed by `systemd` sends its logs to `journald`—a special part of `systemd`. There is a special command-line tool for managing those logs: `journalctl`. In its simplest form, running a `journalctl` command will simply output all system logs, with the newest ones being at the top. While the format of the logs is similar to `syslog`—the traditional tool for gathering logs on Linux—`journald` captures more data. It collects data from the boot process, kernel logs, and more.

The boot logs are transient by default. This means they are not saved between system reboots. There is, however, a possibility to record them permanently. There are two ways to do it, as follows:

- Create a special directory. When `journald` detects it during system boot, it will save logs there: `sudo mkdir -p /var/log/journal`.

- Edit the `journald` configuration file and enable persistent boot logs. Open the `/etc/systemd/journald.conf` file with your favorite editor and, in the `[Journal]` section, edit the `Storage` option to look like this:

```
[Journal]
Storage=persistent
```

`journald` is capable of filtering logs by service by the use of the `-u service.name` option—that is, `journalctl -u httpd.service` will only print logs from the `httpd` daemon. There's a possibility of printing logs from a given timeframe, printing from more than one service, searching logs by **process ID** (**PID**), and more. As usual, we recommend referring to the man page—use the `man journalctl` command for this.

In this section, we have covered the most often used services software in the Linux world—systemd. In the next section, we are going to look at OpenRC—a system used in Alpine Linux, the Linux distribution of choice for containers in the cloud.

OpenRC

Alpine Linux uses another system for managing system services called **OpenRC**. OpenRC is a dependency-based init system originally developed for the use of Gentoo Linux. It is designed to be lightweight, simple, and easy to maintain. OpenRC uses plain text configuration files, which makes it easy to customize and configure. It is also easy to extend with custom scripts and programs. OpenRC is flexible and can be used on a wide variety of systems, from embedded devices to servers.

The following are examples of how OpenRC is used in Alpine Linux:

- **Starting and stopping system services**: OpenRC is used to manage system services such as networking, ssh, and cron. You can use the rc-service command to start, stop, or check the status of a service. For example, to start the ssh service, you can run rc-service ssh start.

- **Customizing system initialization and shutdown**: OpenRC allows you to write custom scripts to customize the behavior of your system during startup or shutdown. These scripts are executed at specific points in the boot process and can be used to set up custom configurations or perform other tasks.

- **Managing system runlevels**: OpenRC uses **runlevels** to define the system's behavior. You can use the rc-update command to add or remove services from different runlevels. For example, to make a service start at boot, you can run rc-update add <service> boot.

To start a service, use the following rc-service command:

```
admin@myhome:~$ rc-service <service> start
```

To stop a service, use the following rc-service command:

```
admin@myhome:~$ rc-service <service> stop
```

To check the status of a service, use the following rc-service command:

```
admin@myhome:~$ rc-service <service> status
```

To enable a service to start at boot, use the following rc-update command:

```
admin@myhome:~$ rc-update add <service> default
```

To disable a service from starting at boot, use the following rc-update command:

```
admin@myhome:~$ rc-update del <service> default
```

Default in this context means a default runlevel of the Alpine Linux system. Runlevels in general are used to define the system's behavior. There are several predefined runlevels common to most Linux distributions, each of which corresponds to a specific set of services that are started or stopped.

In Alpine Linux, these are the default runlevels:

- `default`: This is the default runlevel and is used when the system is booted normally. Services that are started in this runlevel include networking, SSH, and system logging.

- `boot`: This runlevel is used when the system is booting. Services that are started in this runlevel include the system console, the system clock, and the kernel.

- `single`: This runlevel is used when the system is booted into single-user mode. Only a minimal set of services is started in this runlevel, including the system console and the system clock.

- `shutdown`: This runlevel is used when the system is shutting down. Services that are stopped in this runlevel include networking, SSH, and system logging.

OpenRC uses a very similar way of defining service actions to SysV `init` we mentioned earlier in this chapter. Commands such as `start`, `stop`, `restart`, and `status` are defined in a Bash script. Here is a basic example of a service:

```
# Name of the service

name="exampleservice"



# Description of the service

description="This is my example service"



# Start command

start() {

  # Add your start commands

}



# Stop command
```

```
stop() {

    # Add your stop command here

}



# Restart command

restart() {

    stop

    start

}
```

To create a new service, you can copy this file to a new file and modify the `name`, `description`, `start`, `stop`, and `restart` functions as needed. The `start` function should contain the command to start the service, the `stop` function should contain the command to stop the service, and the `restart` function should stop and start the service. Those can be the same as for SysV `init`.

In OpenRC, `init` scripts are typically stored in the `/etc/init.d` directory. These are scripts that are used to start and stop services and manage the system's runlevels.

To create a new `init` script for OpenRC, you can create a new file in the `/etc/init.d` directory and make it executable.

Once you have created your `init` script, you can use the `rc-update` command to add it to the default runlevel, which will cause the service to start at boot. For example, to add the `exampleservice` service to the default runlevel, you can run the following command:

```
admin@myhome:~$ rc-update add exampleservice default
```

In most cases, we'll be using Alpine Linux in a Docker environment where there's little use for OpenRC, but it's still useful to know for some edge-case usage. We will look at Docker in more detail in *Chapter 8* and *Chapter 9*.

In this section, we have looked at OpenRC, software that controls system services in Alpine Linux. In the next section, we are going to introduce, very shortly, an outdated form of SysV `init`, which may come up with older or minimal Linux distributions.

SysV init

As previously mentioned, the `init` process is the most important continuously running process in the system. It's responsible for starting system services when the system boots up or when an administrator requests it, and stopping system services when the system is shutting down or when requested, all in the proper order. It's also responsible for restarting services on request. Since `init` will execute code on behalf of the root user, it is imperative that it's well tested for stability and security risks.

One of the charming properties of the old `init` system was its simplicity. The starting, stopping, and restarting of a service was managed by a script—one that had to be written by either the application author, the distribution package owner for this application, or the system administrator—if the service was not installed from packages. The scripts were simple and easy to understand, write, and debug.

The more the complexity of software grew, however, the more the old `init` system showed its age. Starting up a simple service was okay. Starting up an application consisting of several programs grew more difficult, especially as the dependencies between them became more important. `init` lacks observation of dependencies of startup of the services it takes care of.

Another place where the old `init` system became more and more unfit for modern systems was serial startup: it was unable to start services in parallel, thus negating modern multicore CPU gains. The time was nigh to look for a system more fit to the new times.

A typical `init` system consists of several components, as follows:

- A `/etc/init.d` or `/etc/rc.d/init.d` directory that contains start/stop scripts.
- A `/etc/inittab` file that defines runlevels and sets up a default one.
- A `/etc/rcX.d` directory that contains all the scripts for services that should be started or stopped in the runlevel X, where X is a number from 0 (zero) to 6. We will get to the details in the next paragraph.

The `/etc/init.d/` directory contains shell scripts responsible for starting, stopping, and restarting services. The script accepts a single argument of either `start`, `stop`, or `restart`. Each argument passed to the script executes a proper function (usually named the same as the argument: `start`, `stop`, or `restart`), and the function runs a set of steps to properly start, stop, or restart the given service. The kind of final state of the start process that the system ends up in is called a **runlevel**. The runlevel determines whether services are being started or stopped and which of them are being started if this is the course of action.

To determine the type of action to call, a link to the script would be created in the directory related to the runlevel in question.

Let's assume we want the system to end up in runlevel 3. If we want our service to be started in this runlevel, we would create a link to the `/etc/rc.d/my_service` script pointing to the `/etc/rc3.d/` directory. The name of the link determines the type of action and the order. So, if we wanted the service to be started after numbers 01-49, we would call it /etc/rc.3/S50my_service. The letter S tells the init system to start the service, and the number 50 tells it to start it after all services with lower numbers have been started. Please note that the numbering is more of a framework. It is not guaranteed that there are scripts with all other numbers prior to 50. The same goes for stopping services. After determining which runlevel is the default for stopping the system (usually 0), a proper symlink is created for the service script.

The major problem with the preceding framework is that it's totally unaware of dependencies. The only way to ensure that the services on which your daemon depends are running is to script it around in your start function. In the case of more complex applications comprising many services, this may lead to bigger start/stop scripts.

Another issue with init scripts being raised by admins and developers was that there are multiple standards on how to write them and multiple different tools around init scripts. Basically, every major Linux distribution had its own way of writing those scripts and its own library of helper functions. Let's consider an init script starting the same My Service service on Slackware Linux and Debian/GNU Linux. This is also where the introductory chapter on writing shell scripts comes in handy.

In both cases, Slackware and Debian, we are going to cut some original content out for brevity, leaving only the most important parts. Do not worry as both distributions deliver wholly commented example scripts.

The following init script will work in a Slackware Linux environment. The script starts with a header where we declare the name of the service and some important paths:

```
#!/bin/bash
#
# my_service          Startup script for the My Service daemon
#
# chkconfig: 2345 99 01
# description: My Service is a custom daemon that performs some
important functions.

# Source function library.
. /etc/rc.d/init.d/functions

# Set the service name.
SERVICE_NAME=my_service

# Set the path to the service binary.
SERVICE_BINARY=/usr/local/bin/my_service

# Set the path to the service configuration file.
CONFIG_FILE=/etc/my_service/my_service.conf
```

```
# Set the user that the service should run as.
SERVICE_USER=my_service

# Set the process ID file.
PIDFILE=/var/run/my_service.pid

# Set the log file.
LOGFILE=/var/log/my_service.log
```

The good thing is that an example script delivered with Slackware is very well commented. We have to declare a path to the binary for the daemon. We also have to declare the user and group that the service will run as, effectively deciding filesystem permissions for it.

The next section defines all important actions for the service: start, stop, and restart. There are usually a few others, but we have cut them out for brevity:

```
start() {
  echo -n "Starting $SERVICE_NAME: "
  # Check if the service is already running.
  if [ -f $PIDFILE ]; then
    echo "already running"
    return 1
  fi
  # Start the service as the specified user and group.
  daemon --user $SERVICE_USER --group $SERVICE_GROUP $SERVICE_BINARY
-c $CONFIG_FILE -l $LOGFILE -p $PIDFILE
  # Write a lock file to indicate that the service is running.
  touch $LOCKFILE
  echo "done"
}
```

The start function checks whether the service is running utilizing the pid file. The pid file is a text file that contains the PID of the service. It gives information about both the main process of the service and its status. There is a caveat, however. It is possible that the service isn't running anymore and the pid file still exists. That would prevent the start function from actually starting the service.

After the check determines that the pid file doesn't exist, which means that the service is not running, a special tool called daemon is used to start the process with user and group permissions, pointing to a configuration file, log file, and pid file locations.

The function communicates its actions using the bash echo command, which prints the given text. In case of automatic execution, the output of the echo command will be logged to system logs depending on the logging daemon configuration:

```
stop() {
  echo -n "Stopping $SERVICE_NAME: "
  # Check if the service is running.
  if [ ! -f $PIDFILE ]; then
    echo "not running"
    return 1
  fi
  # Stop the service.
  killproc -p $PIDFILE $SERVICE_BINARY
  # Remove the lock file.
  rm -f $LOCKFILE
  # Remove the PID file.
  rm -f $PIDFILE
  echo "done"
}
```

Again, the stop function checks whether the service is running using the pid file. The probability of the service running without a proper pid file is almost nonexistent. After the check is done, a special command called killproc is used to terminate the process. In the final part of the function, the script does some housekeeping tasks, cleaning the pid file and the lock file:

```
restart() {
  stop
  start
}
```

The restart function is very simple. It reuses the already defined start and stop functions to do exactly what it says: restart the service. This is often useful if the service configuration requires the binary to be loaded again:

```
case "$1" in
  start)
    start
    ;;
  stop)
    stop
    ;;
  restart)
    restart
    ;;
```

```
  *)
    echo "Usage: $0 {start|stop|restart|uninstall}"
esac
```

The final part of the script evaluates the action we want to take—start, stop, or restart the service—and calls the proper function. If we ask it to do something it doesn't recognize, the script will print out usage instructions.

The following script, however, is intended for a Debian Linux environment:

```
#!/bin/bash
#
# my_service        Startup script for the My Service daemon
#
# chkconfig: 2345 99 01
# description: My Service is a custom daemon that performs some
important functions.

# Source function library.
. /lib/lsb/init-functions

# Set the service name.
SERVICE_NAME=my_service

# Set the path to the service binary.
SERVICE_BINARY=/usr/local/bin/my_service

# Set the path to the service configuration file.
CONFIG_FILE=/etc/my_service/my_service.conf

# Set the user that the service should run as.
SERVICE_USER=my_service

# Set the group that the service should run as.
SERVICE_GROUP=my_service
```

Again, the script starts with a header section that defines paths to be later used in the start and stop sections. Normally there are more lines, but we've cut them out for brevity, again:

```
start() {
  log_daemon_msg "Starting $SERVICE_NAME"
  # Check if the service is already running.
  if [ -f $PIDFILE ]; then
    log_failure_msg "$SERVICE_NAME is already running"
    log_end_msg 1
```

```
        return 1
    fi
    # Start the service as the specified user and group.
    start-stop-daemon --start --background --user $SERVICE_USER --group
$SERVICE_GROUP --make-pidfile --pidfile $PIDFILE --startas $SERVICE_
BINARY -- -c $CONFIG_FILE -l $LOGFILE
    # Write a lock file to indicate that the service is running.
    touch $LOCKFILE
    log_end_msg 0
}
```

The start function looks similar to the one from the Slackware version. You'll notice some subtle differences, however, as follows:

- The start-stop-daemon helper function is being used here to manage the running service instead of dacmon and killproc

- There are specialized functions for logging instead of simple echo: log_daemon_msg, log_failure msg, and log_end_msg

- The start-stop-daemon function accepts special flags to determine the action (start and stop) as well as detaching the program from the terminal, making it effectively a system service (--background), as seen here:

```
stop() {
    log_daemon_msg "Stopping $SERVICE_NAME"
    # Check if the service is running.
    if [ ! -f $PIDFILE ]; then
        log_failure_msg "$SERVICE_NAME is not running"
        log_end_msg 1
        return 1
    fi
    # Stop the service.
    start-stop-daemon --stop --pidfile $PIDFILE
    # Remove the lock file.
    rm -f $LOCKFILE
    # Remove the PID file.
    rm -f $PIDFILE
    log_end_msg 0
}
```

The stop function is pretty similar to the stop function from Slackware, with similar differences to those the start function had.

The rest of the script containing the restart function and the task evaluation is not that interesting, and we've cut it out.

As you probably still remember from the section about systemd, it solves some of those problems. SysV is not seen too often in modern systems, so you will most commonly need to deal with systemd instead.

There is, however, another replacement for SysV init called Upstart.

A few words about Upstart, an alternative

Upstart is an event-based replacement for the traditional SysV init system used to manage and control services and daemons on a system. Upstart was introduced in Ubuntu 6.10 and later versions and is designed to improve boot time, simplify system configuration, and provide more flexibility in managing system services. It has now been largely replaced by systemd in most Linux distributions.

Upstart is used to manage the initialization process of the system and to start, stop, and supervise tasks and services. It is designed to be more flexible and efficient than the traditional init daemon and to provide more information about the status of tasks and services.

All that can be managed by systemd and/or cron has become an industry standard, so if you don't have a good reason for using those or already have a system using Upstart, we discourage you from using it as a default.

Summary

In this chapter, we covered system services—or daemons in the Unix and Linux world—and the software most often used to manage them. We explained what those are, how to inspect their state, and how to control them. In the next chapter, we are going to dive into Linux networking.

6

Networking in Linux

Networking is a complex topic no matter the operating system. Linux, in terms of its flexibility, can be very overwhelming regarding the multitude of possibilities of configuration, kernel features, and command-line tools that can help us configure those options. In this chapter, we will lay the foundation for this topic so that you can search for more information on a specific topic in other publications. In this chapter, we are going to cover the following topics:

- Networking in Linux
- ISO/OSI as a networking standard
- Firewalls
- Advanced topics

Networking in Linux

In Linux, networking is implemented in the kernel, which means that it is a part of the operating system. The kernel includes several components that work together to enable networking, including device drivers, protocol implementations, and system calls.

When a user wants to send or receive data over a network, they can do so using any of the networking applications available in Linux, such as `ping`, `traceroute`, `telnet`, or `ssh`. These applications use system calls to communicate with the kernel and request that data be sent or received over the network.

The kernel communicates with the network hardware using device drivers, which are software programs that allow the kernel to access and control the hardware. Different drivers are needed for different types of network hardware, such as Ethernet or Wi-Fi.

The kernel also implements several networking protocols, which are rules and standards that define how data is formatted and transmitted over the network. Common protocols used in Linux include TCP, UDP, and IP (version 4 and version 6).

ISO/OSI as a networking standard

The starting point of any discussion about networks always starts with the reference model defined by the **International Organization for Standardization/Open Systems Interconnection (ISO/OSI)**. The ISO/OSI reference model is a conceptual model that defines a networking framework to implement protocols in seven layers. It is a framework that allows us to view communications between systems (computer or otherwise) as separate from the actual physical and software structure underlying it.

In Linux, the OSI model is implemented through a series of software components that are responsible for performing the functions of each layer. These components work together to enable networking capabilities in Linux.

The seven layers of the OSI model that are implemented in Linux are as follows:

- Physical
- Data link
- Network
- Transport
- Session
- Presentation
- Application

In a system running in the cloud, you will have access to all the layers that are implemented in the Linux kernel. These layers are Network, Transport. Session, Presentation, and Application. To debug network connectivity, check statistics, and find any other possible issues, Linux has a console tool you can use. Let's go through every layer one by one and investigate what command-line tools we can use in Linux to investigate each of them.

To learn more about the OSI model, you can refer to `https://osi-model.com/`. We are going to delve into the layers and explain them in the following subsections.

Physical layer

This layer is responsible for transmitting raw bits over a communication channel and is implemented through device drivers that control the network hardware, such as Ethernet standards such as `10BASE-T`, `10BASE2`, `10BASE5`, `100BASE-TX`, `100BASE-FX`, `100BASE-T`, `1000BASE-T`, and others. We will not look at this layer in more detail here as we will focus on software implementation and how to interact with it on the Linux console. You can find plenty of information about cabling, hardware devices, and networking online.

Data link layer – MAC, VLAN

The **data link layer** is responsible for providing a reliable link between devices on a network. It's divided into the **Logical Link Control** (**LLC**) sublayer and the **Media Access Control** (**MAC**) sublayer.

The data link layer takes the raw data from the network layer (layer 3) and converts it into a format that can be transmitted over the physical link. It also provides error detection and correction, flow control, and MAC functions.

The LLC sublayer provides a consistent interface to the network layer, regardless of the type of physical network being used. It also provides flow control and error correction services.

The MAC sublayer controls access to the physical network and provides addressing services. It uses MAC addresses, which are unique identifiers that are assigned to each device on a network, to ensure that data is delivered to the correct destination.

The data link layer also includes the use of protocols such as Ethernet, PPP, and Frame Relay to provide communication between devices on a network. It also provides a mechanism for flow and error control – for example, it uses a **cyclic redundancy check** (**CRC**) for error detection and a sliding window for flow control.

Several Linux command-line tools can be used to debug data link layer problems. Here are a few examples:

- `ifconfig`: This command can be used to view the status of network interfaces and their associated IP addresses, netmasks, and MAC addresses. It can also be used to configure network interfaces, such as setting an IP address or enabling or disabling an interface.

- `ping`: This command can be used to test the reachability of a host on a network. It sends an **Internet Control Message Protocol** (**ICMP**) echo request packet to the specified host and waits for an echo reply. If the host responds, it indicates that the host is reachable and that the data link layer is functioning properly.

- `traceroute`, `tracepath`, and `mtr`: These commands can be used to trace the route that packets take from the source to the destination. They can also be used to identify any network hops or devices that may be causing problems. Additionally, `tracepath` measures the **Maximum Transmit Unit** (**MTU**) you can use for this connection. `mtr`, on the other hand, provides more information about network health.

- `arp`: This command can be used to view and manipulate the **Address Resolution Protocol** (**ARP**) cache. ARP is used to map an IP address to a MAC address on a local network. This command can be used to verify that the correct IP-MAC address mappings are in the ARP cache.

- `ethtool`: This command can be used to view and configure advanced settings on Ethernet interfaces such as link speed, duplex mode, and auto-negotiation settings.

- `tcpdump`: This command can be used to capture and analyze network packets in real time. It can be used to troubleshoot problems such as packet loss, delayed packets, and network congestion.

We will dive deeper into the preceding tools in the next sections of this chapter as most of them can be used to look into several different OSI layers at once.

Network layer – IPv4 and IPv6

Each device on a public (internet) or private (your office or home) network has a unique address that is used to identify it and connect to it. When you send a request to a website, your device sends a message to the destination server using its address. The server then responds by sending a message back to your device using its address. This process is how devices communicate with each other over the internet. There are two types of addresses we use: IPv6 and IPv4. **IP** stands for **Internet Protocol** and the main difference between v4 (version 4) and v6 (version 6) is the number of addresses you can use.

IPv4 is the most widely used version of IP and there are only about 4.3 billion unique addresses available. However, this is not enough to support the increasing number of devices that are being connected to the internet. To address this issue, a new version of IP called IPv6 was developed to provide a much larger address space.

IPv4 addresses are 32-bit numbers that are usually represented in a dotted decimal notation, with four octets ranging from 0 to 255. For example, 192.168.0.1 or 1.2.3.4 are valid IPv4 addresses.

IPv6 addresses are 128-bit numbers (between zero and FFFF in hexadecimal, which is equal to the decimal value 65535), represented in hexadecimal notation with eight groups of four hexadecimal digits separated by colons. An example of an IPv6 address is 2001:0db8:bad:f00d:0000:dead:beef:7331.

We're mainly going to focus on IPv4 here as it's generally easier to understand, but similar principles work for IPv6, so it's going to be easier to reapply what we'll learn later in this chapter to the IPv6 world.

Subnets, classes, and network masks

Subnets are pieces of a network that are created by dividing a larger network into smaller networks. This is done for several reasons, including security, organization, and the efficient use of IP addresses. In public internet networks, every organization *owns* a piece of the network – a subnet.

When a network is subnetted, the host portion of the IP address (the part of the address that identifies a specific device on the network) is divided into two parts: one that identifies the subnet and one that identifies the host IPs within the subnet. The subnet mask, which is a binary representation that is applied to the IP address, is used to determine which part of the IP address identifies the subnet and which part identifies the host.

Imagine a network that uses the IP address range 192.168.11.0/24 (or 192.168.11.0/255.255.255.0 in decimal form). The /24 part is a **Classless Inter-Domain Routing** (CIDR) notation, which is a decimal representation of the number of bits used for the network (here, this is 24) and host.

This means that this particular network has 24 bits (or three numbers from the full address) for the `network` portion of the IP address and 8 bits (or one, last, number) for the `host` portion. So, for this network, you'll have an available range of `192.168.11.0 - 192.168.11.255`, where `192.168.11.0` is a network address and `192.168.11.255` is a broadcast address.

A **broadcast network address** is a special type of IP address that is used to send a message to all hosts on a particular network or subnet. The broadcast address is the highest in a network or subnet's IP address range, and it is used in conjunction with the subnet mask to identify the broadcast domain. When a host sends a packet to a broadcast address, the packet is delivered to all hosts on the same network or subnet. It's important to note that broadcast packets don't leave the current subnet network – they only work within the local network or subnet.

Before standardizing CIDR, IP addresses were divided into classes (A, B, and C) based on the number of hosts that a network needed to support. These classes were defined by the leading bits of the IP address, and each class had a different number of bits for the network portion and host portion of the IP address. Right now, we're mainly using CIDRs, but some network addressing remains. For example, `10.0.0.0`, `12.0.0.0`, or `15.0.0.0` usually have `255.0.0.0` or `/8` network masks. Here are some other networks you could encounter:

- **Class A**: `10.0.0.0/8`, `12.0.0.0/8`, and `15.0.0.0/8`
- **Class B**: `172.16.0.0/16`, `172.17.0.0/16`, and `172.18.0.0/16`
- **Class C**: `192.168.0.0/24`, `192.168.1.0/24`, and `192.168.2.0/24`

These examples are not actual networks, but just a representation of how to identify the network class by looking at the leading bits of the IP address and the subnet mask – it's not a rule and you could create smaller (or bigger) networks within your infrastructure.

You can find a lot of calculators online that will help you better understand how networks address work. Here are some examples:

- `https://www.calculator.net/ip-subnet-calculator.html`
- `https://mxtoolbox.com/subnetcalculator.aspx`
- `https://www.subnet-calculator.com/cidr.php`

Now that we have learned about subnets, classes, and network masks, let's move on to the next subsection.

Network configuration and console tools

With the knowledge you've gained so far, you can easily check your Linux network configuration using some CLI tools that are available in every modern Linux environment.

Here are some basic console tools to use for network configuration:

- The `iproute2` package, replacing the `ifconfig` and `route` commands
- `ifconfig`
- `route`
- `ip`
- `netplan`

Let's go through the syntax and what's possible while using these tools.

ifconfig

One of the commands that's usually available in Linux is `ifconfig`. This tool is used to configure network interfaces. It can be used to show the status of an interface, assign an IP address to an interface, set the netmask, and set the default gateway. `ifconfig` (from the net-tools package) was replaced in recent years with the `iproute2` set of tools; the most well-known CLI command from this package is `ip`.

Here's some example output of the `ifconfig` command:

```
admin@myhome:~$ ifconfig

eth0: flags=4163<UP,BRODCAST,RUNNING,MULTICAST>  mtu 1500

        inet 172.17.0.2  netmask 255.255.0.0  broadcast 172.17.255.255

        ether 02:42:ac:11:00:02  txqueuelen 0  (Ethernet)

        RX packets 17433  bytes 26244858 (26.2 MB)

        RX errors 0  dropped 0  overruns 0  frame 0

        TX packets 8968  bytes 488744 (488.7 KB)

        TX errors 0  dropped 0 overruns 0  carrier 0  collisions 0

lo: flags=73<UP,LOOPBACK,RUNNING>  mtu 65536

        inet 127.0.0.1  netmask 255.0.0.0

        loop  txqueuelen 1000  (Local Loopback)

        RX packets 0  bytes 0 (0.0 B)
```

```
RX errors 0   dropped 0   overruns 0   frame 0

TX packets 0   bytes 0 (0.0 B)
TX errors 0   dropped 0 overruns 0   carrier 0   collisions 0
```

`ifconfig`, when invoked without any additional options, will list all available interfaces in our system with some basic information such as interface name (here, these are `eth0` and `lo`), the MAC address of the network card (the physical address of your device), network configuration (IP address, network mask, and broadcast address), received and transmitted packets, and other information. This will allow you to check your network status at a glance.

The loopback device (named `lo` in the preceding example) is a virtual network interface that is used to send network packets to the same host that they were sent from. It is also known as the loopback interface and is represented by `lo` or `lo0`.

The primary purpose of the loopback device is to provide a stable and consistent way for the host to communicate with itself over the network stack, without having to rely on any physical network interface.

The loopback interface is typically used for testing, troubleshooting, and some system and application functions, as well as **inter-process communication** (**IPC**) between processes running on the same host.

Using `ifconfig` will also allow you to bring some interfaces up and down and configure their settings. To persist the configuration, you will need to save it to the `/etc/network/interfaces` file:

```
auto lo

iface lo inet loopback

# eth0 network device

auto eth0

iface eth0 inet static

    address 172.17.0.2

    netmask 255.255.0.0

    gateway 172.17.0.1
    dns-nameservers 8.8.8.8 4.4.4.4
auto eth1

allow-hotplug eth1
iface eth1 inet dhcp
```

In the preceding example, we're setting an automatic loopback device with eth0 and eth1. The eth0 interface has a static network configuration and will be also configured, the same as lo, when your system boots. The eth1 interface has a dynamic network configuration that's retrieved by the **Dynamic Host Configuration Protocol** (DHCP) server in the network it's connected to. (DHCP is a widely used and de facto standard way to automatically and dynamically assign a system IP address; it is also possible to pass some other configuration parameters related to networking through DHCP.) Additional allow-hotplug configuration means that this device will be started upon detection by the Linux kernel.

It's useful to know that after editing the /etc/network/interfaces file, you'll need to use either ifup or ifdown tools on Debian Linux or Ubuntu Linux, or ifupdown tools in Alpine Linux. Alternatively, you can restart networking by using the systemctl restart network on Debian Linux, Ubuntu Linux, or RHEL/CentOS. You must use the rc-service networking restart command on Alpine Linux.

To manually configure a device using ifconfig, you'll need to run the following commands:

```
admin@myhome:~$ ifconfig eth0 192.168.1.2 netmask 255.255.255.0
admin@myhome:~$ ifconfig eth0 up
```

This will configure the eth0 interface with a 192.168.1.2 IP address for the network with a netmask of 255.255.255.0 (or /24 in CIDR notation).

Instead of using the ifconfig up and ifconfig down commands, you can also use ifup and ifdown in Debian Linux and Ubuntu Linux systems, respectively, or ifupdown in Arch Linux systems.

route

The route command is used to view and manipulate the IP routing table. It's used to determine where network packets are sent given the destination IP addresses.

Invoking the route command without any options will display the current route table in our system:

```
admin@myhome:~$ route

Kernel IP routing table


Destination     Gateway         Genmask          Flags Metric
Ref     Use Iface

default         172.17.0.1      0.0.0.0          UG    0        0       0
eth0

172.17.0.0      0.0.0.0         255.255.0.0      U     0        0       0
eth0
```

To add a new entry to the routing table, use the following command:

```
admin@myhome:~$ sudo route add default gw 172.17.0.1 eth0
```

There can be only one default route. To add a custom route, do the following:

```
admin@myhome:~$ sudo route add -net 192.168.1.0 netmask 255.255.255.0
gw 192.168.1.1 dev eth0
```

The `del` command deletes an entry from the routing table, similar to the preceding example:

```
admin@myhome:~$ route del -net 192.168.1.0 gw 192.168.1.1 netmask
255.255.255.0 dev eth0
```

Finally, the `flush` command deletes all entries from the routing table, which means you will lose all network connections and if you are connected to a remote machine, you won't be able to work on it anymore.

There are more possibilities while using the `ifconfig` and `route` commands, but as we've already stated, both commands got replaced by the `iproute2` package (a successor of `iproute`), which includes the `ip` command.

iproute2

A more advanced command to manipulate routes and the configuration of network devices is `ip`, which can be used to perform a wider range of tasks, such as creating and deleting interfaces, adding and removing routes, and displaying network statistics.

Let's look at the most common commands you can perform using `iproute2`:

- `ip addr` or `ip a`: This shows information about network interfaces and their IP addresses. It also supports subcommands; for instance, the `ip addr add` command can be used to add an IP address to an interface, and `ip route add` can be used to add a route to the routing table.
- `ip link` or `ip l`: This shows information about network interfaces and their link-layer settings.
- `ip route` or `ip r`: This shows the IP routing table.
- `ip -s link` (or `ip -s l`): This shows statistics about network interfaces.

The following is the output of running the `ip link` and `ip addr` commands:

```
admin@myhome:~$ sudo ip link
1: lo: <LOOPBACK,UP,LOWER_UP> mtu 65536 qdisc noqueue state UNKNOWN
mode DEFAULT group default qlen 1000
    link/loopback 00:00:00:00:00:00 brd 00:00:00:00:00:00
2: tun10@NONE: <NOARP> mtu 1480 qdisc noop state DOWN mode DEFAULT
group default qlen 1000
    link/ipip 0.0.0.0 brd 0.0.0.0
```

```
3: ip6tnl0@NONE: <NOARP> mtu 1452 qdisc noop state DOWN mode DEFAULT
group default qlen 1000
    link/tunnel6 :: brd :: permaddr 4ec4:2248:2903::
47: eth0@if48: <BROADCAST,MULTICAST,UP,LOWER_UP> mtu 1500 qdisc
noqueue state UP mode DEFAULT group default
    link/ether 02:42:ac:11:00:02 brd ff:ff:ff:ff:ff:ff link-netnsid 0
admin@myhome:~$ sudo ip addr
1: lo: <LOOPBACK,UP,LOWER_UP> mtu 65536 qdisc noqueue state UNKNOWN
group default qlen 1000
    link/loopback 00:00:00:00:00:00 brd 00:00:00:00:00:00
    inet 127.0.0.1/8 scope host lo
        valid_lft forever preferred_lft forever
2: tunl0@NONE: <NOARP> mtu 1480 qdisc noop state DOWN group default
qlen 1000
    link/ipip 0.0.0.0 brd 0.0.0.0
3: ip6tnl0@NONE: <NOARP> mtu 1452 qdisc noop state DOWN group default
qlen 1000
    link/tunnel6 :: brd :: permaddr 4ec4:2248:2903::
47: eth0@if48: <BROADCAST,MULTICAST,UP,LOWER_UP> mtu 1500 qdisc
noqueue state UP group default
    link/ether 02:42:ac:11:00:02 brd ff:ff:ff:ff:ff:ff link-netnsid 0
    inet 172.17.0.2/16 brd 172.17.255.255 scope global eth0
        valid_lft forever preferred_lft forever
```

Both commands print out very similar information, but `ip addr`, apart from providing information about physical interfaces, adds information about network configuration. The `ip link` command is being used to control the interface status. Similar to how `ifconfig up eth0` was enabling the interface, `ip link set dev eth0 up` will do the same.

To configure a network interface using `iproute2`, you will need to execute the following commands:

```
admin@myhome:~$ sudo ip addr add 172.17.0.2/255.255.0.0 dev eth0
admin@myhome:~$ sudo ip link set dev eth0 up
```

To set the interface as the default route, use the following command:

```
admin@myhome:~$ sudo ip route add default via 172.17.0.1 dev eth0
```

To check only the `eth0` interface's status, you'd execute the following command:

```
admin@myhome:~$ sudo ip addr show dev eth0
47: eth0@if48: <BROADCAST,MULTICAST,UP,LOWER_UP> mtu 1500 qdisc
noqueue state UP group default
    link/ether 02:42:ac:11:00:02 brd ff:ff:ff:ff:ff:ff link-netnsid
0
    inet 172.17.0.2/16 brd 172.17.255.255 scope global eth0
        valid_lft forever preferred_lft forever
```

netplan

This tool is a new network configuration tool that was introduced in Ubuntu and is supported in Debian. It is a YAML configuration file and can be used to manage network interfaces, IP addresses, and other network settings. It was first introduced in Ubuntu 17.10 as a replacement for the traditional `/etc/network/interfaces` file and has since been adopted by other distributions. such as Debian and Fedora. Ubuntu 18.04 and newer versions have `netplan` installed by default. Other distributions, such as Debian 10 and Fedora 29 onwards, have also included `netplan` by default.

To use Netplan, you will first need to create a configuration file in the `/etc/netplan/` directory. The file should have a `.yaml` extension and should be named something descriptive, such as `01-eth0.yaml` or `homenetwork.yaml`.

The sample configuration looks like this:

```
network:

  version: 2

  renderer: networkd

  ethernets:

    eth0:

      dhcp4: true
```

This configuration defines a single network interface, `eth0`, that uses DHCP to obtain an IP address. The `renderer` key tells netplan which network manager to use (in this case, `networkd`). The `version` key is used to indicate the version of `netplan` being used. `networkd` is a network management daemon that is part of the `systemd` system and service manager.

A configuration with a static IP address for the `eth0` interface would look like this:

```
network:

  version: 2

  renderer: networkd

  ethernets:

    eth0:

      addresses: [192.168.0.2/24]
```

```
    gateway4: 192.168.0.1

    nameservers:

      addresses: [8.8.8.8, 4.4.4.4]
```

This configuration file defines one ethernet interface (eth0) with a static IP address. This interface also has a gateway and DNS servers defined. Note that we've used a CIDR notation instead of a decimal one.

Once you saved your configuration, to apply the changes, you can run the following command:

```
admin@myhome:~$ sudo netplan apply
```

You can also use the following command to check the configuration for syntax errors:

```
admin@myhome:~$ sudo netplan --debug generate
```

You can check the current status of the network interfaces using the following command:

```
admin@myhome:~$ sudo netplan --debug try
```

Finally, if you want to check the network interface's status without applying the configuration, use the following command:

```
admin@myhome:~$ sudo netplan --debug networkd try
```

To view logs regarding errors while bringing up network interfaces, you can use the dmesg command to view kernel messages, including those related to network interfaces. You can use dmesg | grep eth0 to filter logs related to the eth0 interface specifically. Other locations include the /var/log/messages file, journalctl (for example, the journalctl -u systemd-networkd.service command), and /var/log/netplan/, which contains logs generated by netplan.

In day-to-day operations, it's much more likely you'll edit the /etc/network/interfaces file or netplan configuration than configure the interface manually, but it's very useful to know how to do it in case you need to temporarily change something in the network configuration for testing or debugging issues.

Next, we will cover the transport layer.

Transport layer – TCP and UDP

In the **transport layer** of the OSI model, we focus more on **Transmission Control Protocol** (TCP) and IP, which are the foundation of the modern internet. Additionally, we'll look into the **User Datagram Protocol (UDP)**. We talked about IP in the *Network layer – IPv4 and IPv6* subsection of this chapter, so we're only going to deepen our knowledge of this protocol a bit here.

TCP is used for communication and needs reliable two-way communication between services. UDP is a stateless protocol that does not need a constant connection.

A TCP connection is established using a *three-way handshake*:

1. The client sends a SYN (synchronize) packet to the server, to initiate the connection.

2. The server receives the SYN packet and sends back a SYN-ACK (synchronize-acknowledgment) packet to the client to confirm that the connection has been established.

3. The client receives the SYN-ACK packet and sends back an ACK (acknowledgment) packet to the server to complete the three-way handshake.

Once the three-way handshake is completed, the devices can start sending data to each other over the established TCP connection.

The connection for TCP and UDP protocols is initiated by sending packets to a port on the server end of the connection. The port number is included in the IP packet header, along with the IP address, so that the destination device knows which process to send the data to.

A port is an integer ranging from 0 to 65535. It is used to identify a specific process running on a device and to differentiate it from other services. The port number is added to the IP address. Thus, if more than one program listens for a connection on one IP address, the other side can tell exactly which program it wants to talk to. Let's assume that two processes are running and listening: a WWW server and an SSH server. The WWW server will typically listen on ports 80 or 443 and the SSH server will listen on port 22. Let's assume the system has an IP address of 192.168.1.1. To connect to the SSH server, we would pair the IP address with the port address (commonly written as 192.168.1.1:22). The server will know that this incoming connection has to be handled by the SSH process. A UDP connection, unlike TCP, doesn't establish a connection to a server machine. Instead, devices can start sending UDP datagrams (**packets**) to each other on known ports.

When a device sends a UDP datagram, it includes the destination IP address and port number in the packet header. The receiving device checks the destination IP address and port number in the packet header to determine which process to send the data to.

Because UDP is connectionless, there is no guarantee that the data will be received by the destination device or that it will be received in the order it was sent. There is also no error checking or retransmission of lost packets. UDP is typically used for services that require a low overhead and fast communication. The most well-known service that uses UDP for communication is **Domain Name Service (DNS)**.

There are three types of port numbers:

- **Well-known ports**, which you can typically find in your Linux system in the /etc/services file. These are port numbers that are reserved for specific services, such as HTTP traffic on port 80 or DNS traffic on port 53.

- **Privileged ports** are ports that only root users can open a listening service on. Those ports are between 1 and 1024.

- **Ephemeral ports** are port numbers that are used for temporary connections and are assigned dynamically by the operating system, such as for TCP connections.

There are various tools available for viewing details on TCP and UDP traffic on the machine you're working on. You can also set up a firewall so that you can control access to your machine from the network.

netstat

To view all TCP connections made from your machine, you can use a netstat command. It can be used to view a list of open connections, as well as display statistics about the network traffic on your system.

To view a list of all open connections on your system, you can use the netstat -an command. This will display a list of all current connections, including the local and remote IP addresses and ports, as well as the status of the connection (for example, listening, established, and so on).

Using netstat -s, you can view statistics on each connection on your system. This will display a variety of statistics about the network traffic on your system, including the number of packets sent and received, the number of errors, and more.

To view only the UDP connections using netstat, you can use the netstat -an -u command. This will display a list of all current UDP connections, including the local and remote IP addresses and ports, as well as the status of the connection.

Alternatively, you can use the netstat -an -u | grep "udp" | grep "0.0.0.0:*" command to only show UDP connections that are in a listening state. This command filters the output of netstat -an -u to show only lines that contain "UDP" and "0.0.0.0:*", which indicates a listening UDP connection.

You can also use other options, such as -p to show the process ID and name of the process that owns each connection, -r to show the routing table, and -i to show the statistics for a particular interface.

tcpdump

tcpdump is a command-line packet analyzer that allows you to capture and analyze network traffic by displaying packets being transmitted or received over a network to which the computer is attached. tcpdump can be used to troubleshoot network issues, analyze network performance, and monitor network security. You can capture packets on a specific interface, filter packets based on various criteria, and save the captured packets to a file for later analysis.

To capture and display all network traffic on the eth0 interface, you can use the -i option, followed by the interface name. To capture and display all network traffic on the eth0 interface, you would need to run sudo tcpdump -i eth0.

You can also save the captured packets to a file for later analysis by using the -w option – for example, tcpdump -i eth0 -w all_traffic.pcap. This will save all the captured packets on the eth0 interface in the file.

By default, tcpdump captures packets indefinitely.

To capture packets for a specific amount of time, you can use the -c option, followed by the number of packets to capture – for example, sudo tcpdump -i eth0 -c 100. This command captures and displays 100 packets, then exits.

You can also filter the traffic by using filters such as port, ip, host, and so on – for example, sudo tcpdump -i eth0 'src host 192.168.1.2 and (tcp or udp)'. This filter captures all the packets on the eth0 interface with a source IP address of 192.168.1.2 and is either TCP or UDP.

To show more advanced use of tcpdump, let's capture only SYN packets (see all connections being established). You can achieve this by using the tcp[tcpflags] & (tcp-syn) != 0 filter. This filter checks if the SYN flag is set in the TCP header of the packet.

Here is an example command that captures and displays all SYN packets on the eth0 interface:

```
sudo tcpdump -i eth0 'tcp[tcpflags] & (tcp-syn) != 0'
```

You can also save the captured packets to a file for later analysis by using the -w option, like this:

```
sudo tcpdump -i eth0 -w syn_packets.pcap 'tcp[tcpflags] & (tcp-syn) !=
0'
```

This will save all the captured SYN packets in the syn_packets.pcap file.

You can also specify a filter with a more complex filter, like this:

```
sudo tcpdump -i eth0 'tcp[tcpflags] & (tcp-syn) != 0 and src host
192.168.1.2'
```

This filter captures only SYN packets that have a source IP address of 192.168.1.2.

Wireshark

Another popular tool that's similar to tcpdump is Wireshark. It can be used both headless (only in a command line) and via a graphical interface:

- To show all traffic on the eth0 interface using Wireshark, you can use the sudo wireshark -i eth0 command. This will start Wireshark and listen for traffic on the eth0 interface.

You can also use the -k flag to start capturing immediately and the -w flag to write the captured traffic to a file: `sudo wireshark -k -i eth0 -w output_file.pcap`.

- If you'd like to show only SYN packets, as we showed in the `tcpdump` example, you can run the `sudo wireshark -i eth0 -f "tcp.flags.syn == 1"` command. The preceding command uses a filter, `"tcp.flags.syn == 1"`, which says that we want to see only TCP protocol flags marked as SYN. You can also use this filter in the GUI version of Wireshark by going to the **Capture** menu, selecting **Options**, then entering the filter in the **Capture Filter** field before starting the capture.

Alternatively, you can apply this filter after capturing the traffic by clicking on the **filter** button on the top-right corner of the Wireshark window, type `"tcp.flags.syn==1"` in the filter field, and press *Enter*.

ngrep

The next very useful tool that's similar to `tcpdump` and Wireshark is `ngrep`. What differentiates it from other tools we've been talking about is that it's much simpler to use, and it allows you to (similar to `grep`) search for strings inside network packets.

For instance, to monitor GET HTTP requests, you can use the following command:

```
admin@myhome:~$ ngrep -d eth0 -q -W byline "GET" "tcp and port 80"
interface: en0 (192.168.1.0/255.255.255.0)
filter: ( tcp and port 80 ) and ((ip || ip6) || (vlan && (ip || ip6)))
match: GET
# Later output will contain actual GET requests, removed for
readability
```

This command will listen on the eth0 interface and match only TCP packets on port 80 (the default port for HTTP). The -q option tells `ngrep` to be quiet and not display packet summary information, while -W tells `ngrep` to print each packet data in separate lines. At this point, we can move to the session layer.

Session layer

The **session layer**, as we've already mentioned, is responsible for establishing, maintaining, and terminating connections between devices. This means it sets up, coordinates, and terminates conversations, exchanges, or connections between the applications. The session layer ensures that data is transmitted reliably and in the proper sequence by using techniques such as token management and checkpointing. It is also responsible for resolving any conflicts that may arise during a session, such as when two applications attempt to initiate a session at the same time. In a nutshell, the session layer establishes, maintains, and terminates the connections between the devices on a network.

In other words, the session layer is a glue between the lower layers we've already covered and the higher layers we will be covering in the next few sections of this chapter.

The best tools to figure out issues with sessions are logs tied to the service you're working with. If you have a problem with an FTP connection, you might want to look into logs of your client and/or server running on the machine you manage. The tools we covered previously can also help if logs are not enough to understand the issue you're trying to resolve.

Presentation layer – SSL and TLS

The **presentation layer** is the sixth layer of the OSI model, and it is responsible for representing the data in a format that can be understood by the application layer. This includes converting data between different formats, such as the **American Standard Code for Information Interchange (ASCII)** and **Extended Binary Coded Decimal Interchange Code (EBCDIC)**. Both are character encoding standards that represent characters using 7-bit (ASCII) or 8-bit (EBCDIC) integers. We also handle encryption and compression in this layer. The presentation layer also ensures that the data is in the correct format for the application to process. It acts as an intermediary between the application and the data, allowing the application to be independent of the specific format of the data being received.

For this layer, the most common encryption standards, **Secure Sockets Layer (SSL)** and **Transport Layer Security (TLS)** are the ones you might need to debug and fix issues with.

TLS is a widely used protocol for securing communications over a network. It is a successor to **SSL** and is used to encrypt and authenticate data transmitted over a network, such as the internet.

TLS works by establishing a secure *tunnel* between two devices, such as a web server and a web browser. This tunnel is used to transmit data between the two devices in an encrypted format, making it difficult for an attacker to intercept and read the data.

The process of establishing a TLS connection involves several steps:

1. **Handshake**: The client and server exchange information to establish a shared understanding of the encryption method and keys that will be used to secure the connection.

2. **Authentication**: The server authenticates itself to the client by providing a digital certificate that contains information about the server's identity and public key. The client can then use this information to verify the server's identity.

3. **Key exchange**: The client and server exchange public keys to establish a shared secret key that will be used to encrypt and decrypt data.

4. **Data encryption**: Once the shared secret key has been established, the client and server can start encrypting data using symmetric encryption algorithms.

5. **Data transfer**: Data is then transferred over the secure connection, protected by the encryption established during the handshake.

TLS has multiple versions, among which the newest are considered more secure. However, older systems might not support the most recent versions. The available TLS versions are as follows:

- **TLS 1.0**: This was the first version of the protocol, released in 1999.

- **TLS 1.1**: This was released in 2006.

- **TLS 1.2**: This was released in 2008. It added support for new cryptographic algorithms and made several other security enhancements.

- **TLS 1.3**: This was released in 2018 and includes the use of forward secrecy, which makes it harder for attackers to decrypt captured data.

SSL is a security protocol that is widely used to secure communications over a network, such as the internet. It was developed by Netscape in the 1990s and was later succeeded by the TLS protocol.

Like TLS, SSL works by establishing a secure *tunnel* between two devices, such as a web server and a web browser. This tunnel is used to transmit data between the two devices in an encrypted format, making it difficult for an attacker to intercept and read the data. It's not recommended to use SSL anymore and it's better to use the latest version of TLS, but you might still encounter systems using it.

The best tool to debug issues with both SSL and TLS is the openssl command. You can use it to test a connection to a server using SSL. For example, you can use the following command to test a connection to a server on port (usually, it's 443 as it's a common port for HTTPS):

```
admin@myhome:~$ openssl s_client -connect myhome:443
```

You can use the openssl command to check the details of an SSL certificate, including the expiration date, the issuing authority, and the public key. For example, you can use the following command to check the details of a certificate:

```
admin@myhome:~$ openssl s_client -connect myhome:443 -showcerts
```

This will allow you to check if the certificates served by your server are valid and what you expect them to be. This is much faster than using a web browser.

Using the openssl command, you can also check which ciphers are supported by a server. For example, you can use the following command to check the ciphers supported by a server:

```
admin@myhome:~$ openssl s_client -connect myhome:443 -cipher
'ALL:eNULL'
# Cut most of the output for readability
SSL handshake has read 5368 bytes and written 415 bytes
---
New, TLSv1/SSLv3, Cipher is AEAD-AES256-GCM-SHA384
Server public key is 2048 bit
Secure Renegotiation IS NOT supported
Compression: NONE
```

```
Expansion: NONE
No ALPN negotiated
SSL-Session:
    Protocol  : TLSv1.3
    Cipher    : AEAD-AES256-GCM-SHA384
    Session-ID:
    Session-ID-ctx:
    Master-Key:
    Start Time: 1674075742
    Timeout   : 7200 (sec)
    Verify return code: 0 (ok)
---
closed
```

Additionally, `openssl` has a built-in diagnostic that can detect known vulnerabilities that exist in your system:

```
admin@myhome:~$ openssl s_client  connect myhome:443 -tlsextdebug
-status
CONNECTED(00000006)
140704621852864:error:1404B42E:SSL routines:ST_CONNECT:tlsv1 alert
protocol version:/AppleInternal/Library/BuildRoots/aaefcfd1-5c95-
11ed-8734-2e32217d8374/Library/Caches/com.apple.xbs/Sources/libressl/
libressl-3.3/ssl/tls13_lib.c:151:
---
no peer certificate available
---
No client certificate CA names sent
---
SSL handshake has read 5 bytes and written 303 bytes
---
New, (NONE), Cipher is (NONE)
Secure Renegotiation IS NOT supported
Compression: NONE
Expansion: NONE
No ALPN negotiated
SSL-Session:
    Protocol  : TLSv1.3
    Cipher    : 0000
    Session-ID:
    Session-ID-ctx:
    Master-Key:
    Start Time: 1674075828
    Timeout   : 7200 (sec)
    Verify return code: 0 (ok)
---
```

It's also worth noting that both SSL and TLS use public key encryption. With this encryption method, we create two files: a private key and a public key. Together, they form a pair. The public key is designed to encrypt data, while the private key is designed to decrypt it. The private key should be kept private at all times, as its name implies. This method of encryption is based on the mathematical properties of large prime numbers, and it is considered to be very secure.

In the case of TLS, public key encryption is used during the *handshake* phase to establish a secure connection between the client and the server. The process works as follows:

1. The server generates a public key and a private key. The public key is sent to the client as part of the server's digital certificate.

2. The client generates a session key, which is used to encrypt data that is sent to the server. The server's public key is used to encrypt the session key. Once the server receives the encrypted data, it uses its private key to decrypt it and retrieve the session key.

3. Once the session key has been decrypted, it is used to encrypt data that is sent between the client and the server.

The next and final layer is the top layer of the OSI model. In the next section, which covers the application layer, we'll cover protocols such as HTTP and FTP, which are commonly used for browsing the web and sharing files.

Application layer – HTTP and FTP

The **application layer** is the seventh and highest layer of the OSI network model. It provides the interface between the software application and the network, allowing the application to access the network's communication services. The application layer defines protocols and services that are specific to the application, such as file transfer (**File Transfer Protocol (FTP)**), email (**Simple Mail Transfer Protocol (SMTP)** and **Internet Message Access Protocol (IMAP)**), and well-known web services (**Hypertext Transfer Protocol (HTTP)**).

Let's take a closer look at HTTP. This communication protocol is used for transferring data on the World Wide Web. It is based on a client-server model, where a web browser (the client) sends a request to a web server, and the server sends back a response.

When a user enters a **Uniform Resource Locator (URL)** in their web browser, the browser sends an HTTP request to the web server associated with that URL. A URL is something you can see in your browser address input field and is usually located at the top of the browser window – for example, `https://google.com`.

The request includes the method (such as `GET`, `POST`, `PUT`, `PATCH`, or `DELETE`), which indicates the type of action the browser wants the server to perform, as well as any additional information, such as data for a `POST` or `PUT` request.

The web server then processes the request, after which the server sends back an HTTP response, which includes a status code (such as 200 for success or 404 for not found) and any data requested by the browser, such as the HTML and CSS that make up the website.

Once the browser receives the response, it parses the HTML, CSS, and – very often – JavaScript to display the website to the user.

HTTP is a stateless protocol, which means that each request is independent and the server does not retain any information about previous requests. However, many web applications use cookies or other techniques to maintain state across multiple requests.

HTTP version 1.1 introduced new features such as persistent connections, a host header field, and byte-serving which, improved the overall performance of the protocol and made it more suitable for heavy usage scenarios. The most recent version of this protocol is 2.0 and is described in detail by RFC 7540 (https://www.rfc-editor.org/rfc/rfc7540.html), which was released in 2015, and updated by RFC 8740 (https://www.rfc-editor.org/rfc/rfc8740.html), which was released in 2020.

To resolve issues with HTTP, you can use any of the tools for debugging network problems, such as tcpdump or ngrep. There are several console and GUI tools available you can use for debugging HTTP. The most common are wget and curl for the console and Postman or Fiddler for the GUI. There will also be debugging tools built into your browser, such as Firefox or Chrome Developer Tools.

We're going to focus on console tools right now, so let's look into wget first. This tool is intended to download files, but we still can use it to debug HTTP. For the first example, we will show detailed information about the request and response:

```
admin@myhome:~$ wget -d https://google.com
DEBUG output created by Wget 1.21.3 on darwin22.1.0.

Reading HSTS entries from /home/admin/.wget-hsts
URI encoding = 'UTF-8'
Converted file name 'index.html' (UTF-8) -> 'index.html' (UTF-8)
--2023-01-20 14:11:02--  https://google.com/
Resolving google.com (google.com)... 216.58.215.78
Caching google.com => 216.58.215.78
Connecting to google.com (google.com)|216.58.215.78|:443... connected.
Created socket 5.
Releasing 0x0000600000d2dac0 (new refcount 1).
Initiating SSL handshake.
Handshake successful; connected socket 5 to SSL handle
0x00007fca4a008c00
certificate:
  subject: CN=*.google.com
  issuer:  CN=GTS CA 1C3,O=Google Trust Services LLC,C=US
```

```
X509 certificate successfully verified and matches host google.com

---request begin---
GET / HTTP/1.1
Host: google.com
User-Agent: Wget/1.21.3
Accept: */*
Accept-Encoding: identity
Connection: Keep-Alive

---request end---
HTTP request sent, awaiting response...
# Rest of the output omitted
```

You might need to send an HTTP request with specific headers. To do so, you can use the
`--header` option:

```
admin@myhome:~$ wget --header='<header-name>: <header-value>' <url>
```

POST is a special HTTP request – it uses a URL, but it will also require some request data as it's
intended to send some data to your system, be it a username and password or files. To send an HTTP
POST request with prepared data, you can use the `--post-data` option, like this:

```
admin@myhome:~$ wget --post-data=<data> <url>
```

A more powerful tool for debugging HTTP issues is `curl`. To send an HTTP GET request and
display the response, you can use the following command:

```
admin@myhome:~$ curl linux.com
<html>
<head><title>301 Moved Permanently</title></head>
<body>
<center><h1>301 Moved Permanently</h1></center>
<hr><center>nginx</center>
</body>
</html>
```

To send an HTTP POST request and display the response, you can use the `-X  POST` option and
the `-d` option:

```
admin@myhome:~$ curl -X POST -d <data> <url>
```

Using the `-X` option, you can send other types of requests, such as PATCH or DELETE.

To send an HTTP request with specific headers, you can use the -H option:

```
admin@myhome:~$ curl -H '<header-name>: <header-value>' <url>
```

To display the response headers only, use the following code:

```
admin@myhome:~$ curl -I linux.com
HTTP/1.1 301 Moved Permanently
Connection: keep-alive
Content-Length: 162
Content-Type: text/html
Location: https://linux.com/
Server: nginx
X-Pantheon-Styx-Hostname: styx-fe3-b-7f84d5c76-q4hpv
X-Styx-Req-Id: f1409a84-9832-11ed-abcb-7611ac88195f
Cache-Control: public, max-age=86400
Date: Fri, 20 Jan 2023 13:14:42 GMT
X-Served-By: cache-chi-kigq8000084-CHI, cache-fra-eddf8230050-FRA
X-Cache: HIT, HIT
X-Cache-Hits: 37, 1
X-Timer: S1674220483.602573,VS0,VE1
Vary: Cookie, Cookie
Age: 62475
Accept-Ranges: bytes
Via: 1.1 varnish, 1.1 varnish
```

You can also show detailed information about the request and response by using the -v option, like this:

```
admin@myhome:~$ curl -v <url>
```

By sending different types of requests and analyzing the responses, you can use either wget or curl to debug various HTTP issues, such as connectivity problems, response errors, and performance issues. As usual, you can refer to the full documentation of both tools to deepen your understanding of how to use them.

In this section, we introduced several network layers, as defined by the ISO/OSI standard. Each of them standardizes the functions of different elements of networking communication. Next, we are going to discuss firewalls.

Firewalls

A **firewall** is a security measure that controls incoming and outgoing network traffic based on predefined rules and policies. It is typically placed between a protected network and the internet, and its main purpose is to block unauthorized access while allowing authorized communication. Firewalls can be hardware-based or software-based, and they can use a variety of techniques, such as packet filtering, stateful inspection, and application-level filtering, to control network traffic. In this section, we're going to look into a firewall available on Linux systems.

To control a Linux firewall, you will need to use `iptables`, `ufw`, `nftables`, or `firewalld`. Packet filtering is built into the Linux kernel, so those CLI tools will interact with it.

iptables

iptables is the most verbose tool for controlling a firewall, meaning it does not have much abstraction built into it, but it's important to understand the basic concepts so that we can move on to more user-friendly tools.

As mentioned previously, `iptables` allows you to create rules for filtering and manipulating network packets, and it can be used to control incoming and outgoing network traffic based on various criteria, such as IP or MAC addresses, ports, and protocols.

`iptables` uses several concepts to organize rules and divide them into functional parts: tables, chains, rules, and targets. The most general are tables to organize the rules.

There are three tables we can use: `filter`, `nat`, and `mangle`. The `filter` table is used to filter incoming and outgoing packets, the `nat` table is used for **Network Address Translation** (**NAT**), which we will get back to later, and the `mangle` table is used for advanced packet alteration.

Each table contains a set of chains, which are used to organize the rules. The `filter` table, for example, contains three predefined chains: INPUT, OUTPUT, and FORWARD. The INPUT chain is used for incoming packets, the OUTPUT chain is used for outgoing packets, and the FORWARD chain is used for packets that are being forwarded through the network.

Each chain contains a set of rules, which are used to match packets and decide what to do with them. Each rule has a match condition and an action. For example, a rule might match packets coming from a specific IP address and drop them, or it might match packets going to a specific port and accept them.

Each rule has a target, which is the action that should be taken when the rule's match condition is met. The most common targets are ACCEPT, DROP, and REJECT. ACCEPT means to allow the packet through the firewall, DROP means to discard the packet without any feedback to the other end, and REJECT means to refuse the packet actively so that the remote end will know access is rejected to the said port.

The default table of `iptables` will add rules to the filter table and by default, each chain (INPUT, OUTPUT, and FORWARD) has a default policy set to ACCEPT. You can also create additional tables and direct packets to this table for later processing. It's a good practice to set at least the FORWARD and INPUT policies to DROP:

```
admin@myhome:~$ sudo iptables -P INPUT DROP
admin@myhome:~$ sudo iptables -P FORWARD DROP
```

At the same time, we can allow all loopback interface access to ACCEPT:

```
admin@myhome:~$ sudo iptables -A INPUT -i lo -j ACCEPT
```

Additionally, all packets that are in the ESTABLISHED or RELATED state should be accepted; otherwise, we will lose all established connections or connections that are in the process of being established:

```
admin@myhome:~$ sudo iptables -A INPUT -m conntrack --ctstate
ESTABLISHED,RELATED -j ACCEPT
```

To allow HTTP and HTTPS traffic, we can do the following:

```
admin@myhome:~$ sudo iptables -A INPUT -p tcp --dport 80 -j ACCEPT
admin@myhome:~$ sudo iptables -A INPUT -p tcp --dport 443 -j ACCEPT
```

It's a good idea to allow SSH traffic so that we can remotely log into this machine:

```
admin@myhome:~$ sudo iptables -A INPUT -p tcp --dport 22 -j ACCEPT
```

Here are some other more commonly used options you can use in iptables:

- -A or --append: Appends a rule to the end of a chain
- -I or --insert: Inserts a rule at a specific position in a chain
- -D or --delete: Deletes a rule from a chain
- -P or --policy: Sets the default policy for a chain
- -j or --jump: Specifies the target for a rule
- -s or --source: Matches packets based on the source IP address or network
- -d or --destination: Matches packets based on the destination IP address or network
- -p or --protocol: Matches packets based on the protocol (for example, TCP, UDP, or ICMP)
- -i or --in-interface: Matches packets based on the incoming interface
- -o or --out-interface: Matches packets based on the outgoing interface
- --sport or --source-port: Matches packets based on the source port
- --dport or --destination-port: Matches packets based on the destination port
- -m or --match: Adds a match extension, which allows you to match packets based on additional criteria such as connection state, packet length, and more

There are many more features available when dealing with iptables, such as setting up NAT, interface bonding, TCP multipath, and many more. We will cover some of these in the *Advanced topics* section of this chapter.

nftables

nftables is a versatile tool for managing firewall configurations, offering a more streamlined and user-friendly approach compared to `iptables`, which is known for its verbosity and lack of built-in abstraction.

`nftables` employs a logical structure to organize rules that comprises tables, chains, rules, and verdicts. Tables serve as the top-level containers for rules and are instrumental in categorizing them. `nftables` offers several table types: `ip`, `arp`, `ip6`, `bridge`, `inet`, and `netdev`.

Within each table, there are chains, which help organize the rules further between categories: `filter`, `route`, and `nat`.

Each chain comprises individual rules, which serve as the criteria for matching packets and determining subsequent actions. A rule consists of both a matching condition and a verdict. For instance, a rule may match packets originating from a specific IP address and instruct the firewall to drop them, or it might match packets headed for a particular port and dictate acceptance.

Let's set a default policy for incoming and forwarded packets to "drop" (stop processing the packet and don't respond):

```
sudo nft add rule ip filter input drop
sudo nft add rule ip filter forward drop
```

Additionally, it's common practice to allow all loopback interface access:

```
sudo nft add rule ip filter input iifname "lo" accept
```

To ensure established and related connections are permitted, you can run the following command:

```
sudo nft add rule ip filter input ct state established,related accept
```

You can run the following command to allow HTTP and HTTPS traffic:

```
sudo nft add rule ip filter input tcp dport {80, 443} accept
```

Finally, to enable SSH traffic for remote access, you can use the following command:

```
sudo nft add rule ip filter input tcp dport 22 accept
```

These are some frequently used options in `nftables`:

- `add`: Appends a rule to the end of a chain
- `insert`: Inserts a rule at a specific position in a chain
- `delete`: Deletes a rule from a chain
- `chain`: Specifies the target for a rule

- `ip saddr`: Matches packets based on the source IP address or network

- `ip daddr`: Matches packets based on the destination IP address or network

- `ip protocol`: Matches packets based on the protocol (for example, TCP, UDP, or ICMP)

- `iifname`: Matches packets based on the incoming interface

- `oifname`: Matches packets based on the outgoing interface

- `tcp sport`: Matches packets based on the source port

- `tcp dport`: Matches packets based on the destination port

- `ct state`: Adds a match extension, allowing packet matching based on additional criteria such as connection state, packet length, and more

`nftables` is set to be a replacement for `iptables`, but both are frequently used in modern systems. Let's move on to other tools that are more abstract and by this, more user-friendly.

ufw

`ufw` is a frontend for the Linux `iptables` firewall and provides a simple and easy-to-use interface for managing it. `ufw` is designed to be easy to use, and it automatically sets up the `iptables` rules for you based on the configuration options you specify. It's much more user-friendly and easier to use for more common tasks.

Before you start to use `ufw`, you will need to enable it so that all rules you add or remove will persist after system reboot. To do just that, run the following command:

```
admin@myhome:~$ sudo ufw enable
```

To open TCP ports `80` and `443` using `ufw`, you can use the following commands:

```
admin@myhome:~$ sudo ufw allow 80/tcp
admin@myhome:~$ sudo ufw allow 443/tcp
```

Alternatively, you can open both ports with one command:

```
admin@myhome:~$ sudo ufw allow 80,443/tcp
```

Once you have opened the ports, you can verify the changes by checking the status of `ufw`:

```
admin@myhome:~$ sudo ufw status
```

`ufw` is available on all major Linux distributions, including Debian Linux, Ubuntu Linux, Arch Linux, and Fedora Linux. However, in some cases, you will need to install it as it's not part of the default system.

firewalld

Another tool you can use to manage a firewall on Linux is `firewalld`. This is a program that was created to streamline the dynamic configuration of firewalls. One of the big features of `firewalld` is zones, which allow you to declare different levels of trust in interfaces and networks. It is included by default in many popular Linux distributions such as Red Hat Enterprise Linux, Fedora, CentOS, and Debian. Some other Linux distributions, such as Ubuntu, do not include `firewalld` by default but it can be installed and used on those systems as well.

To open TCP ports `80` and `443` using `firewalld`, you can use the `firewall-cmd` command-line tool. Here are the commands to open those ports:

```
admin@myhome:~$ sudo firewall-cmd --add-port=80/tcp --permanent
admin@myhome:~$ sudo firewall-cmd --add-port=443/tcp --permanent
```

You can open both ports with one command:

```
admin@myhome:~$ sudo firewall-cmd --add-port=80/tcp --add-port=443/tcp
--permanent
```

After adding the ports, you need to reload the firewall for the changes to take effect:

```
admin@myhome:~$ sudo firewall-cmd --reload
```

You can also check the status of the ports using the following command:

```
admin@myhome:~$ sudo firewall-cmd --list-ports
```

Whatever tool you use to configure your firewall, it's always a good idea to set default rule policies to `DROP` and only allow the traffic you expect to be handled by your system. There are some topics we don't have much space for in this chapter, but it's useful to know about possibilities when dealing with networking.

Advanced topics

In this section, we will cover more advanced uses of network features. Some are very common (such as port forwarding or NAT), while some are less known. Let's start with well-known features you will most likely encounter very often and then make our way down to more advanced and less-known features.

NAT

Network Address Translation (NAT) is a technique that involves mapping one network to another. The original reason for this was to simplify routing whole network segments without changing the address of every host in the packets.

Source NAT (SNAT) is a type of NAT that changes the source IP address of a packet. It is used to allow hosts on a private network to access the internet using a single public IP address.

Destination NAT (DNAT) is a type of NAT that changes the destination IP address of a packet. It is used to forward incoming traffic to a specific internal host based on the destination IP address. This is often used to allow external clients to access services running on internal hosts using a public IP address.

To set up NAT using `iptables`, you can use the following basic commands:

```
admin@myhome:~$ echo 1 > /proc/sys/net/ipv4/ip_forward
```

This will enable forwarding, which allows your machine to forward packets from one interface to another. This particular command updates the Linux kernel configuration on the fly by using the `proc` filesystem. You can achieve the same thing using the `sysctl` command:

```
admin@myhome:~$ sudo sysctl -w net.ipv4.ip_forward=1
```

To configure NAT for a local network such as 192.168.10.0/24, you will need to run the following commands as `root user`:

```
admin@myhome:~$ sudo iptables -t nat -A POSTROUTING -o eth0 -j
MASQUERADE
admin@myhome:~$ sudo iptables -A FORWARD -i eth0 -o eth1 -m state
--state RELATED,ESTABLISHED -j ACCEPT
admin@myhome:~$ sudo iptables -A FORWARD -i eth1 -o eth0 -j ACCEPT
```

Note that `eth0` is the interface connected to the internet, and `eth1` is the interface connected to the internal network.

The `MASQUERADE` target is used to implement NAT on a Linux router. When packets pass through the router and are sent out to the internet, the `MASQUERADE` target changes the source address of the packets to the router's public IP address. This allows devices on the internal network to communicate with devices on the internet using the router's public IP address as a source address, effectively hiding the internal network from the internet.

The `MASQUERADE` target is typically used in the `POSTROUTING` chain of the `nat` table and is commonly applied to the interface that is connected to the internet. It works only with dynamically assigned IP addresses (DHCP) and it's mostly used in the case of a home router.

Port forwarding

Port forwarding is a technique that's used to direct network traffic from one network address and port to another. This can be useful for directing incoming traffic to a specific service or application running on a computer or network device. This helps with accessing services or applications on a private network from a remote location, or for making services or applications running on a private network accessible to the public.

Essentially, it's another use of NAT as you will change the destination IP (and also port) of the packet that arrived on your machine.

To forward packets coming into TCP port 80 on the eth0 interface to internal IP 192.168.10.101 on port 8080 using iptables, you can use the following commands:

```
admin@myhome:~$ sudo iptables -t nat -A PREROUTING -i eth0 -p tcp
--dport 80 -j DNAT --to-destination 192.168.10.101:8080
admin@myhome:~$ sudo iptables -t nat -A POSTROUTING -j MASQUERADE
```

We need MASQUERADE here as we want to hide the internal IP from the outside.

Interface bonding

Network interface bonding is a feature in Linux that allows multiple network interfaces to be combined into a single *bonded* interface. This can provide increased bandwidth, redundancy, and failover capabilities. Several different bonding modes can be used, each with its benefits and drawbacks. Some common modes include active-backup, balance-rr, and 802.3ad. active-backup mode means that one of the two bonding interfaces is used as a backup when the primary device is down. balance-rr will use both interfaces simultaneously with a round-robin policy. The 802.3ad bonding creates aggregation groups that share the same technical specifications (speed and duplex settings). You can read more about modes and bonding settings on the official Linux kernel website: https://www.kernel.org/doc/Documentation/networking/bonding.txt

TCP multipath

TCP multipath refers to using multiple paths to send and receive data between two points in a network, rather than just one. This can improve the reliability and performance of the network by allowing for failover if one of the paths becomes unavailable, and by allowing for load balancing across multiple paths. This can be achieved using various techniques, such as by using multiple interfaces on a device or by using multiple routing paths through the network.

Configuring multipath is quite easy with the use of the iproute2 package. To configure multipath on Linux using the eth0 and eth1 interfaces, you will need to run the following command:

```
admin@myhome:~$ ip route add default scope global nexthop via
192.168.1.1 dev eth0 nexthop via 192.168.2.1 dev eth1 weight 1
```

This command creates a new default route that uses both the eth0 and eth1 interfaces, with a weight of 1. The IP addresses used in this example (192.168.1.1 and 192.168.2.1) should be replaced with the actual IP addresses of the next hop routers on the eth0 and eth1 interfaces.

When running `ip route show`, you will see that a new multipath route is present. To start using it, you will need to change the default route:

```
admin@myhome:~$ ip route change default scope global nexthop via
192.168.1.1 dev eth0 nexthop via 192.168.2.1 dev eth1 weight 1
```

You can read more about multipath and how to use it at `https://www.multipath-tcp.org/`.

BGP

Border Gateway Protocol (**BGP**) is a routing protocol that's used to distribute routing information within a single **autonomous system** (**AS**) or between multiple autonomous systems on the internet. BGP is used to build routing tables in routers in the internet backbone, as well as in enterprise networks.

BGP routers exchange routing information with their neighbors, which can be other BGP routers in the same AS or a different AS. When a BGP router receives routing information from its neighbors, it uses a set of rules and policies to decide which routes to add to its routing table and advertise to its neighbors. This allows BGP to support multiple paths to a destination, and to choose the best path based on various factors such as distance, cost, or preference.

BGP is considered a path vector protocol because it exchanges information about the complete path to a destination, rather than just the next hop. This allows BGP to support advanced features such as routing policy and traffic engineering.

There are several ways to use BGP on a Linux machine, depending on your specific use case and network environment.

BIRD is a routing daemon for Linux that supports BGP and other routing protocols. You can configure BIRD so that it acts as a BGP speaker and exchange routing information with other BGP routers. BIRD can be installed on most Linux distributions and can be configured using a simple configuration file.

Quagga is another open source routing software suite for Linux that supports BGP, OSPF, and other routing protocols. You can configure Quagga to act as a BGP speaker and exchange routing information with other BGP routers. Quagga can be installed on most Linux distributions and can be configured using a command-line interface or a configuration file.

Free Range Routing (**FRR**) is a routing software suite for Linux that supports BGP, OSPF, and other routing protocols; it's a fork of Quagga.

You can read more about BGP and using BIRD in the following Linux Journal article: `https://www.linuxjournal.com/content/linux-advanced-routing-tutorial`.

Summary

In this chapter, we covered basic networking topics that you will probably encounter during your work in a DevOps team. This is a starting point and a base to help you understand network-related topics when dealing with services running inside a container. You will probably also want to expand your knowledge on this topic by reading about the IPv6 protocol, which is yet to replace IPv4.

In the next chapter, we will switch focus to a **version control system** (**VCS**) that is mainly used in modern organizations: **Git**.

Git, Your Doorway to DevOps

Git is a free and open source **version control system** (**VCS**) that is widely used by software developers and teams to track changes to their code base and collaborate on projects. It allows multiple people to work on the same code base without overwriting each other's changes, and it keeps a record of every change made to the code, making it easy to roll back to a previous version if necessary.

Git was created by Linus Torvalds in 2005 for the development of the Linux kernel, and it has since become the de facto standard for version control in the software industry. It is used by millions of developers worldwide and is supported by a large and active open source community.

In this chapter, we will cover the most commonly used Git commands and how to use them. We will start with the basics of setting up a Git repository and making your first commit, and then move on to more advanced topics such as branching and merging.

The chapter will cover the following topics:

- Basic Git commands
- Local versus remote Git repositories
- GitFlow and GitHub Flow

Technical requirements

In this chapter, you will need a system with a Bash shell. You will have to have the Git command available or be able to install it in this system. We prefer Linux or macOS systems, but it is possible to set up Windows to have functional Bash and Git. The installation of this environment is beyond the scope of this book.

Basic Git commands

There are many commands that you can use with Git, but some of the most commonly used ones include the following:

- `git config`: This is the command used to configure your local Git environment. The configuration can be global; the values will then be kept in your home directory in the `.gitconfig` file. The values can only be set per repository, and then they will be kept within the repository.

- `git init`: This initializes a new Git repository. When you run this command in a directory, it creates a new `.git` directory in the root of the project, which is used to track changes made to the project's files.

- `git clone`: This creates a local copy of a remote Git repository. When you run this command, it creates a new directory with the same name as the repository and clones all of the files and their history into that directory.

- `git add`: This stages files for commit. When you make changes to a file in a Git repository, those changes are not automatically tracked. You must use the `git add` command to tell Git to track changes you have made.

- `git commit`: This saves your changes to the Git repository. When you run this command, it opens a text editor for you to write a commit message, which is a short description of the changes you have made. After you write the commit message and save it, the changes you have made are saved to the repository, and a new commit is created.

- `git push`: This sends your local commits to a remote repository. When you run this command, it pushes all of your local commits to the remote repository, updating the project's history and making your changes visible to other developers working on the project.

- `git pull`: This retrieves updates from a remote repository and merges them into your local repository. When you run this command, it fetches the latest changes from the remote repository and merges them into your local repository, bringing your copy of the project up to date.

- `git branch`: This creates, lists, or deletes branches in a Git repository. Branches allow you to work on multiple versions of a project simultaneously and are often used for feature development or bug fixes.

- `git checkout` or `git switch`: This switches between branches or restores files in your working directory. When you run this command, it switches your working directory to the branch you specify or restores the specified files to their state at a previous commit.

- `git merge`: This merges one branch into another. When you run this command, it combines the changes made in the specified branch into the current branch, creating a new commit that represents the combined changes.

- `git stash`: This temporarily stores changes that you are not ready to commit. When you run this command, it saves your changes to a temporary area and restores your working directory to the state it was in at the last commit. You can later use the `git stash apply` command to restore the stashed changes to your working directory.

Configuring the local Git environment

Before using the Git command, there are at least two options you should set. They are your name and your email address. This is done using the `git config` command. Ahead, I am going to demonstrate setting up the name of the Git user and their email address. We are going to set up the global variables. They will be used by default for every repository cloned locally unless set up specifically for a repository:

```
admin@myhome:~$ git config -global user.name "Damian Wojsław"
admin@myhome:~$ git config -global user.email damian@example.com
```

Now, when we take a look at the `~/.gitconfig` file, we will see this section:

```
# This is Git's per-user configuration file
[user]
    name = Damian Wojsław
    email = damian@example.com
```

There are more configuration options, such as default editor, and so on, but they are out of scope for this section.

Setting up a local Git repository

Before you can start using Git, you need to create a **repository** (also known as a **repo**). A repository is a directory where Git stores all the files and metadata for a project.

To create a new repository, you can use the `git init` command. This creates a new directory with a `.git` subdirectory, which contains all the necessary files for the repository.

For example, to create a new repository in the current directory, you can run the following command:

```
admin@myhome:~$ mkdir git-repository && cd git-repository
admin@myhome:~/git-repository$ git init
```

This creates a new repository in the current directory and sets up the necessary files and metadata. Once you have a Git repository set up, you can start adding and committing files to it.

To add a file to the repository, you can use the `git add` command. This adds the file to the staging area, which is a list of changes that will be included in the next commit.

To add a file called `main.c` to the staging area, you can run the following command:

```
admin@myhome:~/git-repository$ git add main.py
```

You can also add multiple files or directories by separating their names with a space:

```
admin@myhome:~/git-repository$ git add main.py utils.py directory/
```

To commit changes in the staging area, you can use the `git commit` command. This creates a new commit that includes all changes in the staging area.

Every commit needs to have a commit message. This is a short description of the changes you're making. To specify a commit message, you can use the `-m` option followed by the message:

```
admin@myhome:~/git-repository$ git commit -m "Added main.py and
additional utils"
```

You can also use the `git commit` command without the `-m` option to open a text editor where you can write a more detailed commit message.

On every step while working in your Git repository, you can use the `git status` command. This is used to view the current state of a Git repository. It shows which files have been modified, added, or deleted, and whether they have been staged for commit.

There are numerous ways to format a commit message, and almost every project has its own convention. In general, a good practice is to add an ID of `issue` from the ticketing system, followed by a short description that has 72 characters or fewer. The second line should be left empty, and a more detailed description follows on line three. Here's an example of such a commit message:

```
[TICKET-123] Adding main.py and utils.py

main.py contains the root module of the application and is a default
entry point for a Docker image.

utils.py contains helper functions for setting up the environment and
reading configuration from the file.

Co-authors: other-commiter@myproject.example
```

When you run `git status`, it will display a list of modified files, as well as any untracked files that are present in the repository. It will also show the current branch and the status of the staging area.

Here is an example of the output of the `git status` command:

```
admin@myhome:~/git-repository$ git status

On branch main
```

```
Your branch is up to date with 'origin/main'.

Changes not staged for commit:

  (use "git add <file>..." to update what will be committed)

  (use "git restore <file>..." to discard changes in working
directory)

        modified:    main.py

        deleted:     root/__init__.py


Untracked files:

  (use "git add <file>..." to include in what will be committed)

        ui/main.py



  no changes added to commit (use "git add" and/or "git commit -a")
```

In this example, git status shows that two files have been modified (main.py and __init__.py) and one file is untracked (main.py). It also indicates that the current branch is main and that the branch is up to date with the origin/main branch.

git status is a useful tool to understand the current state of your repository and to identify which files have been modified and need to be committed. You can use it to get a quick overview of changes that have been made and to see which files are ready to be committed. You can also use it to see which branch you are currently on, and whether you are up to date with the remote repository.

The git revert command is used to undo changes that have been made to a repository. It does this by creating a new commit that undoes the changes introduced by a previous commit.

For example, imagine that you have made several commits to your repository, and you want to undo the changes introduced by the last commit. You can use the git revert command to do this:

```
admin@myhome:~/git-repository$ git revert HEAD
```

This will create a new commit that undoes the changes introduced by the last commit. The commit history of your repository will now look as if the last commit never happened.

You can also use the `git revert` command to undo the changes introduced by a specific commit. To do this, you need to specify the commit hash of the commit that you want to revert:

```
admin@myhome:~/git-repository$ git revert abc123
```

This will create a new commit that undoes the changes introduced by the commit with the `abc123` hash.

It is important to note that `git revert` does not delete commits or destroy history. Instead, it creates a new commit that undoes the changes introduced by a previous commit. This means that the changes are still present in the repository, but they are no longer visible in the current state of the branch.

If you need to permanently delete commits, you can use the `git reset` command or the `git filter-branch` command. However, these commands can permanently destroy history, so you should use them with extreme caution.

Now that we know how to work on our local repository, we can talk about remote and local copies of the repository.

Local versus remote Git repositories

In Git, a **repository** is a collection of files and their history, as well as configuration files that are used to manage the repository. A repository can be either local or remote.

A **local repository** is a repository that is stored on your local machine. When you initialize a new Git repository using the `git init` command or when you clone an existing repository using the `git clone` command, you are creating a local repository. Local repositories are useful for working on projects when you don't have an internet connection or when you want to keep a copy of the project on your own machine.

A **remote repository** is a repository that is stored on a server and accessed over the internet. When you push commits to a remote repository using the `git push` command, you are updating the remote repository with your local changes. Remote repositories are useful for collaborating with other developers, as they allow multiple people to work on the same project and share their changes with each other.

Git uses a distributed VCS, which means that each developer has a complete copy of the repository on their local machine. This allows developers to work on the project locally and push their changes to a remote repository when they are ready to share them with others. It also allows developers to collaborate on a project even when they are not connected to the internet.

The `git clone` command is used to create a local copy of a remote Git repository. When you run this command, it creates a new directory with the same name as the repository and clones all of the files and their history into that directory.

Here is an example of how you might use the `git clone` command:

```
admin@myhome:~/git-repository$ git clone https://github.com/user/
repo.git
```

This will create a new directory called `repo` and clone the repository located at `https://github. com/user/repo.git` into that directory.

You can also specify a different name for the local directory by adding it as an argument:

```
admin@myhome:~/git-repository$ git clone https://github.com/user/repo.
git my-local-repo
```

This will create a new directory called `my-local-repo` and clone the repository into it.

If you are using a Git hosting service such as GitHub, you can also use a shorthand version of the repository URL:

```
admin@myhome:~/git-repository$ git clone git@github.com:user/repo.
git
```

This will clone the repository using the SSH protocol, which requires you to have an SSH key configured for your account.

The `git clone` command is a useful way to create a local copy of a remote repository, whether you are starting a new project or contributing to an existing one. It allows you to work on a project locally and push your changes back to the remote repository when you are ready.

Interacting with remote repositories

The `git pull` command is used to retrieve updates from a remote repository and merge them into your local repository. It is a combination of the `git fetch` command, which downloads updates from the remote repository, and the `git merge` command, which merges updates into your local repository.

The `git fetch` command downloads updates from the remote repository but does not merge them into your local repository. Instead, it stores the updates in a temporary area called *remote-tracking branches*. You can use the `git fetch` command to update your remote-tracking branches and see which changes are available, but you will need to use the `git merge` command to actually incorporate those changes into your local repository.

Here are some common keywords that you might use when working with `git fetch`:

- `origin`: This is the default name for the remote repository that you cloned from. You can use `origin` to specify the remote repository that you want to fetch updates from. It's also possible to change the default name and add multiple remote repositories.

- main or master: master is the name of the default branch in a Git repository. main is the new default name introduced in the GitHub platform.

- REMOTE_HEAD: This is a special reference that points to the head commit of the branch on the remote repository. You can use REMOTE_HEAD to fetch updates for the branch that is currently checked out on the remote repository.

- HEAD: This is a special reference that points to the head commit of the current branch in your local repository.

Here is an example of using git fetch to update the main branch on the origin repository:

```
admin@myhome:~/git-repository$ git fetch origin main

From gitlab.com:gstlt/git-repository
 * branch              main          -> FETCH_HEAD
```

This will download updates for the master branch on the origin repository and store them in the origin/master remote-tracking branch.

You can then use the git merge command to merge the updates into your local repository:

```
admin@myhome:~/git-repository$ git merge origin/main
Already up to date.
```

This will incorporate the updates from the origin/master remote-tracking branch into your local master branch. In this case, we do not have anything to be merged.

Alternatively, you can use the git pull command to accomplish both steps in a single command:

```
admin@myhome:~/git-repository$ git pull origin main
Already up to date.
```

This will fetch the updates from the origin repository and merge them into your local repository unless you've configured your Git client, as we suggested at the beginning of this chapter—in this case, Git will try to rebase against the main branch in a remote repository.

git rebase and git merge are commands that are both used to integrate changes from one branch into another branch. However, they work in slightly different ways and have different implications for your repository.

git rebase is a command that is used to apply a series of commits from one branch into another branch. When you run git rebase, it looks at commits that are present in the target branch but not in the source branch, and applies them one by one to the source branch. This has the effect of *replaying* the commits from the source branch on top of the target branch.

For example, imagine that you have two branches in your repository: `main` and `develop`. `main` represents the main development branch, and `develop` represents a feature that you are working on. If you make several commits to the `develop` branch and then run `git rebase main`, Git will apply those commits to the `main` branch one by one, as if you had made them directly on the `main` branch.

`git merge`, however, is a command that is used to combine changes from one branch into another branch. When you run `git merge`, it looks at changes that are present in the source branch and applies them to the target branch in a single commit. This has the effect of creating a new commit on the target branch that incorporates all changes from the source branch.

For example, you have the same two branches as in the rebase example: `main` and `develop`. If you make several commits to the `develop` branch and then run `git merge develop`, Git will create a new commit on the `main` branch that incorporates all of the changes from the `develop` branch.

Both `git rebase` and `git merge` are useful tools for integrating changes from one branch into another, but they have different implications for your repository. `git rebase` keeps a linear history and avoids unnecessary additional (merge) commits, but it can also cause conflicts if the target branch has been modified since the source branch was created.

`git merge` is a more straightforward way to combine changes, but it can create a lot of merge commits, which can make the history of the repository more difficult to read, even with graphical tools.

Before merging or rebasing our local repository against a remote version of it, it's useful to check what the differences are between the two. For that purpose, we have the `git diff` command.

What's the git diff command?

The `git diff` command is used to compare changes that have been made to a repository. It shows the differences between two versions of a file or between two branches in the repository (local or remote).

Here are some common uses of the `git diff` command.

Compare the differences between the current state of a file and the most recent commit:

```
admin@myhome:~/git-repository$ git diff path/to/file
```

This will show the differences between the current state of the file and the version that was last committed.

Compare the differences between two commits:

```
admin@myhome:~/git-repository$ git diff hash1 hash2
```

This will show the differences between the two commits you've provided.

Compare the differences between a branch and its upstream branch:

```
admin@myhome:~/git-repository$ git diff branch-name @{u}
```

This will show the differences between the specified branch and its upstream counterpart branch.

Compare the differences between the staging area and the most recent commit:

```
admin@myhome:~/git-repository$ git diff --staged
```

This will show the differences between changes that have been added to the staging area and the most recent commit.

If you've already added some files to be committed but want to double-check before creating a commit, you can use the following command:

```
admin@myhome:~/git-repository$ git diff --cached
```

This will show you all the changes to files you've performed the git add command on.

To compare local and remote branches of the same repository, you will need to refer to the remote branch, which can be done by executing git diff:

```
admin@myhome:~/git-repository$ git diff main origin/main
```

origin/main refers to a remote branch where the remote repository is named as an "origin."

git diff is useful for understanding what changed in the Git repository and how those changes will be incorporated when you commit them. You can use the various options and arguments of the git diff command to specify versions of the files or branches you want to compare.

Since we've covered comparing changes in your repository, there's one particular case where this is a highly useful skill to have: solving conflicts in your repository while rebasing or merging your changes.

Viewing the commit history

Git keeps a record of every commit made to a repository, and you can view the commit history using the git log command. This displays a list of all commits in the repository, along with their commit messages and the date they were made.

Some common use cases for the git log command include the following:

- **Reviewing the history of a repository**: You can use git log to view the entire commit history of a repository, including changes that have been made and the reasons for those changes. This can be helpful when you are working on a project and need to understand how it has evolved over time.

- **Finding the source of a bug**: If you are trying to track down the source of a bug in a repository, you can use `git log` to view the commit history of the affected files and see which commit introduced the bug.

- **Debugging issues**: If you are trying to troubleshoot an issue in a repository, you can use `git log` to view the history of the affected files and see what changes might have caused the issue.

You can use a variety of options with `git log` to customize its output and filter commits that are displayed. For example, you can use the `--author` option to display only commits made by a specific person or the `--since` option to display only commits made in the last month.

Here are some examples of using the `git log` command.

To display the commit history for the current repository, invoke a `git log` command:

```
admin@myhome:~/git-repository$ git log
commit 9cc1536fb04be3422ce18a6271ab83f419840ae3 (HEAD -> main, origin/
main, origin/HEAD)
Author: Grzegorz Adamowicz <grzegorz@devopsfury.com>
Date:   Wed Jan 4 10:13:26 2023 +0100
    Add first version of the table schema

commit fb8a64f0fd7d21c5df7360cac427668c70545c83
Author: Grzegorz Adamowicz <grzegorz@devopsfury.com>
Date:   Tue Jan 3 22:02:25 2023 +0100

    Add testing data, mock Azure class, remove not needed comments

commit ae0ac170f01142dd80bafcf40fd2616cd1b1bc0b
Author: Grzegorz Adamowicz <grzegorz@devopsfury.com>
Date:   Tue Dec 27 14:50:47 2022 +0100

    Initial commit
```

You can check the commit history for a specific file by running the following command:

```
admin@myhome:~/git-repository$ git log path/to/file
```

With this command, you can display the commit history for a specific branch:

```
admin@myhome:~/git-repository$ git log branch-name
```

This will display the commit history for the specified branch, showing only the commits that were made on that branch.

You can display the commit history for a range of commits with this command:

```
admin@myhome:~/git-repository$ git log hash1..hash2
```

This will display the commit history for a range of commits between the `hash1` commit ID and the `hash2` commit ID, showing only the commits that occurred in that range.

`git log --oneline` will display the commit history in a compact format, showing only the commit hash and a message for each commit:

```
admin@myhome:~/git-repository$ git log --oneline
```

With this command, you can display the commit history with diffs:

```
admin@myhome:~/git-repository$ git log -p
```

This will display the commit history, along with the diffs for each commit, showing the changes that were made in each commit.

These are just a few examples of how you can use the `git log` command. You can find more information about the available options in the Git documentation.

In the next section, we are going to look into shortening our Git history before merging our changes back to the `main` branch. This is useful if we have a very long commit history in our local development branch.

Branching

`git branch` and `git switch` are two commands that are used to manage branches in a Git repository.

`git branch` is a command that is used to create, list, or delete branches in a repository. You can use it to create a new branch by specifying the name of the branch as an argument, like so:

```
admin@myhome:~/git-repository$ git branch new-branch
```

This will create a new branch called `new-branch` that is based on the current branch.

You can use the `git branch` command with the `-a` option to list all branches in the repository, including both local branches and remote-tracking branches:

```
admin@myhome:~/git-repository$ git branch -a
```

You can use the `git branch` command with the `-d` option to delete a branch, as in the following example:

```
admin@myhome:~/git-repository$ git branch -d old-branch
```

This will delete the `old branch` but only if it has been already fully merged into the upstream branch. If you wish to remove this branch, you can add a `--force` option or use the `-D` option, which is an alias to the `--delete --force` Git branch.

Finally, to remove a remote branch, we will need to use the `git push` command instead:

```
admin@myhome:~/git-repository$ git push origin --delete old-branch
```

This is a destructive command and can't be undone, so you should act with caution when using it.

`git switch` is a command that is used to switch between branches in a repository. You can use it to switch to a different branch by specifying the name of the branch as an argument, like so:

```
admin@myhome:~/git-repository$ git switch new-branch
```

This will switch to the `new-branch` branch and make it the current branch.

Both `git branch` and `git switch` commands allow you to create new branches, list available branches, and switch between branches as needed. In the next section, we are going to introduce ways of working with Git, called workflows: `git workflow` and `github workflow`.

Squashing commits using an interactive rebase

To squash commits using `git rebase`, you first need to determine the range of commits that you want to squash. This is typically done by specifying the commit hash of the first commit in the range and the commit hash of the last commit in the range. For example, if you want to squash the last three commits in your repository, you could use the following command:

```
admin@myhome:~/git-repository$ git rebase -i HEAD~3
```

This will open an editor window showing a list of the last three commits, along with some instructions. Each commit is represented by a line in the file, and the line begins with the word `pick`. To squash a commit, you need to change the word `pick` to `squash` or `s`.

Here is an example of what the file might look like:

```
pick abc123 Add feature X
pick def456 Fix bug in feature X
pick ghi789 Add test for feature X
```

To squash the second and third commits into the first commit, you would change the file to look like this:

```
pick abc123 Add feature X
squash def456 Fix bug in feature X
squash ghi789 Add test for feature X
```

After making the changes, you can save and close the file. Git will then apply the changes and present you with another editor window, where you can enter a new commit message for the combined commit.

It is important to note that squashing commits using `git rebase` can be a destructive operation, as it permanently alters the commit history of the repository. It is generally recommended to use `git rebase` with caution and to make sure that you have a backup of your repository before using it.

If you want to undo changes made by a `git rebase` operation, you can use the `git rebase --abort` command to discard the changes and restore the repository to its previous state.

After successfully squashing commits, you will be able to push it back to a remote repository, but only by using the `git push --force` command, which will ignore the fact that you just rewrote a commit history of this branch. This is a destructive operation and can't be undone, so again, please double- and triple-check before pushing changes with the `--force` option.

Solving Git conflicts

Conflicts can occur when you are trying to merge or rebase branches that have conflicting changes. This can happen when the same lines of code have been modified in both branches, and Git is unable to automatically resolve the conflicts.

When a conflict occurs during a merge or rebase, Git will mark the conflicting lines in the affected files, and you will need to manually resolve the conflicts before you can continue.

Here is an example of how you might resolve a conflict during a merge.

Run `git merge` to merge the two branches:

```
admin@myhome:~/git-repository$ git merge branch-to-merge
```

Git will detect any conflicts and mark the affected lines in the affected files. The conflicting lines will be surrounded by `<<<<<<<`, `=======`, and `>>>>>>>` markers:

```
<<<<<<< HEAD

something = 'this is in our HEAD';

=======

something = 'this is in our branch-to-merge';

>>>>>>> branch-to-merge
```

Open the affected files and resolve the conflicts by choosing which version of the code you want to keep. You can keep either the version from the current branch (HEAD) or the version from the branch that you are merging (`branch-to-merge`):

```
something = 'this is in our branch-to-merge';
```

Stage the resolved files using `git add`:

```
admin@myhome:~/git-repository$ git add path/to/file
```

Continue the merge or rebase by running `git rebase --continue` or `git merge --continue`:

```
admin@myhome:~/git-repository$ git merge --continue
```

Resolving conflicts during a merge or rebase can be a tedious process, but it is an important part of working with Git. By carefully reviewing conflicting changes and choosing the correct version of the code, you can ensure that your repository remains consistent and free of errors.

It is also a good idea to regularly synchronize your local copy of the repository with the source branch before merging or rebasing, in case you encounter conflicts or other issues that you are unable to resolve. This can help you recover from any mistakes or accidents that might occur during the process.

There are also amazing graphical tools to solve conflicts in a bit more interactive manner. If you're using an **integrated development environment** (IDE), you can use a built-in interface instead of using an external tool. Sometimes, it might not be enough, and you will need a dedicated application just for resolving conflicts, such as `KDiff3`, `WinMerge`, `Meld`, or one of the other options out there.

In this section, we have explained branches, repositories, local and remote repositories, merging, and rebasing. Next, we will look into browsing your repository's historical changes and modifying it.

GitFlow and GitHub Flow

GitFlow is a branching model for Git, created by Vincent Driessen. It aims to provide a consistent and easy-to-use branching model for teams working on software projects. The GitFlow model uses two main branches: `develop` and `master`. The `develop` branch is used to develop new features, while the `master` branch represents the current production-ready state of the code base. There are also several supporting branches, such as `feature`, `release`, and `hotfix`, which are used to manage the process of developing, releasing, and maintaining software.

GitHub Flow is a model that was developed by and for GitHub for use with its hosting and collaboration platform. It is similar to GitFlow, but it is designed to be simpler and easier to use, especially for smaller teams and projects. In the GitHub Flow model, all development is done in branches, and new features are merged into the `master` branch using pull requests. There is no separate `develop` branch, and the `master` branch is always considered to be the current production-ready version of the code base.

Another branching model that is very similar is **GitLab Flow**. It's used to manage the development and maintenance of software projects, and it is specifically designed to be used with GitLab, a web-based Git repository manager that provides **source code management** (SCM), **continuous integration** (CI), and more.

In the GitLab Flow model, all development is done in branches, and new features are merged into the `master` branch using merge requests. There is no separate `develop` branch, and the `master` branch is always considered to be the current production-ready version of the code base. However, GitLab Flow does include some additional features and tools, such as the ability to use protected branches and merge request approvals, which can help teams enforce development best practices and maintain a high level of code quality.

In the next section, we'll look into configuring Git to our needs using configuration files.

Global git configuration – .gitconfig

The `.gitconfig` file is a configuration file that is used to set global options for Git. It is typically located in the user's home directory and can be edited using any text editor.

Some common options that you may want to include in your `.gitconfig` are as follows:

- `user.name` and `user.email`: These options specify the name and email address that will be associated with your Git commits. It is important to set these options correctly, as they will be used to identify you as the author of the commits.

- `color.ui`: This option enables or disables colored output in Git. You can set this option to `auto` to enable colored output when Git is run in a terminal that supports it or to `true` or `false` to always enable or disable colored output, respectively.

- `core.editor`: This option specifies the text editor that Git will use when it needs you to enter a commit message or other input. You can set this option to the command for your preferred text editor, such as `nano`, `vi`, or `emacs`.

- `merge.tool`: This option specifies the tool that Git will use to resolve conflicts when merging branches. You can set this option to the command for a visual merge tool, such as `kdiff3`, `meld`, or `tkdiff`.

- `push.default`: This option specifies the behavior of the `git push` command when you do not specify a branch. You can set this option to `simple`, which pushes the current branch to the same name on the remote, or `upstream`, branch, which pushes the current branch to the remote branch that it is tracking.

- `alias.*`: These options allow you to create aliases for Git commands. For example, you could set `alias.st` to `status`, which would allow you to use the `git st` command instead of `git status`.

The following is an example `.gitconfig` file making use of the preceding options, with some comments after every section:

```
[user]
name = Jane Doe
email = jane@doe.example
```

```
[color]
    ui = always
```

The `user` section defines the user and email that will be used as the author for every commit. The `color` section enables colors for readability purposes. `ui = always` will enable colors always for all output types (machine consumption intended or not). Other possible options are `true`, `auto`, `false`, and `never`.

The `alias` section lets you simplify some long commands while working with Git. It will create an alias defined on the left-hand side for a Git command, which you can add after the equals sign.

Here's an example:

```
[alias]
    ci = commit
```

We're defining the `ci` command as `commit`. After adding this to your `.gitconfig` file in your home directory, you will get another Git command: `git ci`, which will effectively run the `git commit` command. You can add aliases for all common Git commands you use daily. The following command is telling Git where to find a default `.gitignore` file:

```
[core]

    excludesfile = ~/.gitignore
```

The following `push` setting is changing the default push behavior, which will require you to specify to which remote branch you're going to push your local branch:

```
[push]
    default = current
```

By specifying the `current` option, we instruct Git to try to push the local branch to a remote branch with exactly the same name as the local branch. Other options are set out here:

- `nothing`: Don't try to push anything.
- `matching`: Consider all branches remote and local with the same name to be a match and push all matches.
- `upstream`: Push the current branch to an upstream branch.
- `simple`: This is the default option. It will refuse to push to the upstream if the name of the local branch differs from the name of the remote upstream branch.

When running the `git pull` command, Git will try to integrate remote commits into your local branch. By default, it will try to merge upstream, which will result in possibly unwanted merge commits. This will change the default behavior to `rebase`:

```
[pull]

rebase = false
```

See the *Local versus remote Git repositories* section for more information.

Ignoring some files using a .gitignore configuration file

A `.gitignore` file is a configuration file that is used to tell Git which files or directories to ignore when tracking changes in a repository. This can be useful if you have files that are generated by your build process, are specific to your local environment, or are otherwise not relevant to the project and do not need to be tracked by Git.

Here are some examples of the types of files and directories that you might include in your `.gitignore` file:

- **Temporary files**: These are created by your editor or operating system and are not necessary for the project. Examples might include `*.tmp`, `*.bak`, or `*.swp` files.

- **Build artifacts**: These are files that are generated by your build process and are not necessary for the project. Examples might include `*.exe`, `*.jar`, `*.war` files and `bin/`, `obj/`, or `dist/` directories.

- **Dependencies**: These are directories that contain dependencies from your package manager that you don't want to include in the Git repository. For example, you might want to ignore the `node_modules/` directory if you are using npm, or the `vendor/` directory if you are using Composer.

- **IDE-specific files**: The IDE creates some cache and configuration directories in the same directory as you have your local copy of the Git repository. You might want to ignore files that are specific to that IDE—for example, the `idea/` directory if you are using JetBrains IDEs, or the `.vscode/` directory if you are using Visual Studio Code.

- **Sensitive information**: Files containing passwords or API keys. We strongly advise taking special care to ignore those files to prevent them from being committed to the repository. This will save you a lot of headaches and unnecessary risk. These days, bots such as GitHub's *Dependabot* will raise an alert (or even block the commit) if some of those sensitive files get into your repositories, but it's better to catch these earlier in the development process.

Here is an example `.gitignore` file that ignores some common types of files:

```
# Temporary files or logs

*.tmp

*.bak

*.swp

*.log



# Build artifacts

bin/

obj/

dist/

venv/

*.jar

*.war

*.exe

# Dependencies

node_modules/

vendor/

requirements.txt

# IDE-specific files

.idea/
.vscode/
```

```
# Sensitive information

secrets.txt

api_keys.json
```

It's also possible to use wildcards and **regular expressions** (**regexes**) to specify patterns of files to ignore. You can also put `.gitignore` files inside directories where you need a special rule that it is not desired to be there globally.

Finally, you can put your site-wide `.gitignore` file inside your home directory to make sure you won't commit any files needed for your local development.

Summary

Git is such a powerful tool that it's hard to fit all its features into a single book. In this chapter, we've learned the most basic tasks you will need to know in your day-to-day work as a DevOps engineer, and you are well equipped to handle most of the issues you will encounter.

In the next chapter, we will focus on Docker and containerization, where we will put all the skills you've gained so far to the test.

8

Docker Basics

In this chapter, we introduce one of the building blocks of the DevOps toolkit – containers. We are going to explain the differences between virtualization and containers, and then present the advantages and disadvantages of both solutions. Additionally, we are going to present a way to choose between both solutions for a given workload.

The main topics covered in this chapter are as follows:

- Virtualization versus containerization
- Anatomy of Docker
- Docker commands
- Dockerfile
- Docker image registries
- Docker networking

Technical requirements

For this chapter, you will need a Linux system with an installed Docker Engine. We are not going to cover the installation steps here. Different Linux distributions provide Docker in different ways. We are going to use Docker Engine 20.10.23 here. Since in this chapter all examples are very basic, older versions of Docker will most probably work. Still, if you run into issues with following our examples, updating Docker to our version should be your first step in troubleshooting.

Virtualization versus containerization

In this section, we are going to explain what virtualization and containerization are and what the major differences between them are.

Virtualization

Virtualization is a technique of running a complete simulated computer within another computer. Complete means that it mirrors everything a physical computer would have: motherboard, BIOS, processor, hard drives, USB ports, and so on. Simulated means that it is entirely a product of software. This computer does not exist physically, thus it is called virtual. To exist, the **virtual machine** (**VM**), as simulated computers are often called, needs a real, physical one to emulate it. The physical machine is called a host or hypervisor.

So, I have a physical computer. It is powerful. Why would I want to run a VM in it? For obvious reasons, the VM will be less powerful than the host: after all, the host requires RAM, CPU, and hard drive space for itself. There is also some small drop in performance (since we are actually running a program that emulates full hardware) when compared to the physical machine.

The reasons can vary depending on the use case, but there are a lot of them.

You may want to run a full operating system different from your own to test some software, to run software unavailable for your operating system, or to dutifully recreate a development environment for your application. You may want to recreate, as closely as possible, a production environment for your application. All those are valid and pretty popular reasons for using VMs.

Let us see what the advantages of virtualization are:

- **Isolation**: VMs present themselves, as mentioned, as fully functional computers. To the running operating system, they create the illusion of being separate physical machines. Whatever we run in them shouldn't be able to access the host computer (unless specifically allowed to) and, indeed, barring a few incidents where such things were possible (as a result of a programming error), VMs have provided secure environments. This isolation is a very good solution in malware analysis, running workloads that require separate servers, and so on. As an example, if a VM runs a single WWW server, the security vulnerability in the server may grant the attacker access to the operating system, thus allowing them a free run. But since other components of infrastructure, for example, databases, are run in separate VMs, the incident can be contained to the WWW server only.

- **Tuning**: With a sufficiently powerful host, it is possible to partition its resources so that each running VM has guaranteed RAM, hard disk space, and CPU.

- **Operating system simplification**: When running various workloads, such as databases, WWW servers, and mail servers, the complexity of maintaining a single server running them all grows pretty fast. Every installed software requires additional software to be installed (such as libraries and helper programs). Libraries required by various programs may introduce incompatibilities (especially if we install software not distributed by operating system developers, so-called third-party programs). On rare occasions, even the software included in the distribution may be incompatible with each other to a degree that makes it impossible or very difficult to install them on one operating system. Maintenance of such a system can become troublesome and require a lot of detective work. Modern hypervisor software alleviates many system administration hurdles by means of clones, snapshots, golden images, and so on.

- **Automation**: Modern virtualization software provides a lot of features that promote the automation of system management on many levels. Snapshots – a point-in-time capture of the whole system – allow a rollback at any given moment to a previous system state. This allows it to easily back out of unwanted changes to the last known good state. Clones let us provision new VMs based on another, already running and configured. Golden images are archived images of VMs that we can easily and quickly import and start – completely omitting installation and limiting configuration to the absolute minimum. This also allows for reliable environment recreation.

- **Speed up**: Properly setting up a workflow utilizing VMs allows us to start a new operating system complete with its own server or desktop hardware in a matter of minutes instead of hours. This opens new possibilities for testing environments, remote desktop solutions, and so on.

The preceding list is not exhaustive but should easily demonstrate why virtualization became a darling of data centers and hosting companies. The availability of a wide variety of servers we can cheaply rent is a direct result of virtualization allowing companies to partition the hardware.

As great as the solution is, it is not a panacea and not everything should be virtualized. More so, running a separate VM for every piece of software easily leads to resource utilization overhead. 100 virtual servers not only will use the CPU and RAM provided to the operating system in it but also, some percent will be used for housekeeping on the host machine. Each of those servers will utilize the disk space required by the operating system within, even though it probably will be an exact copy of the 99 other servers on the same server – a waste of space, RAM, and CPU. Also, bringing up a new VM will take some time. Granted, if you have everything configured and automated properly, it is going to be shorter than setting up a new hardware machine, but still.

Before virtualization became widely available, operating system developers were trying to provide techniques that would allow system operators to isolate various workloads. The main targets were data centers, where one physical server was too much for one workload (a database or a WWW server), but running more than one posed a risk (security, stability, etc.). After the virtualization became widespread, it became obvious that using it was sometimes like using cannons against sparrows. It doesn't make sense to procure a whole new server (even virtual) with the whole operating system when you want to run a small program. Thus, through a natural progression of innovating new features that allow for better process isolation, containers arose.

Containerization

Containers are lightweight virtual environments that allow us to run a single process in an isolated manner. An ideally prepared container consists of only software and libraries required to run the application. The host operating system takes care of operating hardware, managing memory, and all other peripheral tasks. The main assumption of the container is that it doesn't emulate being a separate operating system or a separate server. Processes or users within the container can easily find out they are enclosed there. The downside is that containers won't emulate hardware. You cannot use them to test new drivers, for instance. The upside is that a single container can take as little as just a few megabytes of hard drive space and only the memory required for the process to run.

As a consequence, starting up a container takes only the time that the application needs to start. The bootup time – the BIOS, hardware tests, and operating system boot time – is all shaved off. All the software not required by the application can, and should, be omitted. Given the small size of container images, their redistribution times became almost negligible, their start times almost instantaneous, and the build time and process largely simplified. This has led to much easier recreation of the environment. This, in turn, has led to easier test environment setup and very often deployments of new versions of the software – as often as several thousand times a day. The scaling of applications has become much easier and faster.

The preceding brought another change in the approach to running applications. The logical consequence of the preceding change is that containers are not maintained the same way the operating system is. You do not upgrade software within the container – you deploy a container with a new version of software in place of the obsolete one. This leads to an assumption that you don't keep data within a container but in a filesystem that you attach to the container during the run.

The poster child of Linux containerization is **Docker**. One thing that Docker did that has probably helped to bring the revolution is creating an ecosystem for easy sharing of container images.

A container image is, in the case of Docker, a simple archive consisting of all the binaries and libraries that are required for the application and a few files with configuration information. Since the size tends to be rather small and the image never has any data within, it is logical to allow people to share the image they have built. Docker has an image hub (called Docker Hub) with a nice WWW UI, and command-line tools for searching, downloading, and uploading images. The hub allows rating images and giving comments and feedback to the authors.

Now that we know what containerization is, we can look deeper into how Docker works internally and what makes it tick. Let's look into the anatomy of Docker.

Anatomy of Docker

Docker comprises several components:

- Command-line utility – Docker
- Host
- Objects
- Registries

The Docker CLI tool – `docker` – is the main means of managing containers and images. It is used to build images, pull them from the registry, upload them to the registry, run containers, interact with them, set runtime options, and, finally, destroy them. It is a command-line tool that communicates with Docker hosts using an API. By default, it is assumed that the `docker` command is being invoked on the host, but it is not strictly necessary. One `docker` CLI tool can manage more than one host.

The host is more interesting. The host runs `dockerd` – a daemon responsible for actually performing the actions ordered via the `docker` tool. It is here that container images are stored. The host also provides resources such as networking, storage, and the containers themselves.

The `dockerd` daemon is the beating heart of the containers. It's the background process that runs on a host machine and manages the containers. `dockerd` manages creating and managing containers, providing an API for interacting with the daemon, managing volumes, networks, and image distribution, providing an interface to manage images and containers, and storing and managing metadata for containers and images. It also manages communication between other processes in Docker Swarm mode.

OverlayFS

OverlayFS was first released as a part of the Linux kernel version 3.18 in August 2014. It was initially developed as a means to provide a more efficient and flexible way to handle container storage in comparison to the previous storage driver, **Another UnionFS** (**AUFS**). OverlayFS was considered the next generation of UnionFS, which was the storage driver used by Docker at that time.

This filesystem was included as a built-in storage driver in Docker starting from version 1.9.0. Since then, OverlayFS has become the default storage driver for Docker on most Linux distributions, and it is widely used in various container orchestration platforms such as Kubernetes and OpenShift.

OverlayFS is a filesystem for Linux that allows for the overlay of one directory on top of another. It allows for the creation of a *virtual* filesystem that is composed of two different directories: a lower directory and an upper directory. The upper directory contains the files that are visible to the user, while the lower directory contains the *base* files that are hidden.

When a file or directory is accessed in the upper directory, OverlayFS first looks for it in the upper directory, and if it doesn't find it, it looks in the lower directory. If the file or directory is found in the upper directory, that version is used. If it is found in the lower directory, that version is used.

This mechanism allows for the creation of *overlay* filesystems, where the upper directory can be used to add, modify, or delete files and directories in the lower directory, without modifying the lower directory itself. This is useful in scenarios such as containerization, where the upper layer can be used to store the changes made in a container, while the lower layer contains the base image for the container.

What is an image?

A **Docker image** is a pre-built package that contains all of the files and settings needed to run a piece of software in a container. It includes your application code or a binary, runtime, system tools, libraries, and all needed configuration files. Once an image is built, it can be used to start one or more containers, which are isolated environments that provide a consistent way to run the software.

When starting a container, you must select a program to run as a primary process of the container. If this process quits, the whole container will be terminated as well.

Building a Docker image typically involves creating a Dockerfile, which is a script that contains instructions for building the image. The Dockerfile specifies the base image to use, any additional software to be installed, any files to be added to the image, and any configuration settings to be applied.

When building an image, Docker reads the instructions in the Dockerfile and performs the steps we've prepared inside the Dockerfile.

Once the image is built, it can be saved and used to start one or more containers. The process of building an image can also be automated using a tool such as Jenkins, GitHub, or GitLab actions, which can automatically build and test new images whenever changes are made to the code base.

The resulting image consists of a unique ID (SHA-256 hash), which is a hash of the image's content and metadata, and it also can have a tag, which is a human-readable string that can be used to refer to a specific version of the image. UnionFS takes care of merging all content when running a container.

To inspect the metadata and content parts of an image, you can run the following commands:

```
admin@myhome:~$ docker pull ubuntu
admin@myhome:~$ docker inspect ubuntu
[
    {
        "Id": "sha256:6b7dfa7e8fdbe18ad425dd965a1049d984f31cf0ad-
57fa6d5377cca355e65f03",
        "RepoTags": [
            "ubuntu:latest"
        ],
        "RepoDigests": [
            "ubuntu@sha256:27cb6e6ccef575a4698b66f5de06c7ecd-
61589132d5a91d098f7f3f9285415a9"
        ],
        "Created": "2022-12-09T01:20:31.321639501Z",
        "Container":
"8bf713004e88c9bc4d60fe0527a509636598e73e3ad1e71a9c9123c863c17c31",
        "Image": sha256:070606cf58d59117ddc1c48c0af233d6761addbcd-
4bf9e8e39fd10eef13c1bb7",
        "GraphDriver": {
            "Data": {
```

```
            "MergedDir": "/var/lib/docker/overlay2/
f2c75e37be7af790f0823f6e576ec511396582ba71defe5a3ad0f661a632f11e/
merged",
            "UpperDir": "/var/lib/docker/overlay2/
f2c75e37be7af790f0823f6e576ec511396582ba71defe5a3ad0f661a632f11e/
diff",
            "WorkDir": "/var/lib/docker/overlay2/
f2c75e37be7af790f0823f6e576ec511396582ba71defe5a3ad0f661a632f11e/work"
          },
          "Name": "overlay2"
      },
      "RootFS": {
          "Type": "layers",
          "Layers": [
"sha256:6515074984c6f8bb1b8a9962c8fb5f310fc85e70b04c88442a3939c026db-
fad3"
          ]
      },
   }
]
```

There's a lot of information here, so we've stripped the output and left the information we want to focus on. You can see an image ID, all merged directories under the GraphDriver section, and the RootFS sha256 layer. RootFS contains the whole filesystem created by UnionFS when we start a process within the container.

What is a container runtime?

A **container runtime** (or **container engine**) is a software component that runs containers on your system. Container runtimes load container images from a Docker registry, monitoring system resources, allocating system resources for a container, and managing its life cycle.

There are a number of runtime containers that are being used. The most known and used on laptops – you probably have it installed on your system already – is containerd. It's a high-performance container runtime that is designed to be embedded in a larger system. It is used by many cloud providers and is also the default runtime for Kubernetes.

LXC is a runtime that uses Linux namespaces and cgroups to provide isolation for containers. It is considered to be more lightweight and efficient than Docker (containerd). It's also harder to use.

Another interesting runtime is **Container Runtime Interface for OCI (CRI-O)**. CRI-O is fully compliant with the **Open Container Initiative** (**OCI**) specification, which means that it can run any OCI-compliant container image. Also, it's designed to work natively with Kubernetes Pods, which allows it to provide better integration with Kubernetes than other runtimes.

Rocket (**rkt**) is an alternative container runtime that is designed to be more secure and efficient than Docker. It uses the **App Container** (**appc**) image format and has a simpler architecture than Docker. It's also not used very often.

Other container engines worth noting are **run Open Container** (**runC**), a low-level container engine that provides the basic functionality for creating and managing containers, and Firecracker developed by AWS.

cgroups

Linux **cgroups** (short for **control groups**) are a Linux kernel feature that allows for the management and isolation of system resources for groups of processes. Cgroups allow the system administrator to allocate resources such as CPU, memory, and network bandwidth to specific groups of processes, and to monitor and control their usage.

This can be used to limit the resources used by a particular application or user, or for isolating different types of workloads on a shared system.

Docker by default doesn't limit either CPU or memory consumption for an application inside a container. It's quite easy to enable without any direct interaction with cgroups or kernel settings – the Docker daemon will do it for us.

You can limit the amount of memory a Docker container can use by using the `--memory` or `-m` option with the `docker run` command.

For example, use the following to run the `alpine` image with a memory limit of 500 MB:

```
admin@myhome:~$ docker run --memory 500m alpine /bin/sh
```

You can specify the memory limit in bytes, kilobytes, megabytes, or gigabytes by using the appropriate suffix (b, k, m, or g).

When you limit the memory for a container, Docker will also limit the amount of memory swap that a container can use. By default, the memory swap limit is twice the value of the memory limit. It's also possible to limit the memory swap by using the `--memory-swap` or `--memory-swappiness` option.

Limiting usage of the CPU time that an application inside a Docker container can use can be done by using the CPU shares limit (`--cpus` or `-c` option). CPU shares are a relative measure of CPU time that a container can use. By default, a container is allocated a certain number of CPU shares, which it can use to consume CPU time proportional to its share. For example, if a container has 0.5 CPU shares, it can use up to 50% of the CPU time if no other containers are consuming CPU.

Other options available are the following:

- `--cpuset-cpus`: This allows you to specify the range of CPU cores that the container can use, for example, 0-1 to use the first two cores, or 0,2 to use the first and third core.

- `--cpu-shares`: This allows you to set a CPU time limit for a Docker container. It specifies the amount of CPU time, in microseconds, that the container can use in a given period of time. The period of time is specified by the `--cpu-period` option.

- `--cpu-quota` and `--cpu-period`: `--cpu-quota` is the CPU time limit in microseconds and `--cpu-period` is the length of the CPU time period in microseconds.

The `--cpu-quota` and `--cpu-period` options allow you to specify a more precise CPU time limit for a container compared to the `--cpus` and `--cpuset-cpus` options. It is useful if you need to limit the CPU time for a container more precisely to prevent performance issues or ensure that your application runs reliably.

In this section, we went through the container runtime and how it works. Next, we will be looking into the command-line interface for the `containerd` daemon to interact with all Docker components in an easy and robust way.

Docker commands

The **Docker command-line interface** is a tool that allows users to interact with containers. It provides a set of commands that you can use to build Docker images and create and manage containers, images, networks, and volumes. It interacts with the `containerd` daemon using a socket file or network.

The most common commands you can use are the following:

- `build`: This allows you to build a new Docker image using a Dockerfile

- `run`: This starts a new container

- `start`: This restarts one or more stopped containers

- `stop`: This will stop one or more running containers

- `login`: This is used to gain access to private registries

- `pull`: This downloads an image or a repository from a registry

- `push`: This uploads an image or a repository to a registry

- `build`: This helps create an image from a provided Dockerfile

- `images`: This lists all images on your machine

- `ps`: This lists all running containers

- `exec`: This executes a command in a running container

- `logs`: This shows the logs of a container

- `rm`: This removes one or more containers

- `rmi`: This removes one or more images

- `network`: This is used to manage Docker networks
- `volume`: This is used to manage volumes

docker build

The `docker build` command is used to build a Docker image from a Dockerfile. The basic syntax is as follows:

```
docker build [OPTIONS] PATH | URL | -
```

`PATH` is the path to the directory containing the Dockerfile.

`URL` is the URL to a Git repository containing the Dockerfile.

`-` (a dash) is used to build an image from the contents of `stdin`, so you could pipe Dockerfile content to it from the output of some previous command that would build a Dockerfile, for example, generate it from a template.

To build an image from a Dockerfile located in the current directory, you would run the following command:

```
admin@myhome:~$ docker build .
```

You can also use a specific tag for the build image, as in the following example:

```
admin@myhome:~$ docker build -t my-image:1.0 .
```

You can also pass `--build-arg` to your build command to pass build-time variables to the Dockerfile:

```
admin@myhome:~$ docker build --build-arg VAR1=value1 -t my-image:1.0 .
```

docker run

When you're running a container, you're essentially taking a Docker image and executing a process within that environment. An image is a blueprint or a snapshot of a container; it's a read-only template with instructions for creating a container. A container that is running is an instance of that image but with its own state.

The `docker run` command is used to start a container from a Docker image. For example, to start a container from the `myimage` image and run `/bin/bash`, you would run the following command:

```
admin@myhome:~$ docker run myimage /bin/bash
```

You can also pass options to the `run` command, as in the following example:

```
admin@myhome:~$ docker run -d -p 8080:80 --name containername myimage
```

This command starts the container in detached mode (the -d option; it puts a container in the background), maps port 80 in the container to port 8080 on the host (-p 8080:80), and assigns the name containername to the container (--name containername).

You can also pass environment variables to the container:

```
admin@myhome:~$ docker run -e VAR1=value1 -e VAR2=value2
myimage:latest
```

Docker containers don't store data after they are killed. To make data persistent you would use storage external to the docker container itself. In the simplest setup, this would be a directory or a file on filesystem outside the container.

There are two ways to do it: create a Docker volume:

```
admin@myhome:~$ docker volume create myvolume
```

To use this volume for data persistence you'd mount it when starting the container:

```
admin@myhome:~$ docker run -v myvolume:/mnt/volume myimage:latest
```

You can also bind mount local folder (**bind mount** is a Docker term for this operation) as a volume inside the container using the -v option. In this case you don't run the Docker volume create command:

```
admin@myhome:~$ docker run -v /host/path:/mnt/volume myimage:latest
```

You can also specify the working directory inside the container using the -w option:

```
admin@myhome:~$ docker run -w /opt/srv my-image
```

Other options that are useful are as follows:

- --rm: This option will remove the container after it is stopped
- -P, --publish-all: This option will publish all exposed ports (EXPOSE option in Dockerfile) to a random local port
- --network: This option will connect the container to the Docker network specified

You can find more available options to use by invoking the docker run --help command.

docker start

The docker start command is used to start one or more stopped Docker containers. For example, to start a container, you would run the following command:

```
admin@myhome:~$ docker start mycontainer
```

`mycontainer` is the name or ID of the container you want to start. You can check all running and stopped containers using the `docker ps` command; we will get into it a bit later. You can also start multiple containers at once. To do that, you can list their names or IDs separated by spaces.

To start multiple containers, you would run the following command:

```
admin@myhome:~$ docker start mycontainer othercontainer lastcontainer
```

To attach to the container's process so that you can see its output, use the `-a` option while starting the container:

```
admin@myhome:~$ docker start -a mycontainer
```

docker stop

This command is used to stop containers running in the background. The syntax for the command is the same as for starting the container. The difference lies in the available options you can use.

To stop multiple containers at once, you can list their names or IDs separated by spaces.

```
admin@myhome:~$ docker stop mycontainer
```

To stop multiple containers, you would run the following command:

```
admin@myhome:~$ docker stop mycontainer othercontainer lastcontainer
```

You can also use the `-t` option to specify the amount of time (in seconds) to wait for the container to stop before sending a `SIGKILL` signal. For example, to wait for `10` seconds before stopping a container, run the following:

```
admin@myhome:~$ docker stop -t 10 mycontainer
```

You can also use `--time` or `-t` to specify the amount of time to wait for the container to stop before sending a `SIGKILL` signal.

By default, the `docker stop` command sends a `SIGTERM` signal to the container, which gives the process running in the container a chance to cleanly shut down. If the container does not stop after the default 10-second timeout, a `SIGKILL` signal will be sent to force it to stop.

docker ps

This command is used to list the running or stopped containers. When you run the `docker ps` command without any options, it will show you the list of running containers along with their container ID, names, image, command, created time, and status:

```
admin@myhome:~$ docker ps
```

It's possible to view all containers with the -a option:

```
admin@myhome:~$ docker ps -a
CONTAINER ID           IMAGE                          COMMAND
STATUS                       PORTS                 NAMES
a1e83e89948e   ubuntu:latest                         "/
bin/bash -c 'while…"   29 seconds ago        Up 29
seconds                                 pedantic_mayer
0e17e9729e9c   ubuntu:lat-
est                                    "bash"                 About a
minute ago   Exited (0) About a minute ago
angry_stonebraker
aa3665f022a5   ecr.aws.com/pgsql/server:latest   "/opt/
pgsql/bin/nonr…"   5 weeks ago            Exited (255) 8 days
ago           0.0.0.0:1433->1433/tcp    db-1
```

You can use the --quiet or -q option to display only the container IDs, which might be useful for scripting:

```
admin@myhome:~$
docker ps --quiet
```

You can use the --filter or -f option to filter the output based on certain criteria:

```
admin@myhome:~$
docker ps --filter "name=ubuntu"
```

To inspect how much disk space the container is utilizing, you will need to use the -s option:

```
admin@myhome:~$ docker ps -s
CONTAINER ID       IMAGE         COMMAND     CREATED
STATUS             PORTS     NAMES         SIZE
a1e83e89948e   ubuntu:latest   "/bin/bash -c 'while…"   14 seconds
ago    Up 13 seconds              pedantic_mayer    0B (virtual 77.8MB)
```

This container doesn't use additional space on the disk, as it didn't save any data.

docker login

The docker login command is used to log in to a Docker registry. A registry is a place where you can store and distribute Docker images. The most commonly used registry is Docker Hub, but you can also use other registries, such as AWS **Elastic Container Registry** (**ECR**), Project Quay, or Google Container Registry.

By default, docker login will connect to the Docker Hub registry. If you want to log in to a different registry, you can specify the server URL as an argument:

```
admin@myhome:~$ docker login quay.io
```

When you run the `docker login` command, it will prompt you for your username and password. If you don't have an account on the registry, you can create one by visiting the registry's website.

Once you are logged in, you will be able to push and pull images from the registry.

You can also use the `--username` or `-u` option to specify your username, and `--password` or `-p` to specify the password on the command line, but it is not recommended due to security reasons.

You can also use the `--password-stdin` or `-P` option to pass your password via `stdin`:

```
admin@myhome:~$ echo "mypassword" | docker login --username myusername
--password-stdin
```

It can be an output for any command. For example, to log in to AWS ECR, you would use the following command:

```
admin@myhome:~$ aws ecr get-login-password | docker login --username
AWS --password-stdin 1234567890.dkr.ecr.region.amazonaws.com
```

You can also use the `--token` or `-t` option to specify your token:

```
admin@myhome:~$ docker login --token usertokenwithrandomcharacters
```

Once you are logged in, you will be able to push and pull images from the registry.

docker pull

To pull a Docker image, you can use the `docker pull` command followed by the image name and a tag. By default, `pull` will pull a tag `latest` (latest is the name of the tag, or the version of the image).

For example, use the following to pull the latest version of the `alpine` image from Docker Hub:

```
admin@myhome:~$ docker pull alpine
```

Use the following to pull a specific version of the `alpine` image, such as version 3.12:

```
admin@myhome:~$ docker pull alpine:3.12
```

You can also pull an image from a different registry by specifying the registry URL in the image name.

For example, use the following to pull an image from a private registry:

```
admin@myhome:~$ docker pull myregistry.com/myimage:latest
```

docker push

After you build a new image, you can push that image to the image registry. By default, push will try to upload it to the Docker Hub registry:

```
admin@myhome:~$ docker push myimage
```

To push a specific version of the image, such as version 1.0, you will need to tag the image locally as version 1.0 and then push it to the registry:

```
admin@myhome:~$ docker tag myimage:latest myimage:1.0
admin@myhome:~$ docker push myimage:1.0
```

You can also push an image to a different registry by specifying the registry URL in the image name:

```
admin@myhome:~$ docker push myregistry.com/myimage:latest
```

You need to be logged in to the registry to which you are pushing the image using the docker login command before pushing an image.

docker image

This is a command to manage images. The common use cases for docker image are shown in the following examples.

To list available images on your machine, you can use the docker image ls command:

```
admin@myhome:~$ docker image ls
REPOSITORY              TAG              IMAGE ID         CREATED         SIZE
ubuntu                  latest           6b7dfa7e8fdb     7 weeks ago     77.8MB
mcr.microsoft.com/mssql/server           2017-latest
                        a03c94c3147d     4 months ago     1.33GB
mcr.microsoft.com/azure-functions/python    3.0.15066-python3.9-
buildenv     b4f18abb38f7    2 years ago      940MB
```

You can also use the docker images command to do the same action:

```
admin@myhome:~$ docker images
REPOSITORY       TAG              IMAGE ID         CREATED         SIZE
ubuntu           latest           6b7dfa7e8fdb     7 weeks ago     77.8MB
mcr.microsoft.com/mssql/server           2017-
latest                    a03c94c3147d     4 months ago     1.33GB
mcr.microsoft.com/azure-functions/python    3.0.15066-python3.9-
buildenv     b4f18abb38f7    2 years ago      940MB
```

To pull an image from the Docker registry, use the following:

```
admin@myhome:~$ docker image pull ubuntu
```

When you specify an image name, you can use either the full repository name (for example, `docker.io/library/alpine`) or just the image name (for example, `alpine`), if the image is in the default repository (Docker Hub). See also the `docker pull` command discussed in an earlier section.

It's also possible to build an image:

```
admin@myhome:~$ docker image build -t <image_name> .
```

See also the section on the `docker build` command for more details on building images.

To create a tag for an image, you should run the following:

```
admin@myhome:~$ docker image tag <image> <new_image_name>
```

Finally, you can remove an image:

```
admin@myhome:~$ docker image rm <image>
```

See the section on the `docker rmi` command, which is an alias of this command.

Another option to remove images is the `docker image prune` – command. This command will remove all unused images (dangling images).

docker exec

`docker exec` allows you to run a command in a running Docker container. The basic syntax is as follows:

```
docker exec CONTAINER COMMAND ARGUMENTS
```

In the preceding example, the terms have the following meanings:

- `CONTAINER` is the name or ID of the container to run the command in
- `COMMAND` is the command to run in the container
- `ARGUMENTS` represents any additional arguments for the command (this is optional)

For example, to run the `ls` command in the container named `my_container`, you can use the following command:

```
admin@myhome:~$ docker exec mycontainer ls
```

docker logs

`docker logs` is used to fetch logs generated by a Docker container:

```
docker logs CONTAINER_NAME_OR_ID
```

Additional options you can pass to the command are as follows:

- `--details, -a`: Show extra details provided to logs
- `--follow, -f`: Follow log output
- `--since, -t`: Only display logs since a certain date (e.g., 2013-01-02T13:23:37)
- `--tail, -t`: Number of lines to show from the end of the logs (default `all`)

An example of its use is as follows:

```
admin@myhome:~$ docker logs CONTAINER_ID
```

docker rm

`docker rm` is used to remove one or more Docker containers:

```
docker rm CONTAINER_NAME_OR_ID
```

An example of its use is as follows:

```
admin@myhome:~$ docker rm CONTAINER_ID
```

To list all containers, use the `docker ps -a` command.

docker rmi

Images that are pulled or built locally can take up a lot of space on your disk, so it's useful to check and remove unused ones. `docker rmi` is used to remove one or more Docker images.

The following is its usage:

```
docker rmi IMAGE
```

An example of its use is as follows:

```
admin@myhome:~$ docker rmi IMAGE_ID
```

docker network

The `docker network` command is used to manage Docker networks. Apart from the usual actions (create, delete, and list), it's possible to attach (and disconnect) a running Docker container to a different network.

It's also possible to extend Docker with a specialized network plugin of your choice. There are multiple options here, so we will just list some of the network plugins with a short description. Plugins can also be used with more advanced setups, such as Kubernetes clusters:

- **Contiv-VPP** (`https://contivpp.io/`) uses the **Vector Packet Processing** (**VPP**) technology to provide an efficient, scalable, and programmable networking solution for containers, suitable for use in enterprise and service provider environments, where high-performance and scalable networking are a requirement.

- **Weave Net** (`https://www.weave.works/docs/net/latest/overview/`) allows containers to communicate with each other, regardless of which host they are running on. Weave Net creates a virtual network that spans multiple hosts, making it possible to deploy containers in a highly available, redundant, and load-balanced manner.

- **Calico** (`https://www.tigera.io/tigera-products/calico/`) is one of the most recognizable plugins out there. It uses a pure IP-based approach to networking, providing simplicity and scalability. Calico allows administrators to define and enforce network policies, such as allowing or denying specific traffic flows based on the source, destination, and port. Calico is designed for large-scale deployments and supports both virtual and physical networks.

The following are common operations when using the `docker network` command:

- Creating a new network:

```
admin@myhome:~$ docker network create mynetwork
```

- Inspecting a network:

```
admin@myhome:~$ docker network inspect mynetwork
```

- Removing a network:

```
admin@myhome:~$ docker network rm mynetwork
```

- Listing networks:

```
admin@myhome:~$ docker network ls
```

- Connecting a container to a network:

```
admin@myhome:~$ docker network connect mynetwork
    CONTAINER_NAME_OR_ID
```

- Disconnecting a container from a network:

```
admin@myhome:~$ docker network disconnect mynetwork
    CONTAINER_NAME_OR_ID
```

docker volume

The `docker volume` command is used to manage volumes in Docker. With this single command, you're able to list available volumes, clean unused ones, or create one for later use.

Docker supports multiple volume drivers, including the following:

- `local`: The default driver; stores data on the local filesystem using UnionFS

- `awslogs`: Makes it possible to store logs generated by your applications in Amazon CloudWatch Logs

- `cifs`: Allows you to mount an SMB/CIFS (Windows) share as a Docker volume

- `GlusterFS`: Mounts the GlusterFS distributed filesystem as a Docker volume

- NFS: Mounts the **Network File System** (**NFS**) as a Docker volume

Many more drivers are available. The list of available drivers can be found in the official Docker documentation: `https://docs.docker.com/engine/extend/legacy_plugins/`.

The following are examples of its use:

- **List volumes**: `docker volume ls`

- **Create a volume**: `docker volume create <volume-name>`

- **Inspect a volume**: `docker volume inspect <volume-name>`

- **Remove a volume**: `docker volume rm <volume-name>`

- **Mount a volume** to a container, which can be done when starting a new container: `docker volume create myvolume`

  `docker run -v myvolume:/opt/data alpine`

In this section, we've learned how to interact with all Docker components using a command-line interface. Up to now, we've been using publicly available Docker images, but it's time to learn how to build your own images.

Dockerfile

A Dockerfile is essentially a text file with a predetermined structure that contains a set of instructions for building a Docker image. The instructions in the Dockerfile specify what base image to start with (for example, Ubuntu 20.04), what software to install, and how to configure the image. The purpose of a Dockerfile is to automate the process of building a Docker image so that the image can be easily reproduced and distributed.

The structure of a Dockerfile is a list of commands (one per line) that Docker (`containerd` to be exact) uses to build an image. Each command creates a new layer in the image in UnionFS, and the resulting image is the union of all the layers. The fewer layers we manage to create, the smaller the resulting image.

The most frequently used commands in a Dockerfile are the following:

- `FROM`
- `COPY`
- `ADD`
- `EXPOSE`
- `CMD`
- `ENTRYPOINT`
- `RUN`
- `LABEL`
- `ENV`
- `ARG`
- `VOLUME`
- `USER`
- `WORKDIR`

You can find a complete list of commands on the official Docker documentation website: `https://docs.docker.com/engine/reference/builder/`.

Let's go through the preceding list to understand which command does what and when it's best to use it.

FROM

A Dockerfile starts with a `FROM` command, which specifies the base image to start with:

```
FROM ubuntu:20.04
```

You can also name this build using `as` keyword followed by a custom name:

```
FROM ubuntu:20.04 as builder1
```

`docker build` will try to download Docker images from the public Docker Hub registry, but it's also possible to use other registries out there, or a private one.

COPY and ADD

The `COPY` command is used to copy files or directories from the host machine to the container file system. Take the following example:

```
COPY . /var/www/html
```

You can also use the `ADD` command to add files or directories to your Docker image. `ADD` has additional functionality beyond `COPY`. It can extract a TAR archive file automatically and check for the presence of a URL in the source field, and if it finds one, it will download the file from the URL. Finally, the `ADD` command has a `--chown` option to set the ownership of the files in the destination. In general, it is recommended to use `COPY` in most cases, and only use `ADD` when the additional functionality it provides is needed.

EXPOSE

The `EXPOSE` command in a Dockerfile informs Docker that the container listens on the specified network ports at runtime. It does not actually publish the ports. It is used to provide information to the user about which ports are intended to be published by the container.

For example, if a container runs a web server on port 80, you would include the following line in your Dockerfile:

```
EXPOSE 80
```

You can specify whether the port listens on TCP or UDP – after specifying the port number, add a slash and a TCP or UDP keyword (for example, `EXPOSE 80/udp`). The default is TCP if you specify only a port number.

The `EXPOSE` command does not publish the ports. To make ports available, you will need to publish them with the use of the `-p` or `--publish` option when running the `docker run` command:

```
admin@myhome:~$ docker run -p 8080:80 thedockerimagename:tag
```

This will map port 8080 on the host machine to port 80 in the container so that any incoming traffic on port 8080 will be forwarded to the web server running in the container on port 80.

Regardless of the `EXPOSE` command, you can publish different ports when running a container. `EXPOSE` is used to inform the user about which ports are intended to be published by the container.

ENTRYPOINT and CMD

Next on our list is the ENTRYPOINT command, which in a Dockerfile specifies the command that should always be run when the container starts. It cannot be overridden by any command-line options passed to the docker run command.

The ENTRYPOINT command is used to configure the container as an executable. It is similar to the CMD command, but it is used to configure the container to run as an executable. It is typically used to specify the command that should be run when the container starts, such as a command-line tool or a script.

For example, if you have a container that runs a web server, you might use the ENTRYPOINT command to specify the command that starts the web server:

```
ENTRYPOINT ["nginx", "-g", "daemon off;"]
```

If you want to run a container with different arguments, you can use the CMD command to set default arguments that can be overridden when the container is started:

```
ENTRYPOINT ["/usr/bin/python"]
CMD ["main.py","arg1","arg2"]
```

CMD is used to specify the command that should be run when a container is started from the image. Take the following example:

```
CMD ["nginx", "-g", "daemon off;"]
```

The rule of thumb here is that if you want your application to take custom arguments, you are free to use ENTRYPOINT to launch a process and CMD to pass arguments to it. This way, you can be flexible with what your process will do by passing different options via the command line.

RUN

The RUN command in a Dockerfile is used to execute commands inside the container. It creates a new layer in the image each time it is executed.

The RUN command is used to install software packages, create directories, set environment variables, and perform any other actions that are required to set up the environment inside the container.

For example, you can use the RUN command to install a package:

```
RUN apt-get update && apt-get install -y python3 python3-dev
```

You can use the RUN command to create a directory:

```
RUN mkdir /var/www
```

You can use the RUN command to set environment variables:

```
RUN echo "export JAVA_HOME=/usr/lib/jvm/java-8-openjdk-amd64" >>
~/.bashrc
```

It's worth noting that the order of the RUN commands in the Dockerfile is important, as each command creates a new layer in the image, and the resulting image is the union of all the layers. So, if you're expecting some packages to be installed later in the process, you need to do it before using them.

LABEL

The LABEL command is used to add metadata to the image. It basically adds the key-value pairs of data to the image. Those can be used to store information such as the image's version, maintainer, and other relevant information you might need in your organization. The following is an example of the command:

```
LABEL maintainer-"Chris Carter <chcarter@your.comain.tld>"
```

You can also add multiple labels in one line:

```
LABEL maintainer="Chris Carter <chcarter@your.comain.tld>"
version="0.2"
```

The labels added to the image can be viewed using the docker inspect command:

```
admin@myhome:~$ docker inspect --format='{{json .Config.Labels}}'
<image_name_or_ID>
```

The use of LABEL commands to add metadata to the image can help users understand what the image's purpose is or who they should ask about the details, and help to manage the images.

ENV and ARG

The ENV command is used to set environment variables in the following format:

```
ENV <key>=<value>
```

The ARG command, on the other hand, is used to define build-time variables. These variables can be passed to the docker build command using the --build-arg flag and their values can be used in the Dockerfile.

The ARG command is used to define build-time variables similar to the ENV format:

```
ARG <name>[=<default value>]
```

The ARG command creates a variable that is only accessible during the build process, whereas the ENV command creates an environment variable that is accessible to all processes running inside the container.

In the next chapter, we'll get into more detail about the build process of the Docker image, where ARG and ENV are used together to persist ENV variables across build stages.

VOLUME

Another command is VOLUME. With it, you can configure a container to create a mount point for a volume at a specific location. Volumes are a way to store data outside of the container's filesystem, which means the data can persist even if the container is deleted or recreated. The following is the command:

```
VOLUME /opt/postgresql_data
```

Use the following to specify multiple directories:

```
VOLUME /opt/postgresql_data /opt/postgresql_xferlog
```

Or, the following is also valid:

```
VOLUME ["/opt/postgresql_data", "/opt/postgresql_xferlog"]
```

If there is any data in the directory marked as volume, when running a Docker using the docker run command, a new volume will be created with the content of this directory. That way, you can ensure that data created while this Docker container is running won't be lost when the container gets killed or stopped otherwise. It's especially important for databases, as we suggested in the preceding example.

USER

The USER command in a Dockerfile is used to set the default user that the container runs as. By default, the container runs as the root user; it is recommended to run the container as a custom user without root capabilities.

The USER command is used to set the user and optionally the group that the container runs as. For example, you can use the USER command to run the container as the webserver user:

```
USER webserver
```

You can also specify the user and group:

```
USER webserver:webserver
```

It's also possible to set the user ID and group ID instead of the name:

```
USER 1001:1001
```

The USER command only sets the default user for the container, but you can override it when running the container:

```
admin@myhome:~$ docker run --user=root webserver-image
```

Running your application as a non-root user is a best practice for security reasons. It limits the potential damage that can be done if an attacker gains access to the container, as the process running with full permissions is also running with the same UID (here: root) on the host you are running the image.

WORKDIR

The WORKDIR command in a Dockerfile is used to set the current working directory for the container. The working directory is the location in the container's filesystem where all the subsequent RUN, CMD, ENTRYPOINT, COPY, and ADD commands will be executed.

You can use the WORKDIR command to set the working directory to /usr/local/app:

```
WORKDIR /usr/local/app
```

When using WORKDIR, you won't need to set full paths for the files while using any other commands and you could parameterize your application location (with ARG or ENV).

Now that we've familiarized ourselves with Dockerfiles and how to build Docker images, it's useful to know how to store this new image in some way. Docker image registries are used for that exact purpose. We will look into registries in the next section.

Docker image registries

A **Docker image registry** hosts Docker images. Docker images are organized by tags that can be accessed and downloaded by users. These images can be used to create and run containers on a host machine. Image repositories can be hosted either locally or on a remote server, such as on Docker Hub, which is a public repository provided by Docker. You can also create your own private image repositories to share and distribute your images within your organization.

When you pull an image from a Docker image repository, the image is composed of multiple layers. Each layer represents an instruction in the Dockerfile that was used to build the image. These layers are stacked on top of each other to create the final image. Each layer is read-only and has a unique ID.

Thanks to UnionFS, the Docker registry shares common layers between multiple images and containers, reducing the amount of disk space required. When a container modifies a file, it creates a new layer on top of the base image, rather than modifying the files in the base image. This allows for easy rollback to previous states of the container and makes the images highly portable.

There are multiple image repositories you could use depending on a cloud solution you're using (ECR for AWS or Google Container Registry for GCP, for instance) or SaaS solutions (Docker Hub is the most popular – `https://hub.docker.com`). There are also a number of open source licensed solutions available:

- **Harbor**: This is available at `https://goharbor.io/`
- **Portus**: This is available at `http://port.us.org/`
- **Docker Registry**: This is available at `https://docs.docker.com/registry/`
- **Project Quay**: This is available at `https://quay.io` and on GitHub at `https://github.com/quay/quay`

In this section, we've familiarized ourselves with Docker registries to store our Docker images in a remote location. In the next section, we will look more into Docker networking and its extensions.

Docker networking

There are four types of Docker networking: none, bridge, host, and overlay.

Bridge is the default network mode in Docker. Containers in the same bridge network can communicate with each other. Shortly, it creates a virtual network, in which containers are assigned IP addresses and can cummunicate using them, while anything outside of that network cannot reach any of those addresses. In the *Host* network, the container uses the host's network stack. This means that the container shares your machine's IP address and network interfaces.

Overlay mode allows you to create a virtual network that spans multiple Docker hosts. Containers in different hosts can communicate with each other as if they are on the same host. It's useful when running Docker Swarm.

Using the Docker command line, you are able to create a custom network of any of those types.

None network

A **none network** in Docker is a special type of network mode that disables all networking for a container. When a container is run in *none* network mode, it does not have access to any network resources and cannot communicate with other containers or the host machine.

To run a container in *none* network mode, you can use the `--network none` option when running the `docker run` command.

For example, to start a container running the `nginx` image in *none* network mode, you would run the following command:

```
admin@myhome:~$ docker run --network none -d ubuntu:20
```

The *none* network is useful for running workloads that aren't supposed to use any network connections, for example, for processing data in a connected volume.

Bridge mode

When using **bridge mode** on a container being created, a virtual interface is also created and attached to the virtual network. Each container is then assigned a unique IP address on the virtual network, allowing it to communicate with other containers and the host machine.

The host machine acts as a gateway for the containers, routing traffic between the containers and the outside network. When a container wants to communicate with another container or the host machine, it sends the packet to the virtual network interface. The virtual network interface then routes the packet to the correct destination.

By default, it's a 172.17.0.0/16 network and it's connected to a bridge device, docker0, in your machine. Within this network, all traffic between containers and the host machine is allowed.

All containers are attached to the default bridge network if no network was selected using the --network option when executing the docker run command.

You can list all available networks using the following command:

```
admin@myhome:~$ docker network ls
NETWORK ID      NAME            DRIVER    SCOPE
6c898bde2c0c    bridge          bridge    local
926b731b94c9    host            host      local
b9f266305e10    none            null      local
```

To get more information about the network, you can use the following command:

```
admin@myhome:~$  docker inspect bridge
[
    {
        "Name": "bridge",
        "Id":
"6c898bde2c0c660cd96c3017286635c943adcb152c415543373469afa0aff13a",
        "Created": "2023-01-26T16:51:30.720499274Z",
        "Scope": "local",
        "Driver": "bridge",
        "EnableIPv6": false,
        "IPAM": {
            "Driver": "default",
            "Options": null,
            "Config": [
                {
                    "Subnet": "172.17.0.0/16",
```

```
                    "Gateway": "172.17.0.1"
                }
            ]
        },
        "Internal": false,
        "Attachable": false,
        "Ingress": false,
        "ConfigFrom": {
            "Network": ""
        },
        "ConfigOnly": false,
        "Containers": {},
        "Options": {
            "com.docker.network.bridge.default_bridge": "true",
            "com.docker.network.bridge.enable_icc": "true",
            "com.docker.network.bridge.enable_ip_masquerade": "true",
            "com.docker.network.bridge.host_binding_ipv4": "0.0.0.0",
            "com.docker.network.bridge.name": "docker0",
            "com.docker.network.driver.mtu": "1500"
        },
        "Labels": {}
    }
]
```

Let's move on to HOST mode.

Host mode

In **host networking mode**, the container shares the host's network stack and network interfaces. This means that the container uses your machine's IP address and network settings, and can directly access the same network resources as the machine it runs on, including other containers.

Containers running in host networking mode can also directly listen on a port of the host machine (bind to it).

One of the main advantages of host networking mode is that it provides better performance as the container doesn't have to go through an additional network stack.

This mode is less secure than the other networking mode as the container has direct access to the host's network resources and can listen to connections on the host's interface.

Overlay

An **overlay network** is created by a manager node, which is responsible for maintaining the network configuration and managing the membership of worker nodes. The manager node creates a virtual network switch and assigns IP addresses to each container on the network.

Each worker node runs Docker Engine and a container network driver, which is responsible for connecting the containers on that host to the virtual network switch. The container network driver also ensures that packets are properly encapsulated and routed to the correct destination.

When a container on one host wants to communicate with a container on another host, it sends the packet to the virtual network switch. The switch then routes the packet to the correct host, where the container network driver decapsulates the packet and delivers it to the destination container.

The overlay network uses the **Virtual eXtensible Local Area Network** (**VXLAN**) protocol to encapsulate IP packets and make it possible to create a Layer 2 network between multiple hosts.

Summary

In this chapter, we have introduced one of the major building blocks of modern DevOps-led infrastructure, that is, containers. We described the most prominent container technology – Docker. We have also introduced the basics of running Docker containers and building your own. In the next chapter, we are going to build on this knowledge and introduce more advanced Docker topics.

A Deep Dive into Docker

The advent of Docker has revolutionized the way we run, deploy, and maintain our applications. With the rise of containerization, we've been able to abstract away much of the underlying infrastructure and dependencies that applications rely on, making it easier than ever to deploy and manage them across different environments. However, with great power comes great responsibility, and we must understand the internals of Docker and establish good practices to ensure that our applications are secure, reliable, and performant.

In this chapter, we'll delve into the nitty-gritty of Docker, exploring its architecture, components, and key features. We'll also examine some of the helper projects that have emerged on top of Docker, such as Docker Compose and Kubernetes, and learn how to use them to build more complex and scalable applications. Throughout, we'll emphasize best practices for working with Docker, such as creating efficient Docker images, managing containers, and optimizing performance. By the end of this chapter, you'll be well-equipped to confidently run your applications inside Docker and leverage its full power to build robust and scalable systems.

This chapter covers the following topics:

- Docker advanced use cases
- Docker Compose
- Advanced Dockerfile techniques
- Docker orchestration

Docker advanced use cases

While using Docker and its CLI, there are a lot of things we will need to take care of in terms of the life cycle of the container, build process, volumes, and networking. Some of those things you can automate by using other tools, but it's still useful to know what's going on underneath.

Running public images

A lot of public images you can find on Docker Hub (`https://hub.docker.com`) have initialization scripts available that take configuration from environment variables or the mounted files to a predefined directory.

The most commonly used image that uses both techniques is images with databases. Let's look for an official **Docker PostgreSQL** image. You can find the one we'll be using here: `https://hub.docker.com/_/postgres`.

To run the official PostgreSQL Docker image, you can use the following command:

```
admin@myhome:~$ docker run --name some-postgres -e POSTGRES_
PASSWORD=mysecretpassword -d postgres
```

In this command, we have the following:

- `--name some-postgres` gives the container a name of *some-postgres*
- `-e POSTGRES_PASSWORD=mysecretpassword` sets the password for the default PostgreSQL user (*postgres*)
- `-d` runs the container in the background; `postgres` specifies the image to use

It's also possible to override the default user (*postgres*) by adding a `POSTGRES_USER` environment variable. Other configuration environment variables are listed in the documentation.

A very useful feature you can use when working with the official PostgreSQL image is database pre-population using SQL scripts. To achieve this, you will need to bind mount a local directory with scripts to `/docker-entrypoint-initdb.d` inside the container. There are two things you will need to take care of: empty data directory and making sure all scripts are finished with success. An empty data directory is necessary as this will act as the entry point where you can load your SQL or shell scripts; it also prevents data loss. If any of the scripts finish with an error, the database won't be started.

Similar features are provided for other Docker images running any other database available in Docker Hub.

Another useful official image you could use is **nginx**: it's probably much simpler to use as you already have a configured web server inside and you will need to provide either content for it to serve (HTML files, JavaScript, or CSS) or override the default configuration.

Here is an example of mounting a static HTML website to a container:

```
admin@myhome:~$ docker run -p 8080:80 -v /your/webpage:/usr/share/
nginx/html:ro -d nginx
```

In this command, we have the following:

- `-p 8080:80`: This option maps port `8080` on the host machine to port `80` inside the container. This means that when someone accesses port `8080` on the host machine, it will be redirected to port `80` in the container.

- `-v /your/webpage:/usr/share/nginx/html:ro`: This option mounts the `/your/webpage` directory on the host machine to the `/usr/share/nginx/html` directory inside the container. The `ro` option means that the mount is read-only, which means that the container cannot modify the files in the `/your/webpage` directory.

- `-d`: This option tells Docker to run the container in detached mode, which means that it will run in the background.

- `nginx`: This is the name of the Docker image that will be used to run the container. In this case, it's the official nginx image from Docker Hub.

We can override the default nginx configuration like so:

```
admin@myhome:~$ docker run -p 8080:80 -v ./config/nginx.conf:/etc/
nginx/nginx.conf:ro -d nginx
```

In this command, most of the previous options repeat themselves except one: `-v ./config/nginx.conf:/etc/nginx/nginx.conf:ro`. This option mounts the `./config/nginx.conf` file on the host machine to the `/etc/nginx/nginx.conf` file inside the container. The `ro` option means that the mount is read-only, which means that the container cannot modify the `nginx.conf` file.

Running a debugging container

Containers running in production usually have very few tools that are useful for troubleshooting installed. On top of that, those containers aren't running as root users and have multiple security mechanisms to prevent tampering. With that in mind, how do we get into the Docker network to debug if something is not working?

The answer to that question is just running another container we could get into. It would have some tools pre-installed or would allow us to install whatever we need while running. There are multiple techniques we can use to achieve this.

First, we will need a process that will run indefinitely until we stop it manually. While this process is running, we could step in and use the `docker exec` command to get *inside* the running Docker.

Knowing Bash scripting, the easiest way to run this process is to create a `while` loop:

```
admin@myhome:~$ docker run -d ubuntu while true; do sleep 1; done
```

Another method is to use the `sleep` program:

```
admin@myhome:~$ docker run -d ubuntu sleep infinity
```

Alternatively, you could just try to *read* a special device, `/dev/null`, that is outputting nothing and the `tail` command:

```
admin@myhome:~$ docker run -d ubuntu tail -f /dev/null
```

Finally, when one of these commands is running inside the network you're trying to troubleshoot, you can run a command inside it, and effectively be able to run commands from within the environment you need to investigate:

```
admin@myhome:~$ docker exec -it container_name /bin/bash
```

Let us now look at cleaning up unused containers.

Cleaning up unused containers

Docker images can accumulate over time, especially when you frequently build and experiment with containers. Some of these images may no longer be needed, and they can take up valuable disk space. To clean up these unused images, you can use the `docker image prune` command. This command removes all images that are not associated with a container, also known as **dangling images**.

In addition to unused images, there may also be leftover containers that were not removed properly. These containers can be identified using the `docker ps -a` command. To remove a specific container, you can use the `docker rm <container_id>` command, where `<container_id>` is the identifier of the container you want to remove. If you want to remove all stopped containers, you can use the `docker container prune` command.

It's good practice to regularly perform image and container cleanup to maintain a healthy Docker environment. This not only saves disk space but also helps prevent potential security vulnerabilities associated with unused resources. It is also best practice to remove all sensitive information, such as passwords and keys, from the containers and images.

Here's an example of using the `docker image prune` command to remove dangling images:

```
admin@myhome:~$ docker image prune
Deleted Images:
deleted:
sha256:5f70bf18a086007016e948b04aed3b82103a36bea41755b6cddfa-
f10ace3c6ef
deleted:sha256:c937c4dd0c2eaf57972d4f80f55058b3685f87420a9a9fb9ef0d-
fe3c7c3e60bc
Total reclaimed space: 65.03MB
```

Here's an example of using the `docker container prune` command to remove all stopped containers:

```
admin@myhome:~$ docker container prune
WARNING! This will remove all stopped containers.

Are you sure you want to continue? [y/N] y

Deleted Containers:
8c922b2d9708fcee6959af04c8f29d2d6850b3a3b3f3b966b0c360f6f30ed6c8
6d90b6cbc47dd99b2f78a086e959a1d18b7e0cf4778b823d6b0c6b0f6b64b6c64
Total reclaimed space: 0B
```

To automate these tasks, you can use `crontab` to schedule regular cleanups. To edit your `crontab` file, you can use the `crontab -e` command. Here's an example of scheduling a daily cleanup of dangling images at 3 A.M.:

```
0 3 * * * docker image prune -f
```

This line is made up of five fields separated by spaces. These fields represent the minutes, hours, days of the month, months, and days of the week when the command will be executed. In this example, the command will be executed at 3 A.M. every day. Let's look at each element in detail:

- The first field, 0, represents the minutes. In this case, we want the command to be executed at exactly 0 minutes past the hour.
- The second field, 3, represents the hours. In this case, we want the command to be executed at 3 A.M.
- The third field, *, represents the days of the month. The asterisk means "any" day of the month.
- The fourth field, *, represents the months. The asterisk means "any" month of the year.
- The fifth field, *, represents the days of the week. The asterisk means "any" day of the week. 1 represents Monday, 2 represents Tuesday, and so on until 7, which represents Sunday.

Here's an example of scheduling a weekly cleanup of stopped containers at 4 A.M. on Sundays:

```
0 4 * * 7 docker container prune -f
```

The `-f` flag is used to force the removal of the images or containers without confirming this with the user.

To list all existing cron jobs for your user, you can use the `crontab -l` command. More about `crontab` can be found online or by using the `man crontab` command. A great how-to article about using it can be found in the Ubuntu Linux knowledge base: `https://help.ubuntu.com/community/CronHowto`.

Docker volumes and bind mounts

As mentioned in the previous chapter, Docker volumes and bind mounts are two different ways to persist data generated by a Docker container. Volumes are managed by Docker and exist outside the container's filesystem. They can be shared and reused between containers, and they persist even if the original container is deleted. On the other hand, bind mounts link a file or directory on the host system to a file or directory in the container. The data in bind mounts is directly accessible from the host and the container and persists for as long as the host file or directory remains.

To use a Docker volume, you can use the `-v` or `--mount` flag when you run the `docker run` command and specify the host source and container destination. For example, to create a volume and mount it to the container at `/app/data`, you can run the following command:

```
admin@myhome:~$ docker run -v my_data:/app/data <image_name>
```

To use a bind mount, you can use the same flags and specify the host source and container destination, just like with a volume. However, instead of using a volume name, you need to use the host file or directory path. For example, to bind mount the `/host/data` host directory to the container at `/app/data`, you can run the following command:

```
admin@myhome:~$ docker run -v /host/data:/app/data <image_name>
```

When using bind mounts in Docker, you may encounter permission problems with the files and directories within the bind mount. This is because the **user IDs (UIDs)** and **group IDs (GIDs)** of the host and container may not match, leading to issues with accessing or modifying the data in the bind mount.

For example, if the host file or directory is owned by a user with UID 1000, and the corresponding UID in the container is different, the container may not be able to access or modify the data in the bind mount. Similarly, if the group IDs do not match, the container may not be able to access or modify the data due to group permissions.

To prevent these permission problems, you can specify the UID and GID of the host file or directory when you run the `docker run` command. For example, to run a container with a bind mount as the UID and GID 1000, you can run the following command:

```
admin@myhome:~$ docker run -v /local/data:/app/data:ro,Z --user
1000:1000 <image_name>
```

In this example, the `:ro` flag specifies that the bind mount should be read-only, and the `,Z` flag tells Docker to label the bind mount with a private label so that it cannot interact with other containers. The `--user` flag sets the UID and GID of the process running inside the container to `1000`.

By specifying the UID and GID of the host file or directory in the container, you can prevent permission problems with bind mounts in Docker and ensure that the container can access and modify the data as expected.

Docker networking advanced use cases

Docker provides a convenient way to manage the networking of containers in a user-defined network. By using Docker networks, you can easily control the communication between containers and isolate them from the host network.

Docker bridge networking is a default network configuration that enables communication between containers running on the same host. It works by creating a virtual network interface on the host system that acts as a bridge between the containers and the host network. Each container on the bridge network is assigned a unique IP address that allows it to communicate with other containers and the host.

Bridge networks are isolated from each other, meaning that containers connected to different bridge networks cannot communicate with each other directly. To achieve communication between containers on different networks, you can use the Docker service discovery mechanism, such as connecting to a specific container IP address or using a load balancer.

To use bridge networking in practice, you can create a new bridge network using the Docker CLI. For example, you can use the `docker network create --driver bridge production-network` command to create a new bridge network named *production-network*. After the network is created, you can then connect your containers to the network by using the `--network` option in the `docker run` command. You can use the `docker run --network production-network my-image` command to run a container from the *my-image* image on the *production-network* network.

In addition to creating a new bridge network, you can connect containers to the default *bridge* network that is automatically created when you install Docker. To connect a container to the default network, you do not need to specify the `--network` option in the `docker run` command. The container will automatically be connected to the default *bridge* network and assigned an IP address from the bridge network subnet.

Now, if you create multiple networks, by default, they will be separated and no communication will be allowed between them. To allow communication between two bridge networks, such as *production-network* and *shared-network*, you will need to create a network connection between the two networks by connecting a container of your choosing to those two networks or allowing all communication between the two networks. The latter option, if possible, is not supported.

The final option is to use **Docker Swarm** mode and overlay network mode, which we will get into a bit later in this chapter in the *Docker orchestration* section.

The following is an example of how to connect a container to two networks at the same time. First, let's create a production and shared network:

```
admin@myhome:~$ docker network create production-network
fd1144b9a9fb8cc2d9f65c913cef343ebd20a6b30c4b3ba94fdb1fb50aa1333c
admin@myhome:~$ docker network create shared-network
a544f0d39196b95d772e591b9071be38bafbfe49c0fdf54282c55e0a6ebe05ad
```

Now, we can start a container connected to `production-network`:

```
admin@myhome:~$ docker run -itd --name prod-container --network
production-network alpine sh
Unable to find image 'alpine:latest' locally
latest: Pulling from library/alpine
8921db27df28: Pull complete
Digest:sha256:f271e74b17ced29b915d351685fd4644785c6d1559dd1f-
2d4189a5e851ef753a
Status: Downloaded newer image for alpine:latest
287be9c1c76bd8aa058aded284124f666d7ee76c73204f9c73136aa0329d6bb8
```

Let's do the same for `shared-network`:

```
admin@myhome:~$ docker run -itd --name shared-container --network
shared-network alpine sh
38974225686ebe9c0049147801e5bc777e552541a9f7b2eb2e681d5da9b8060b
```

Let's investigate if both containers are running:

```
admin@myhome:~$ docker ps
CONTAINER ID    IMAGE      COMMAND     CREATED        STATUS     PORTS     NAMES
38974225686e    alpine     "sh"        4 seconds ago  Up 3
seconds                    shared-container
287be9c1c76b    alpine     "sh"        16 seconds ago Up 14
seconds                    prod-container
```

Finally, let's also connect `prod-container` to a shared network:

```
admin@myhome:~$ docker network connect shared-network prod-container
```

After that, we can get a shell inside `prod-container` and ping `shared-container`:

```
admin@myhome:~$ docker exec -it prod-container /bin/sh
/ # ping shared-container
PING shared-container (172.24.0.2): 56 data bytes
64 bytes from 172.24.0.2: seq=0 ttl=64 time=0.267 ms
64 bytes from 172.24.0.2: seq=1 ttl=64 time=0.171 ms
^C
--- shared-container ping statistics ---
2 packets transmitted, 2 packets received, 0% packet loss
round-trip min/avg/max = 0.171/0.219/0.267 ms
/ # ping prod-container
PING prod-container (172.23.0.2): 56 data bytes
64 bytes from 172.23.0.2: seq=0 ttl=64 time=0.167 ms
64 bytes from 172.23.0.2: seq=1 ttl=64 time=0.410 ms
64 bytes from 172.23.0.2: seq=2 ttl=64 time=0.108 ms
```

```
^C
5 packets transmitted, 5 packets received, 0% packet loss
round-trip min/avg/max = 0.108/0.188/0.410 ms
```

You can learn more about networking in Docker on the official website: `https://docs.docker.com/network/`.

Security features of Docker

Docker, at its core, wasn't meant to be a security tool. This was built in at a later stage with the support of the Linux Kernel features that are still being developed, and more security features are being added.

There's a lot to cover regarding this topic, but we're going to focus on the four most frequently used security features:

- Namespaces
- **Security computing mode (seccomp)**
- Rootless mode
- Docker signed images

Linux kernel namespaces

Kernel namespaces are an important component of Docker security as they provide isolation between containers and the host system. They allow each container to have a view of the system resources, such as the filesystem, network, and process table, without affecting the host or other containers. This means that a process running inside a container cannot access or modify the host filesystem, network configuration, or processes, which helps secure the host system from malicious or rogue containers.

Docker uses several Linux kernel namespaces to provide isolated environments for containers. These namespaces are used to create isolated environments for processes, networks, mount points, and more.

The USER namespace for the Docker daemon will ensure that the root inside the Docker container is running in a separate context from the host context. It's needed to ensure that the root user inside the container is not equal to the root on the host.

The PID namespace isolates the process IDs between containers. Each container sees its own set of processes, isolated from other containers and the host.

The NET namespace's function is to isolate the network stack of each container so that each container has a virtual network stack, with its own network devices, IP addresses, and routes.

The IPC namespace deals with the **inter-process communication** (IPC) resources between containers. Each container has its own private IPC resources, such as System V IPC objects, semaphores, and message queues.

The UTS namespace is about hostname and domain name isolation for each container. Here, each container has its own hostname and domain name that does not affect other containers or the host.

Finally, the MNT namespace isolates the mount points of each container. This means that each container has a private filesystem hierarchy, with its own root filesystem, mounted filesystems, and bind mounts.

By using these namespaces, Docker containers are isolated from each other and from the host, which helps ensure the security of containers and the host system.

The most confusing to use is the USER namespace as it requires a special UID and GID mapping configuration. It's not enabled by default as sharing PID or NET namespaces with the host (-pid=host or -network=host) isn't possible. Also, using the -privileged mode flag on docker run will not be possible without specifying -userns=host (thus disabling the USER namespace separation). Other namespaces listed previously are in effect mostly without any other special configuration.

Seccomp

Seccomp, short for **secure computing mode**, is a Linux kernel feature that allows a process to specify the system calls it is allowed to make. This makes it possible to restrict the types of system calls that can be made by a container, which can help improve the security of the host system by reducing the risk of container escape or privilege escalation.

When a process specifies its seccomp profile, the Linux kernel filters incoming system calls and only allows those that are specified in the profile. This means that even if an attacker were to gain access to a container, they would be limited in the types of actions they could perform, reducing the impact of the attack.

To create a seccomp profile for a container, you can use the seccomp configuration option in the docker run command. This allows you to specify the seccomp profile to use when starting the container.

There are two main ways to create a seccomp profile: using a predefined profile or creating a custom profile. Predefined profiles are available for common use cases and can be easily specified in the docker run command. For example, the default profile allows all system calls, while the restricted profile only allows a limited set of system calls that are considered safe for most use cases.

To create a custom seccomp profile, you can use the **Podman** (https://podman.io/blogs/2019/10/15/generate-seccomp-profiles.html) or **seccomp-gen** (https://github.com/blacktop/seccomp-gen) tools. Both tools automate figuring out which calls are being made by the container you intend to use in production and generate a JSON file that can be used as the seccomp profile.

Seccomp does not guarantee security. It is important to understand the system calls that are required for your application and ensure that they are allowed in the seccomp profile.

The following is an example of a seccomp profile that allows a limited set of system calls for a container running a web server application:

```
{
    "defaultAction": "SCMP_ACT_ALLOW",
    "syscalls": [
        {
            "name": "accept",
            "action": "SCMP_ACT_ALLOW"
        },
        {
            "name": "bind",
            "action": "SCMP_ACT_ALLOW"
        },
        {
            "name": "connect",
            "action": "SCMP_ACT_ALLOW"
        },
        {
            "name": "listen",
            "action": "SCMP_ACT_ALLOW"
        },
        {
            "name": "sendto",
            "action": "SCMP_ACT_ALLOW"
        },
        {
            "name": "recvfrom",
            "action": "SCMP_ACT_ALLOW"
        },
        {
            "name": "read",
            "action": "SCMP_ACT_ALLOW"
        },
        {
            "name": "write",
            "action": "SCMP_ACT_ALLOW"
        }
    ]
}
```

In this example, `defaultAction` is set to `SCMP_ACT_ALLOW`, which means that all system calls not specifically listed in the `syscalls` array will be allowed. To block all not-defined calls, you can use `SCMP_ACT_ERRNO` as a default action. All available actions are described in the online manual for the `seccomp_rule_add` filter specification: `https://man7.org/linux/man-pages/man3/seccomp_rule_add.3.html`.

The `syscalls` array lists the system calls that should be allowed for the container and specifies the action to take for each call (in this case, all calls are allowed). This profile only allows the system calls necessary for a web server to function and blocks all other system calls, improving the security of the container.

More information about system calls is available here: `https://docs.docker.com/engine/security/seccomp/`.

Rootless mode

Docker Rootless mode is a feature that allows users to run Docker containers without having to run the Docker daemon as the root user. This mode provides an additional layer of security by reducing the attack surface of the host system and minimizing the risk of privilege escalation.

Let's set up a rootless Docker daemon on Ubuntu Linux or Debian Linux. First, make sure you've installed Docker from the official Docker package repository instead of the Ubuntu/Debian package:

```
admin@myhome:~$ sudo apt-get install -y -qq apt-transport-https
ca-certificates curl
admin@myhome:~$ sudo mkdir -p /etc/apt/keyrings && sudo chmod -R 0755
/etc/apt/keyrings
admin@myhome:~$ curl -fsSL "https://download.docker.com/linux/ubuntu/
gpg" | sudo gpg --dearmor --yes -o /etc/apt/keyrings/docker.gpg
admin@myhome:~$ sudo chmod a+r /etc/apt/keyrings/docker.gpg
admin@myhome:~$ echo "deb [arch=amd64 signed-by=/etc/apt/keyrings/
docker.gpg] https://download.docker.com/linux/ubuntu jammy stable" |
sudo tee /etc/apt/sources.list.d/docker.list
admin@myhome:~$ sudo apt-get update
admin@myhome:~$ sudo apt-get install -y docker-ce docker-ce-cli
containerd.io docker-scan-plugin docker-compose-plugin docker-ce-
rootless-extras docker-buildx-plugin
```

`docker-ce-rootless-extras` will install a shell script in your `/usr/bin` directory named `dockerd-rootless-setuptool.sh`, which will automate the whole process:

```
admin@myhome~$ dockerd-rootless-setuptool.sh  --help
Usage: /usr/bin/dockerd-rootless-setuptool.sh [OPTIONS] COMMAND
A setup tool for Rootless Docker (dockerd-rootless.sh).
Documentation: https://docs.docker.com/go/rootless/
Options:
  -f, --force                 Ignore rootful Docker (/var/run/docker.
```

```
sock)
      --skip-iptables          Ignore missing iptables


Commands:
   check        Check prerequisites
   install      Install systemd unit (if systemd is available) and show
how to manage the service
   uninstall    Uninstall systemd unit
```

To run this script, we will need a non-root user with a configured environment to be able to run the Docker daemon. Let's create a dockeruser user first:

```
admin@myhome~$ sudo adduser dockeruser
Adding user `dockeruser' ...
Adding new group `dockeruser' (1001) ...
Adding new user `dockeruser' (1001) with group `dockeruser' ...
Creating home directory `/home/dockeruser' ...
Copying files from `/etc/skel' ...
New password:
Retype new password:
passwd: password updated successfully
Changing the user information for dockeruser
Enter the new value, or press ENTER for the default
      Full Name []:
      Room Number []:
      Work Phone []:
      Home Phone []:
      Other []:
Is the information correct? [Y/n] y
```

Let's also create a UID map configuration before we proceed. To do that, we will need to install the uidmap package and create the /etc/subuid and /etc/subgid configuration files:

```
admin@myhome~$ sudo apt install -y uidmap
admin@myhome~$ echo "dockeruser:100000:65536" | sudo tee /etc/subuid
admin@myhome~$ echo "dockeruser:100000:65536" | sudo tee /etc/subgid
```

Log in as dockeruser and run the dockerd-rootless-setuptool.sh script:

```
admin@myhome~$ sudo -i -u dockeruser
```

Make sure environment XDG_RUNTIME_DIR is set and systemd can read environment variables from dockeruser:

```
$ export XDG_RUNTIME_DIR=/run/user/$UID
$ echo 'export XDG_RUNTIME_DIR=/run/user/$UID' >> ~/.bashrc
$ systemctl --user show-environment
HOME=/home/dockeruser
LANG=en_US.UTF-8
LOGNAME=dockeruser
PATH=/usr/local/sbin:/usr/local/bin:/usr/sbin:/usr/bin:/sbin:/bin:/
usr/games:/usr/local/games:/snap/bin:/snap/bin
SHELL=/bin/bash
SYSTEMD_EXEC_PID=720
USER=dockeruser
XDG_RUNTIME_DIR=/run/user/1001
XDG_DATA_DIRS=/usr/local/share/:/usr/share/:/var/lib/snapd/desktop
DBUS_SESSION_BUS_ADDRESS=unix:path=/run/user/1001/bus
```

Now, you can install rootless Docker using the dockerd-rootless-setuptool.sh script (some output has been truncated for readability):

```
$ dockerd-rootless-setuptool.sh install
[INFO] Creating  [condensed for brevity]
     Active: active (running) since Fri 2023-02-17 14:19:04 UTC; 3s
ago
+ DOCKER_HOST=unix:///run/user/1001/docker.sock /usr/bin/docker
version
Client: Docker Engine - Community
 Version:          23.0.1
[condensed for brevity]
Server: Docker Engine - Community
 Engine:
  Version:         23.0.1
 [condensed for brevity]
 rootlesskit:
  Version:         1.1.0
 [condensed for brevity]
+ systemctl --user enable docker.service
Created symlink /home/dockeruser/.config/systemd/user/default.target.
wants/docker.service → /home/dockeruser/.config/systemd/user/docker.
service.
[INFO] Installed docker.service successfully.
```

Now, let's verify if we can use the Docker rootless daemon:

```
dockeruser@vagrant:~$ export DOCKER_HOST=unix:///run/user/1001/docker.
sock
dockeruser@vagrant:~$ docker ps
CONTAINER ID    IMAGE    COMMAND    CREATED    STATUS    PORTS    NAMES
```

At this point, we have a Docker daemon running as a *dockeruser* system user instead of root. We will be able to run all services we need the same way we would in a standard configuration. There are some exceptions, such as a Docker in Docker setup, which require further configuration.

More detailed information about rootless mode can be found at `https://docs.docker.com/engine/security/rootless/`.

Docker signed images

Docker signed images are a security measure that assures users that a Docker image has come from a trusted source and has not been tampered with. Docker uses a digital signature to sign images, which can be verified by Docker Engine to ensure that the image is exactly as it was when it was signed by the publisher.

Docker signed images can be verified by checking the public key of the signer from a trusted registry (such as Docker Hub). If the image is valid, Docker will allow you to pull and run the image locally.

The first step in signing a Docker image is to generate a signing key. A **signing key** is a pair of keys – a public key and a private key – that can be used to sign and verify Docker images. The private key should be kept safe and not shared with anyone, while the public key can be distributed to users who need to verify the signed images. To generate a signing key, you can use the `docker trust key generate` command:

```
admin@myhome:~/$ docker trust key generate devops
Generating key for devops...
Enter passphrase for new devops key with ID 6b6b768:
Repeat passphrase for new devops key with ID 6b6b768:
Successfully generated and loaded private key. Corresponding public
key available: /home/admin/devops.pub
```

Remember to use a strong password to secure the key from access. The private key will be saved in your home directory – for example, `~/.docker/trust/private`. The name of the file will be hashed.

Once you have generated the signing key, the next step is to initialize the trust metadata for the image. The trust metadata contains information about the image, including the list of keys that are authorized to sign the image. To initialize the trust metadata, you can use the `docker trust signer add` command. Note that you need to be logged into the Docker registry you're using (via the `docker login` command):

```
admin@myhome:~/$ docker trust signer add --key devops.pub private-
registry.mycompany.tld/registries/pythonapps
Adding signer "devops" to private-registry.mycompany.tld/registries/
pythonapps/my-python-app...
Initializing signed repository for private-registry.mycompany.tld/
registries/pythonapps/my-python-app...
You are about to create a new root signing key passphrase. This
passphrase
will be used to protect the most sensitive key in your signing system.
Please
choose a long, complex passphrase and be careful to keep the password
and the
key file itself secure and backed up. It is highly recommended that
you use a
password manager to generate the passphrase and keep it safe. There
will be no
way to recover this key. You can find the key in your config
directory.
Enter passphrase for new root key with ID a23d653:
Repeat passphrase for new root key with ID a23d653:
Enter passphrase for new repository key with ID de78215:
Repeat passphrase for new repository key with ID de78215:
Successfully initialized "private-registry.mycompany.tld/registries/
pythonapps/my-python-app"
Successfully added signer: devops to private-registry.mycompany.tld/
registries/pythonapps/my-python-app
```

You can sign the image by using the docker trust sign command after a successful Docker image build and tagging it with your registry name. This command signs the image using the authorized keys in the trust metadata and pushes this information, along with your Docker image, to the registry:

```
admin@myhome:~/$ docker trust sign private-registry.mycompany.tld/
registries/pythonapps/my-python-app:2.9BETA
Signing and pushing trust data for local image private-registry.
mycompany.tld/registries/pythonapps/my-python-app:2.9BETA, may
overwrite remote trust data
The push refers to repository [private-registry.mycompany.tld/
registries/pythonapps]
c5ff2d88f679: Mounted from library/ubuntu
latest: digest:sha256:41c1003bfccce22a81a49062ddb088ea6478eabe-
a1457430e6235828298593e6 size: 529
Signing and pushing trust metadata
Enter passphrase for devops key with ID 6b6b768:
Successfully signed private-registry.mycompany.tld/registries/
pythonapps/my-python-app:2.9BETA
```

To verify that your Docker image has been signed and with which key, you can use the `docker trust inspect` command:

```
admin@myhome:~/$ docker trust inspect --pretty private-registry.
mycompany.tld/registries/pythonapps/my-python-app:2.9BETA
Signatures for private-registry.mycompany.tld/registries/pythonapps/
my-python-app:2.9BETA


SIGNED TAG     DIGEST                     SIGNERS
latest         41c1003bfccce22a81a49062ddb088ea6478eabe-
a1457430e6235828298593e6    devops


List of signers and their keys for private-registry.mycompany.tld/
registries/pythonapps/my-python-app:2.9BETA


SIGNER     KEYS
devops     6b6b7688a444


Administrative keys for private-registry.mycompany.tld/registries/
pythonapps/my-python-app:2.9BETA


  Repository Key:
de782153295086a2f13e432db342c879d8d8d9fdd55a77f685b79075a44a5c37
  Root Key:
c6d5d339c75b77121439a97e893bc68a804368a48d4fd167d4d9ba0114a7336b
```

Docker Content Trust (**DCT**) is disabled by default in the Docker CLI, but you can enable it by setting the DOCKER_CONTENT_TRUST environment variable to 1. This will prevent Docker from downloading non-signed and verified images to local storage.

More detailed information about DCT can be found on an official website: `https://docs.docker.com/engine/security/trust/`.

Docker for CI/CD pipeline integration

Continuous integration (**CI**) and **continuous deployment** (**CD**) are popular software development practices that aim to ensure that the software development process is streamlined and the quality of code is maintained.

CI refers to the practice of automatically building and testing code changes in a shared repository. CD is the next step after CI, where the code changes are automatically deployed to production or a staging environment.

Docker is a popular tool that's used in CI/CD pipelines as it provides an efficient way to package and distribute applications. In this subsection, we will show you how to build and push a Docker image to AWS **Elastic Container Registry (ECR)** using GitHub Actions.

Let's look at an example of how to set up a GitHub Action to build and push a Docker image to AWS ECR.

Create a new GitHub Actions workflow by creating a new file named main.yml in the .github/ workflows directory of your repository. After adding and pushing it to the main branch, it'll be available and triggered after any new push to this branch.

In the main.yml file, define the steps for the workflow, like so:

```yaml
name: Build and Push Docker Image
on:
  push:
    branches:
      - main
env:
  AWS_REGION: eu-central-1
jobs:
  build:
    runs-on: ubuntu-latest
    steps:
    - name: Checkout code
      uses: actions/checkout@v2
    - name: Configure AWS credentials
      uses: aws-actions/configure-aws-credentials@v1
      with:
        aws-access-key-id: ${{ secrets.AWS_ACCESS_KEY_ID }}
        aws-secret-access-key: ${{ secrets.AWS_SECRET_ACCESS_KEY }}
        aws-region: ${{ env.AWS_REGION }}
    - name: Build Docker image
      uses: docker/build-push-action@v2
      with:
        push: true
        tags: ${{ env.AWS_REGION }}/my-image:${{ env.GITHUB_SHA }}
    - name: Push Docker image to AWS ECR
      uses: aws-actions/amazon-ecr-push-action@v1
      with:
        region: ${{ env.AWS_REGION }}
        registry-url: ${{ env.AWS_REGISTRY_URL }}
        tags: ${{ env.AWS_REGION }}/my-image:${{ env.GITHUB_SHA }}
```

Replace the AWS_REGION and AWS_REGISTRY_URL environment variables with your specific values. You should also replace my-image with the name of your Docker image.

In your GitHub repository settings, create two secrets named `AWS_ACCESS_KEY_ID` and `AWS_SECRET_ACCESS_KEY` with the AWS credentials that have the necessary permissions to push to AWS ECR. Alternatively, you could use your own runner and AWS IAM role attached to the runner or GitHub OIDC, which will authenticate itself with the AWS account. You can find the relevant documentation here: `https://docs.github.com/en/actions/deployment/security-hardening-your-deployments/configuring-openid-connect-in-amazon-web-services`.

With these steps in place, your GitHub Action will now automatically build and push your Docker image to AWS ECR every time you push code changes to the main branch. After the push, you could trigger another process on the server side to evaluate and deploy a new Docker image to one of your environments without further manual interaction. This helps streamline your CI/CD pipeline and ensures that your code changes are deployed to production with confidence.

It's also possible to integrate the same pipeline with GitLab or other CI/CD tools in a similar manner.

In this section, we've learned about some not-so-common use cases for containers, such as rootless mode, secure computing mode, networking advanced use cases, and how to start a debugging container. In the next section, we will focus on automating the process of setting up Docker containers even further and how to orchestrate it a bit better than manually starting containers one by one.

Docker Compose

Docker Compose is a console tool for running multiple containers using one command. It provides an easy way to manage and coordinate multiple containers, making it easier to build, test, and deploy complex applications. With Docker Compose, you can define your application's services, networks, and volumes in a single YAML file, and then start and stop all services from the command line.

To use Docker Compose, you first need to define your application's services in a `docker-compose.yml` file. This file should include information about the services you want to run, their configuration, and how they are connected. The file should also specify which Docker images to use for each service.

The `docker-compose.yaml` file is a central configuration file that's used by Docker Compose to manage the deployment and running of applications. It is written in YAML syntax.

The structure of the `docker-compose.yaml` file is divided into several sections, each of which defines a different aspect of the deployment. The first section, `version`, specifies the version of the Docker Compose file format being used. The second section, `services`, defines the services that make up the application, including their image names, environment variables, ports, and other configuration options.

The `services` section is the most important part of the `docker-compose.yaml` file as it defines how the application is built, run, and connected. Each service is defined by its own set of key-value pairs, which specify its configuration options. For example, the `image` key is used to specify the name of the Docker image to be used for the service, while the `ports` key is used to specify the port mappings for the service.

The docker-compose.yaml file can also include other sections, such as volumes and networks, which allow you to define shared data storage and network configurations for your application. Overall, the docker-compose.yaml file provides a centralized, declarative way to define, configure, and run multi-container applications with Docker Compose. With its simple syntax and powerful features, it is a key tool for streamlining the development and deployment of complex applications.

Environment variables are key-value pairs that allow you to pass configuration information to your services when they are run. In the docker-compose.yaml file, environment variables can be specified using the environment key within the service definition.

One way to specify environment variables in the docker-compose.yaml file is to simply list them as key-value pairs within the environment section. Here's an example:

```
version: '3'
services:
  db:
    image: mariadb
    environment:
      MYSQL_ROOT_PASSWORD: example
      MYSQL_DATABASE: example_db
```

In addition to specifying environment variables directly in the docker-compose.yaml file, you can store them in an external file and reference that file within the docker-compose.yaml file using the env_file key. Here's an example:

```
version: '3'
services:
  db:
    image: mariadb
    env_file:
      - db.env
```

The contents of the db.env file might look like this:

```
MYSQL_ROOT_PASSWORD=example
MYSQL_DATABASE=example_db
```

By using an external env_file key, you can keep sensitive information separate from your docker-compose.yaml file and easily manage environment variables across different environments.

As an example, consider a MariaDB Docker image. The MariaDB image requires several environment variables to be set to configure the database, such as MYSQL_ROOT_PASSWORD for the root password, MYSQL_DATABASE for the name of the default database, and others. These environment variables can be defined in the docker-compose.yaml file to configure the MariaDB service.

Let's look at an example of using Docker Compose to set up a nginx container, a PHP-FPM container, a WordPress container, and a MySQL container. We'll start by defining our services in the `docker-compose.yml` file and break it down into smaller blocks with comments:

```
version: '3'
```

The preceding line defines the version of the Docker Compose file syntax. Next, we will define all Docker images to be run and interact with each other:

```
services:
  web:
    image: nginx:latest
    depends_on:
      - wordpress
    ports:
      - 80:80
    volumes:
      - ./nginx.conf:/etc/nginx/conf.d/default.conf
      - wordpress:/var/www/html
    networks:
      - wordpress
```

This defines a component of our application stack named web. It will use a Docker image from Docker Hub named nginx with the latest tag. Here are some other important settings:

- depends_on: This tells Docker Compose to start this component after the wordpress service.

- ports: This forwards your host port to a Docker port; in this case, it'll open port 80 on your computer and forward all incoming traffic to the same port inside the Docker image, the same way the -p setting does when starting a single Docker container using the command line.

- volumes: This setting is equivalent to the -v option in the Docker command-line tool, so it'll mount a nginx.conf file from the local directory to the /etc/nginx/conf.d/default.conf file inside a Docker image.

- wordpress:/var/www/html: This line will mount a Docker volume named wordpress to the directory inside the Docker image. The volume will be defined ahead.

- networks: Here, we're connecting this service to a Docker network named wordpress, which is defined as follows:

```
wordpress:
  image: wordpress:php8.2-fpm-alpine
  depends_on:
    - db
  environment:
    WORDPRESS_DB_HOST: db:3306
```

```
      WORDPRESS_DB_USER: example_user
      WORDPRESS_DB_NAME: example_database
      WORDPRESS_DB_PASSWORD: example_password
    restart: unless-stopped
    volumes:
      - wordpress:/var/www/html
    networks:
      - wordpress
```

The preceding service is very similar to a *web* service, with the following additions:

- environment: This defines environment variables present inside the Docker image.

- restart: This configures the service so that it's automatically restarted if the process stops working for some reason. Docker Compose will not attempt to restart this service if we've manually stopped it.

- depends_on: This server will only be started after the db service is up.

Let's look at the db service:

```
db:
  image: mariadb:10.4
  environment:
    MYSQL_ROOT_PASSWORD: example_password
    MYSQL_DATABASE: example_database
    MYSQL_USER: example_user
    MYSQL_PASSWORD: example_password
  restart: unless-stopped
  volumes:
    - dbdata:/var/lib/mysql
  networks:
    - wordpress
```

This service is setting up the MariaDB database so that it can store WordPress data. Note that all environment variables we can use for MariaDB or WordPress images are documented on their respective Docker Hub pages:

```
volumes:
  wordpress:
  dbdata:
```

Here, we're defining the Docker volumes we are using for WordPress and MariaDB. These are regular Docker volumes that are stored locally, but by installing Docker Engine plugins, those could be distributed filesystems, such as GlusterFS or MooseFS:

```
networks:
  wordpress:
    name: wordpress
    driver: bridge
```

Finally, we're defining a `wordpress` network with a `bridge` driver that allows communication between all preceding services with isolation from the Docker images running on a machine you will run it on.

In the preceding example, in addition to the options already covered in this section, we have a services dependency (`depends_on`), which will allow us to force the order in which containers will be started.

The two volumes we're defining (`wordpress` and `dbdata`) are used for data persistence. The `wordpress` volume is being used to host all WordPress files and it's also mounted to the web container that is running the nginx web server. That way, the web server will be able to serve static files such as CSS, images, and JavaScript, as well as forward requests to the PHP-FPM server.

Here's the nginx configuration, which uses `fastcgi` to connect to the WordPress container running the PHP-FPM daemon:

```
server {
    listen 80;
    listen [::]:80;
    index index.php index.htm index.html;
    server_name _;
    error_log  /dev/stderr;
    access_log /dev/stdout;
    root /var/www/html;

    location / {
        try_files $uri $uri/ /index.php;
    }

    location ~ \.php$ {
      include fastcgi_params;
      fastcgi_intercept_errors on;
      fastcgi_pass wordpress:9000;
      fastcgi_param  SCRIPT_FILENAME $document_root$fastcgi_script_
name;
    }
```

```
    location ~* \.(js|css|png|jpg|jpeg|gif|ico)$ {
      expires max;
      log_not_found off;
    }
  }
```

With this `docker-compose.yml` file, you can start and stop all the services defined in the file by using the `docker-compose up` and `docker-compose down` commands, respectively. When you run `docker-compose up`, Docker will download the necessary images and start the containers, and you'll be able to access your WordPress website at `http://localhost`.

`docker-compose` is a very useful tool for running applications that require multiple services in an easy and repeatable way. It's most commonly used when running applications locally for development, but some organizations decide to use `docker-compose` in production systems where it serves its purpose.

It's extremely rare if you can use a ready-made Docker image for local development or production. Using public images as a base for your customization is a practice applied in, dare we claim it, all organizations using Docker. With that in mind, in the next section, we will learn how to build Docker images using multi-stage builds and how to use each Dockerfile command properly.

Advanced Dockerfile techniques

Dockerfiles are used to define how an application should be built inside a Docker container. We covered most of the available commands in *Chapter 8*. Here, we will introduce more advanced techniques, such as multi-stage builds or not-so-common ADD command uses.

Multi-stage build

Multi-stage builds are a feature of Docker that allows you to use multiple Docker images to create a single final image. By creating multiple stages, you can separate the build process into distinct steps and reduce the size of the final image. Multi-stage builds are particularly useful when building complex applications that require multiple dependencies as they allow developers to keep the necessary dependencies in one stage and the application in another.

One example of using multi-stage builds with a Golang application involves creating two stages: one for building the application and one for running it. In the first stage, the Dockerfile pulls in the necessary dependencies and compiles the application code. In the second stage, only the compiled binary is copied over from the first stage, reducing the size of the final image. Here's an example Dockerfile for a Golang application:

```
# Build stage
FROM golang:alpine AS build
RUN apk add --no-cache git
```

```
WORKDIR /app
COPY . .
RUN go mod download
RUN go build -o /go/bin/app

# Final stage
FROM alpine
COPY --from=build /go/bin/app /go/bin/app
EXPOSE 8080
ENTRYPOINT ["/go/bin/app"]
```

In the preceding example, the Dockerfile creates two stages. The first stage uses the golang:alpine image and installs the necessary dependencies. Then, it compiles the application and places the binary in the /go/bin/app directory. The second stage uses the smaller Alpine image and copies the binary from the first stage into the /go/bin/app directory. Finally, it sets the entry point to /go/bin/app.

ADD command use cases

The ADD command in a Docker file is used to add files or directories to the Docker image. It works in the same way as COPY but with some additional features. We've talked about basic uses before, but there are other use cases too.

The second use case allows you to unpack files compressed with ZIP or TAR and gzip tools on the fly. While adding a compressed file to the image, the file will be uncompressed and all the files inside it will be extracted to the destination folder. Here's an example:

```
ADD my-tar-file.tar.gz /app
```

The third way of using the ADD command is to copy a remote file from a URL to the Docker image. For example, to download a file named file.txt from a URL, https://yourdomain.tld/configurations/nginx.conf, and copy it to the nginx configuration directory, /etc/nginx, inside the Docker image, you can use the following ADD command:

```
ADD https://yourdomain.tld/configurations/nginx.conf /etc/nginx/nginx.
conf
```

You can also use a Git repository to add your code:

```
ADD --keep-git-dir=true https://github.com/your-user-or-organization/
some-repo.git#main /app/code
```

To clone a Git repository over SSH, you will need to allow the ssh command inside Docker to access a private key with access to the repository you're trying to access. You can achieve this by adding a private key in a multi-stage build and removing it at the end of the stage where you're cloning a repository. This is generally not recommended if you have a choice. You can do this more securely by using Docker secrets and mounting the secret while running the build.

Here's an example of using ARG with a private key:

```
ARG SSH_PRIVATE_KEY
RUN mkdir /root/.ssh/
RUN echo "${SSH_PRIVATE_KEY}" > /root/.ssh/id_rsa
# make sure your domain is accepted
RUN touch /root/.ssh/known_hosts
RUN ssh-keyscan gitlab.com >> /root/.ssh/known_hosts
RUN git clone git@gitlab.com:your-user/your-repo.git
```

Here's an example of using a Docker secret and a mount:

```
FROM python:3.9-alpine
WORKDIR /app

RUN --mount=type=secret,id=ssh_id_rsa,dst=~/id_rsa chmod 400 ~/id_rsa \
    && ssh-agent bash -c 'ssh-add ~/id_rsa; git clone git@gitlab.com:your-user/your-repo.git' && rm -f ~/id_rsa
    # Rest of the build process follows…
```

In the preceding example, we're assuming your private key isn't protected by the password and your key is being saved in the ssh_id_rsa file.

The final way of using the ADD command is to extract a TAR archive from the host machine and copy its contents to the Docker image. For example, to extract a TAR archive named data.tar.gz from the host machine and copy its contents to the /data directory inside the Docker image, you can use the following ADD command:

```
ADD data.tar.gz /data/
```

Secrets management

Docker secrets management is an important aspect of building secure and reliable containerized applications.

Secrets are sensitive pieces of information that an application needs to function, but they should not be exposed to unauthorized users or processes. Examples of secrets include passwords, API keys, SSL certificates, and other authentication or authorization tokens. These secrets are often required by applications at runtime, but storing them in plaintext in code or configuration files can be a security risk.

Securing secrets is crucial to ensuring the security and reliability of applications. Leaking secrets can lead to data breaches, service disruptions, and other security incidents.

In the basic Docker setup, it's only possible to provide secrets to a Docker image using environment variables, as we covered in *Chapter 8*. Docker also provides a built-in secrets management mechanism that allows you to securely store and manage secrets. However, it's only available when Swarm mode needs to be enabled (we will get back to Swarm later in this chapter in the *Docker orchestration* section).

To make secrets available to applications running inside Docker, you can use the `docker secret create` command. For example, to create a secret for a MySQL database password, you can use the following command:

```
admin@myhome:~/$ echo "mysecretpassword" | docker secret create mysql_
password -
```

This command creates a secret named `mysql_password` with a value of `mysecretpassword`.

To use a secret in a service, you need to define the secret in the service configuration file. For example, to use the `mysql_password` secret in a service, you can define it in the `docker-compose.yml` file, as follows:

```
version: '3'
services:
  db.
    image: mysql
    environment:
      MYSQL_ROOT_PASSWORD_FILE: /run/secrets/mysql_password
    secrets:
      - mysql_password
    volumes:
      - db_data:/var/lib/mysql
secrets:
  mysql_password:
    external: true
volumes:
  db_data:
```

In this configuration file, the `mysql_password` secret is defined in the `secrets` section, and the `MYSQL_ROOT_PASSWORD_FILE` environment variable is set to the path of the secret file, which is `/run/secrets/mysql_password`.

To deploy the service, you can use the `docker stack deploy` command. For example, to deploy the service defined in the `docker-compose.yml` file, you can use the following command:

```
admin@myhome:~/$ docker stack deploy -c docker-compose.yml myapp
```

Handling secrets with care is extremely important from a security perspective. The most common mistake is putting a secret directly inside a Docker image, environment file, or application configuration file that is committed into the Git repository. There are existing contingencies that prevent users from doing that (such as Dependabot in GitHub), but if they should fail, it's extremely hard to remove them from the Git history afterward.

In this section, we covered how to handle different aspects of building a container and advanced build techniques. With this knowledge and with the use of Docker Compose, you will be able to build and run your application with a decent dose of automation. What if you have 10 of those applications? 100? Or even more?

In the next section, we will dig into clusters, which will automate things further and deploy your applications to multiple hosts simultaneously.

Docker orchestration

In the world of containerization, **orchestration** is the process of automating deployment and managing and scaling your applications across multiple hosts. Orchestration solutions help simplify the management of containerized applications, increase availability, and improve scalability by providing a layer of abstraction that allows you to manage containers at a higher level, instead of manually managing individual containers.

Docker Swarm is a Docker-native clustering and orchestration tool that allows you to create and manage a cluster of Docker nodes, allowing users to deploy and manage Docker containers across a large number of hosts. Docker Swarm is an easy-to-use solution that comes built-in with Docker, making it a popular choice for those who are already familiar with Docker.

Kubernetes is an open source container orchestration platform that was originally developed by Google. Kubernetes allows you to deploy, scale, and manage containerized applications across multiple hosts, while also providing advanced features such as self-healing, automated rollouts, and rollbacks. Kubernetes is one of the most popular orchestration solutions in use today and is widely used in production environments.

OpenShift is a container application platform that is built on top of Kubernetes and it's developed by Red Hat. This platform provides a complete solution for deploying, managing, and scaling containerized applications, with additional features such as built-in CI/CD pipelines, integrated monitoring, and automatic scaling. OpenShift is designed to be enterprise-grade, with features such as multi-tenancy and **role-based access control** (**RBAC**), making it a popular choice for large organizations that need to manage complex containerized environments.

There is a wide variety of orchestration solutions available, each with its strengths and weaknesses. The choice of which solution to use ultimately depends on your specific needs, but Docker Swarm, Kubernetes, and OpenShift are all popular choices that provide robust and reliable orchestration capabilities for containerized applications.

Docker Swarm

Docker Swarm is a native clustering and orchestration solution for Docker containers. It provides a simple yet powerful way to manage and scale Dockerized applications across a cluster of hosts. With Docker Swarm, users can create and manage a swarm of Docker nodes that act as a single virtual system.

The base components of Docker Swarm are as follows:

- **Nodes**: These are the Docker hosts that form the Swarm. Nodes can be physical or virtual machines running the Docker daemon, and they can join or leave the Swarm as needed.

- **Services**: These are the applications that run on the Swarm. A service is a scalable unit of work that defines how many replicas of the application should run, and how to deploy and manage them across the Swarm.

- **Managers**: These are the nodes responsible for managing the Swarm state and orchestrating the deployment of services. Managers are in charge of maintaining the desired state of the services and ensuring they are running as intended.

- **Workers**: These are the nodes that run the actual containers. Workers receive instructions from the managers and run the desired replicas of the service.

- **Overlay networks**: These are the networks that allow the services to communicate with each other, regardless of the node they are running on. Overlay networks provide a transparent network that spans the entire Swarm.

Docker Swarm provides a simple and easy-to-use way to manage containerized applications. It is tightly integrated with the Docker ecosystem and provides a familiar interface for Docker users. With its built-in features for service discovery, load balancing, rolling updates, and scaling, Docker Swarm is a popular choice for organizations that are just starting with container orchestration.

To initialize Docker Swarm mode and add two workers to the cluster, you will need to initialize Swarm mode:

```
admin@myhome:~/$ docker swarm init
Swarm initialized: current node (i050z7b0tjoew7hxlz419cd81) is now a
manager.

To add a worker to this swarm, run the following command:

    docker swarm join --token SWMTKN-1-0hu2dmht259tb4skyetrpzl2qhxgedd
ij3bc1wof3jxh7febmd-6pzkhrh4ak345m8022hauviil 10.0.2.15:2377

To add a manager to this swarm, run 'docker swarm join-token manager'
and follow the instructions.
```

This will create a new Swarm and make the current node the Swarm manager.

Once the Swarm has been initialized, you can add worker nodes to the cluster. To do this, you need to run the following command on each worker node:

```
admin@myhome:~/$ docker swarm join --token <token> <manager-ip>
```

Here, `<token>` is the token generated by the `docker swarm init` command output, which you can find in the preceding code block, and `<manager-ip>` is the IP address of the Swarm manager.

For example, if the token is `SWMTKN-1-0hu2dmht259tb4skyetrpzl2qhxgeddij3bc1w of3jxh7febmd-6pzkhrh4ak345m8022hauviil` and the manager IP is `10.0.2.15`, the command would be as follows:

```
admin@myhome:~/$ docker swarm join --token SWMTKN-1-0hu2dmht259tb4skye
trpzl2qhxgeddij3bc1wof3jxh7febmd-6pzkhrh4ak345m8022hauviil 10.0.2.15
```

After running the `docker swarm join` command on each worker node, you can verify that they have joined the Swarm by running the following command on the Swarm manager node:

```
admin@myhome:~/$ docker node ls
ID                        HOSTNAME    STATUS    AVAILABILITY    MANAGER
STATUS    ENGINE VERSION
i050z7b0tjoew7hxlz419cd81
*    myhome    Ready    Active         Leader              23.0.1
```

This will show a list of all the nodes in the Swarm, including the manager and any workers you have added.

After that, you can add more nodes and start deploying applications to Docker Swarm. It's possible to reuse any Docker Compose you're using or Kubernetes manifests.

To deploy a sample application, we can reuse a Docker Compose template by deploying a `wordpress` service, but we will need to update it slightly by using MySQL user and password files in the environment variables:

```
wordpress:
  image: wordpress:php8.2-fpm-alpine
  depends_on:
    - db
  environment:
    WORDPRESS_DB_HOST: db:3306
    WORDPRESS_DB_USER_FILE: /run/secrets/mysql_user
    WORDPRESS_DB_NAME: example_database
    WORDPRESS_DB_PASSWORD_FILE: /run/secrets/mysql_password
```

Here's an example of adding secrets to both the `wordpress` and `db` services:

```
  secrets:
    - mysql_user
    - mysql_password
db:
```

```
        image: mariadb:10.4
        environment:
          MYSQL_ROOT_PASSWORD: example_password
          MYSQL_DATABASE: example_database
          MYSQL_USER_FILE: /run/secrets/mysql_user
          MYSQL_PASSWORD_FILE: /run/secrets/mysql_password
```

Here's an example of adding a secrets definition at the bottom of `docker-compose.yml:secrets`:

```
    mysql_user:
      external: true
    mysql_password:
      external: true
```

The `external: true` setting is telling `docker-compose` that secrets are already present and that it should not try to update or recreate them on its own.

In this version of the Compose file, we use secrets to store the MySQL user and password for both the wordpress and db services.

To deploy this file to Docker Swarm, we can use the following commands:

```
admin@myhome:~/$ echo "root" | docker secret create mysql_user -
vhjhswo2qg3bug9w7id08y34f
echo "mysqlpwd" | docker secret create mysql_password -
oy9hsbzmzrh0jrgjo6bgsydol
```

Then, we can deploy the stack:

```
admin@myhome:~/$ docker stack deploy -c docker-compose.yml wordpress
Ignoring unsupported options: restart

Creating network wordpress_wordpress
Creating service wordpress_web
Creating service wordpress_wordpress
Creating service wordpress_db
```

Here, `docker-compose.yaml` is the name of the Compose file and `my-stack-name` is the name of the Docker stack.

Once the stack has been deployed, the `wordpress`, `web`, and `db` services will be running with the MySQL user and password specified in the secrets. You can verify this by listing stacks and checking if containers are running:

```
admin@myhome:~/$ root@vagrant:~# docker stack ls
NAME            SERVICES
```

```
wordpress    3
root@vagrant:~# docker ps
CONTAINER ID     IMAGE        CREATED        STATUS        PORTS       NAMES
7ea803c289b0    mariadb:10.4                    "docker-
entrypoint.s…"    28 seconds ago  Up 27 seconds    3306/
tcp    wordpress_db.1.dogyh5rf52zzsiq0t95nrhuhc
ed25de3273a2    wordpress:php8.2-fpm-alpine     "docker-
entrypoint.s…"    33 seconds ago  Up 31 seconds    9000/
tcp    wordpress_wordpress.1.xmmljnd640ff9xs1249jpym45
```

Docker Swarm is a great project to start your adventure with Docker orchestration methods. It's possible to use it with a production-grade system by using various plugins that will extend its default functionality.

Kubernetes and OpenShift

Two of the most popular tools for orchestrating Docker containers are Kubernetes and OpenShift. Although they share some similarities, they also have some significant differences. Here are the main differences between Kubernetes and OpenShift:

- **Architecture**: Kubernetes is a standalone orchestration platform that is designed to work with multiple container runtimes, including Docker. OpenShift, on the other hand, is a platform that is built on top of Kubernetes. It provides additional features and tools, such as source code management, continuous integration, and deployment. These additional features make OpenShift a more comprehensive solution for enterprises that require end-to-end DevOps capabilities.

- **Ease of use**: Kubernetes is a powerful orchestration tool that requires a high level of technical expertise to set up and operate. OpenShift, on the other hand, is designed to be more user-friendly and accessible to developers with varying levels of technical knowledge. OpenShift provides a web-based interface for managing applications and can be integrated with various development tools, making it easier for developers to work with.

- **Cost**: Kubernetes is an open source project that is free to use, but enterprises may need to invest in additional tools and resources to set it up and operate it. OpenShift is an enterprise platform that requires a subscription for full access to its features and support. The cost of OpenShift may be higher than Kubernetes, but it provides additional features and support that may be worth the investment for enterprises that require advanced DevOps capabilities.

Both solutions are powerful Docker orchestration tools that offer different benefits and trade-offs. Kubernetes is highly customizable and suitable for more technical users. OpenShift, on the other hand, provides a more comprehensive solution with additional features and a user-friendly interface but comes at a higher cost. You should consider specific needs in your organization when choosing between these two tools, keeping in mind that Docker Swarm is also an option. Cloud providers also developed their own solutions, with Elastic Container Service being one of them.

Summary

In this chapter, we covered more advanced topics around Docker, only touching topics around orchestration. Kubernetes, OpenShift, and SaaS solutions provided by cloud operators are driving the creation of new tools that will further ease Docker's use in modern applications.

Docker has had a profound impact on the world of software development and deployment, enabling us to build, ship, and run applications more efficiently and reliably than ever before. By understanding the internals of Docker and following best practices for working with it, we can ensure that our applications are secure, performant, and scalable across a wide range of environments.

In the next chapter, we will look into challenges on how to monitor and gather logs in a distributed environment built on top of Docker containers.

Part 3: DevOps Cloud Toolkit

This last part of the book will focus more on automation using **Configuration as Code (CaC)** and **Infrastructure as Code (IaC)**. We will also talk about monitoring and tracing as a crucial part of modern application development and maintenance. In the last chapter, we will talk about DevOps pitfalls we've experienced in many projects we've been involved with.

This part has the following chapters:

- *Chapter 10, Monitoring, Tracing, and Distributed Logging*
- *Chapter 11, Using Ansible for Configuration as Code*
- *Chapter 12, Leveraging Infrastructure as Code*
- *Chapter 13, CI/CD with Terraform, GitHub, and Atlantis*
- *Chapter 14, Avoiding Pitfalls in DevOps*

10

Monitoring, Tracing, and Distributed Logging

Applications developed nowadays tend to be running inside Docker containers or as a serverless application stack. Traditionally, applications were built as a monolithic entity—one process running on a server. All logs were stored on a disk. It made it easy to get to the right information quickly. To diagnose a problem with your application, you had to log in to a server and search through logs or stack traces to get to the bottom of the problem. But when you run your application inside a Kubernetes cluster in multiple containers that are executed on different servers, things get complicated.

This also makes it very difficult to store logs, let alone view them. In fact, while running applications inside a container, it's not advisable to save any files inside it. Oftentimes, we run those containers in a read-only filesystem. This is understandable as you should treat a running container as an ephemeral identity that can be killed at any time.

We face an identical situation when running serverless applications; on **Amazon Web Services** (**AWS**) Lambda, the process starts when you get a request, data inside that request gets processed, and the application dies after it finishes its job. If you happen to save anything to disk, it will get deleted once processing is concluded.

The most logical solution is, of course, sending all logs to some external system that will save, catalog, and make your logs searchable. There are multiple solutions, including **Software as a Service** (**SaaS**) and cloud-specific applications.

Incidentally, sending logs to an external system is also beneficial for bare-metal servers—for analysis and alerting, or diagnosing if you happen to lose access to the server or it stops responding.

Along with system and application logs, we can also send application-tracing metrics. Tracing is a more in-depth form of metrics where the application will provide you with more insights into system performance and how it behaves in given circumstances. Examples of trace data are the time in which a given request was handled by your application, how many CPU cycles it took, and how long your application was waiting for a database to respond.

In this chapter, you will learn about the following:

- What are monitoring, tracing, and logging?
- How to choose and configure one of the cloud-ready logging solutions
- Self-hosted solutions and how to choose them
- SaaS solutions and how to evaluate which will be most useful for your organization

Additionally, we will be covering the following topics:

- Differences between monitoring, tracing, and logging
- Cloud solutions
- Open source solutions for self-hosting
- SaaS solutions
- Log and metrics retention

So, let's jump right into it!

Differences between monitoring, tracing, and logging

You will hear these terms being used interchangeably depending on the context and person you're talking to, but there's a subtle and very important difference between them.

Monitoring refers to instrumenting your servers and applications and gathering data about them for processing, identifying problems, and, in the end, bringing results in front of interested parties. This also includes alerting.

Tracing, on the other hand, is more specific, as we already mentioned. Trace data can tell you a lot about how your system is performing. With tracing, you can observe statistics that are very useful to developers (such as how long a function ran and whether the SQL query is fast or bottleneck), DevOps engineers (how long we were waiting for a database or network), or even the business (what was the experience of the user with our application?). So, you can see that when it's used right, it can be a very powerful tool under your belt.

The purpose of **logging** is to bring actionable information in a centralized way, which commonly is just saving all messages to a file (it's called a log file). These messages typically consist of the success or failure of a given operation with configurable verbosity. Logging is primarily used by system administrators or DevOps engineers to provide a better view of what's going on in the operating system or with any given application.

With that cleared up, we can jump into specific implementations of the distributed monitoring solutions in the cloud, DIY solutions, or as a SaaS application.

Cloud solutions

Every cloud provider out there is fully aware of the need for proper monitoring and distributed logging, so they will have built their own native solutions. Sometimes it's worth using native solutions, but not always. Let's take a look at the major cloud providers and what they have to offer.

One of the first services available in AWS was **CloudWatch**. At first, it would just collect all kinds of metrics and allow you to create dashboards to better understand system performance and easily spot issues or simply a denial-of-service attack, which in turn allowed you to quickly react to them.

Another function of CloudWatch is alerting, but it's limited to sending out emails using another Amazon service, **Simple Email Service**. Alerting and metrics could also trigger other actions inside your AWS account, such as scaling up or down the number of running instances.

As of the time of writing this book, CloudWatch can do so much more than monitoring. The developers of this service have added the ability to collect and search through logs (**CloudWatch Logs Insights**), monitor changes in AWS resources itself, and trigger actions. We're also able to detect anomalies within our applications using **CloudWatch anomaly detection**.

As for tracing, AWS has prepared a service called **AWS X-Ray**, which is an advanced distributed tracing system that can give you information about how your application is working in the production environment in almost real time. Unfortunately, its capabilities are limited to only a couple of languages out there: Node.js, Java, and .NET. So, you're out of luck if your application is written in Python.

Looking at other popular cloud solutions, there's Google. The **Google Cloud Platform** (**GCP**) consists of a smart solution for gathering logs, querying, and error reporting, and it's called… **Cloud Logging**. If using this service within GCP, similarly to CloudWatch Logs, you will be able to send your application logs, store them, search for data you need (IP addresses, query strings, debug data, and so on), and analyze your logs by using SQL-like queries.

The similarities end here, though, as Google went a couple of steps further with additional features such as the ability to create log dashboards with visualizations of errors reported by your application, or creating log-based metrics.

In GCP, monitoring is carried out by another service entirely–Google Cloud Monitoring. It's focused on gathering data about your application, creating **Service-Level Objectives** (**SLOs**), extensive metrics collection from Kubernetes (**Google Kubernetes Engine**, or **GKE**), and third-party integrations, for example, with a well-known service such as **Prometheus**.

Looking at the Microsoft Cloud Platform—Azure—you will find **Azure Monitor Service**, which consists of several parts that cover the requirements for full-scope application monitoring and tracing. There is **Azure Monitor Logs** for gathering logs, obviously. There is also **Azure Monitor Metrics** for monitoring and visualizing all metrics you push there. You can also analyze, query, and set alerts like you would be able to in GCP or AWS. Tracing is being done by **Azure Application Insights**. It is being promoted by Microsoft as an **Application Performance Management** (**APM**) solution and is part of **Azure Monitor**. It offers a visual map of the application, real-time metrics, code analysis, usage data, and many more features. The implementation, obviously, differs between all cloud providers and their solutions. You will have to refer to the documentation on how to instrument and configure each of those services.

We will focus on AWS services. We will create a log group for our application and gather metrics from an EC2 instance. We will also talk about tracing with AWS X-Ray in Python, which we could use for our application running inside AWS infrastructure no matter the underlying service.

CloudWatch Logs and metrics

CloudWatch Logs is a log management service provided by AWS that enables you to centralize, search, and monitor log data from various sources in a single place. It allows you to troubleshoot operational problems and security incidents, as well as monitor resource utilization and performance.

CloudWatch metrics are a monitoring service provided by AWS that allows you to collect, track, and monitor various metrics for your AWS resources and applications.

CloudWatch metrics provide users with a detailed view of how their AWS resources are performing, by collecting and displaying key metrics, such as CPU utilization, network traffic, and disk I/O, and other metrics related to AWS resources, such as EC2 instances, RDS instances, S3 buckets, and Lambda functions.

Users can use CloudWatch metrics to set alarms that will notify them when certain thresholds are exceeded, as well as to create custom dashboards that display important metrics in near real time. CloudWatch metrics also allow users to retrieve and analyze historical data, which can be used to identify trends and optimize resource usage.

To be able to send logs and metrics to CloudWatch, we will need the following:

- An IAM policy that grants permissions to send logs to CloudWatch Logs. Additionally, we will allow pushing metrics data along with the logs.

- To create an IAM role with the previously created policy attached to it. This role then can be assumed by EC2 instances, Lambda functions, or any other AWS services that require the ability to send logs to CloudWatch Logs.

- To attach the role to a resource that we want to send logs to CloudWatch Logs. For our purpose, we will attach the role to an EC2 instance.

An example of an IAM policy is as follows:

```
{
    "Version": "2012-10-17",
    "Statement": [
        {
            "Sid": "CloudWatchLogsPermissions",
```

```
                    "Effect": "Allow",
                    "Action": [
                        "logs:CreateLogStream",
                        "logs:CreateLogGroup",
                        "logs:PutLogEvents"
                    ],
                    "Resource": "arn:aws:logs:*:*:*"
                },
                {

                    "Sid": "CloudWatchMetricsPermissions",
                    "Effect": "Allow",
                    "Action": [
                        "cloudwatch:PutMetricData"
                    ],
                    "Resource": "*"
                }
        ]
}
```

In this policy, the `logs:CreateLogStream` and `logs:PutLogEvents` actions are allowed for all CloudWatch Logs resources (`arn:aws:logs:*:*:*`), and the `cloudwatch:PutMetricData` action is allowed for all CloudWatch metric resources (`*`).

We will also need a trust policy allowing EC2 to assume a role we're going to create for it, in order to be able to send metrics and logs. The trust policy looks like this:

```
{
  "Version": "2012-10-17",
  "Statement": [
    {
      "Effect": "Allow",
      "Principal": { "Service": "ec2.amazonaws.com"},
      "Action": "sts:AssumeRole"
    }
  ]
}
```

Save this to a `trust-policy.json` file, which we will use in a moment.

Using the AWS CLI tool, to create an instance profile and attach the preceding policy to it, you will need to run the following commands:

```
admin@myhome:~$ aws iam create-instance-profile --instance-profile-
name DefaultInstanceProfile
{
```

```
    "InstanceProfile": {
        "Path": "/",
        "InstanceProfileName": "DefaultInstanceProfile",
        "InstanceProfileId": "AIPAZZUIKRXR3HEDBS72R",
        "Arn": "arn:aws:iam::673522028003:instance-profile/
DefaultInstanceProfile",
        "CreateDate": "2023-03-07T10:59:01+00:00",
        "Roles": []
    }
}
```

We will also need a role with a trust policy attached to it:

```
admin@myhome:~$ aws iam create-role --role-name DefaultInstanceProfile
--assume-role-policy-document file://trust-policy.json
{
    "Role": {
        "Path": "/",
        "RoleName": "DefaultInstanceProfile",
        "RoleId": "AROAZZUIKRXRYB6HO35BL",
        "Arn": "arn:aws:iam::673522028003:role/
DefaultInstanceProfile",
        "CreateDate": "2023-03-07T11:13:54+00:00",
        "AssumeRolePolicyDocument": {
            "Version": "2012-10-17",
            "Statement": [
                {
                    "Effect": "Allow",
                    "Principal": {
                        "Service": "ec2.amazonaws.com"
                    },
                    "Action": "sts:AssumeRole"
                }
            ]
        }
    }
}
```

Now, we can attach the role we just created to the instance profile, so we can use it in the EC2 instance:

```
admin@myhome:~$ aws iam add-role-to-instance-profile --role-name
DefaultInstanceProfile --instance-profile-name DefaultInstanceProfile
```

And now, let's attach a policy to use against the EC2 service:

```
admin@myhome:~$ aws iam put-role-policy --policy-name
DefaultInstanceProfilePolicy --role-name DefaultInstanceProfile
--policy-document file://policy.json
```

The `policy.json` file is the file where you've saved the policy.

An instance profile, as the name suggests, will work only with EC2 instances. To use the same policy for a Lambda function, we will need to create an IAM role instead and attach the newly created role to a function.

Let's create a new instance using the AWS CLI as well, and attach the instance profile we've just created. This particular instance will be placed in a default VPC and in a public subnet. This will cause the instance to get a public IP address and will be available from the public internet.

To create an EC2 instance in a public subnet of the default VPC using `DefaultInstanceProfile`, you can follow these steps:

1. Get the ID of the default VPC:

    ```
    admin@myhome:~$ aws ec2 describe-vpcs --filters
    "Name=isDefault,Values=true" --query "Vpcs[0].VpcId" --output text
    vpc-0030a3a495df38a0e
    ```

    This command will return the ID of the default VPC. We will need it in the following steps.

2. Get the ID of a public subnet in the default VPC and save it for later use:

    ```
    admin@myhome:~$ aws ec2 describe-subnets --filters
    "Name=vpc-id,Values=vpc-0030a3a495df38a0e"
    "Name=map-public-ip-on-launch,Values=true" --query
    "Subnets[0].SubnetId" --output text
    subnet-0704b611fe8a6a169
    ```

    To launch an EC2 instance, we will need an instance template called an **Amazon Machine Image** (**AMI**) and an SSH key that we will use to access this instance. To get an ID of an Ubuntu image, we can also use the AWS CLI tool.

3. We will filter out the most recent AMI ID of Ubuntu 20.04 with the following command:

    ```
    admin@myhome:~$ aws ec2 describe-images --owners
    099720109477 --filters "Name=name,Values=*ubuntu/images/
    hvm-ssd/ubuntu-focal-20.04*" "Name=state,Values=available"
    "Name=architecture,Values=x86_64" --query "reverse(sort_by(Images,
    &CreationDate))[:1].ImageId" --output text
    ami-0a3823a4502bba678
    ```

    This command will list all available Ubuntu 20.04 images owned by Canonical (`099720109477`) and filter them by name (`ubuntu-focal-20.04-*`), architecture (we need `x86_64`, not ARM), and whether they are available for use (state is available). It will also sort them by creation date in descending order and return the most recent (first on the list) image ID.

4. Now, to create an SSH key, you will need to generate one for yourself or use the key you have
    already on your machine. We will need to upload a public part of our key to AWS. You can
    simply run another CLI command to achieve this:

```
admin@myhome:~$ aws ec2 import-key-pair --key-name admin-key --public-
key-material fileb://home/admin/.ssh/admin-key.pub
{
    "KeyFingerprint": "12:97:23:0f:d6:2f:2b:28:4d:a0:ad:62:a7:20:e
3:f8",
    "KeyName": "admin-key",
    "KeyPairId": "key-0831b2bc5c2a08d82"
}
```

With all that, finally, we're ready to launch a new instance in the public subnet
with DefaultInstanceProfile:

```
admin@myhome:~$ aws ec2 run-instances --image-id ami-0abbe417ed83c0b29
--count 1 --instance-type t2.micro --key-name admin-key --subnet-id
subnet-0704b611fe8a6a169 --associate-public-ip-address --iam-instance-
profile Name=DefaultInstanceProfile
{
    "Groups": [],
    "Instances": [
        {
            "AmiLaunchIndex": 0,
            "ImageId": "ami-0abbe417ed83c0b29",
            "InstanceId": "i-06f35cbb39f6e5cdb",
            "InstanceType": "t2.micro",
            "KeyName": "admin-key",
            "LaunchTime": "2023-03-08T14:12:00+00:00",
            "Monitoring": {
                "State": "disabled"
            },
            "Placement": {
                "AvailabilityZone": "eu-central-1a",
                "GroupName": "",
                "Tenancy": "default"
            },
            "PrivateDnsName": "ip-172-31-17-127.eu-central-1.compute.
internal",
            "PrivateIpAddress": "172.31.17.127",
            "ProductCodes": [],
            "PublicDnsName": "",
            "State": {
                "Code": 0,
                "Name": "pending"
```

```
        },
        "StateTransitionReason": "",
        "SubnetId": "subnet-0704b611fe8a6a169",
        "VpcId": "vpc-0030a3a495df38a0e",
        "Architecture": "x86_64",
        "BlockDeviceMappings": [],
        "ClientToken": "5e4a0dd0-665b-4878-b852-0a6ff21c09d3",
        "EbsOptimized": false,
        "EnaSupport": true,
        "Hypervisor": "xen",
        "IamInstanceProfile": {
                "Arn": "arn:aws:iam::673522028003:instance-profile/
DefaultInstanceProfile",
                "Id": "AIPAZZUIKRXR3HEDBS72R"
        },
# output cut for readability
```

The output of the preceding command is information about newly launched instances you could use for scripting purposes or simply to save the instance IP address for later use.

At this point, you won't be able to connect to the machine yet as, by default, all ports are closed. To open the SSH port (22), we will need to create a new security group.

5. Use the following command to achieve that:

```
admin@myhome:~$ aws ec2 create-security-group --group-name
ssh-access-sg --description "Security group for SSH access" --vpc-id
vpc-0030a3a495df38a0e
{
    "GroupId": "sg-076f8fad4e60192d8"
}
```

The VPC ID we used is the one we saved earlier in the process, and the output is the ID of our new security group. We will need to add an ingress rule to it and connect it to our EC2 instance. See the InstanceID value in the long output once the machine is created (i-06f35cbb39f6e5cdb).

6. Use the following command to add an inbound rule to the security group that allows SSH access from 0.0.0.0/0:

```
admin@myhome:~$ aws ec2 authorize-security-group-ingress --group-id
sg-076f8fad4e60192d8 --protocol tcp --port 22 --cidr 0.0.0.0/0
{
    "Return": true,
    "SecurityGroupRules": [
        {
                "SecurityGroupRuleId": "sgr-0f3b4be7d2b01a7f6",
                "GroupId": "sg-076f8fad4e60192d8",
```

```
            "GroupOwnerId": "673522028003",
            "IsEgress": false,
            "IpProtocol": "tcp",
            "FromPort": 22,
            "ToPort": 22,
            "CidrIpv4": "0.0.0.0/0"
        }
    ]
}
```

We've used the ID of the security group that we created in a previous step.

This command added a new inbound rule to the security group that allows TCP traffic on port 22 (SSH) from any IP address (0.0.0.0/0). Instead of allowing full internet access to your new EC2 instance, you could choose to use your own public IP address instead.

7. Now, we can attach this security group to an instance:

```
admin@myhome:~$ aws ec2 modify-instance-attribute --instance-id
i-06f35cbb39f6e5cdb --groups  sg-076f8fad4e60192d8
```

At this point, port 22 should be open and ready to receive connections.

Let's stop here for a moment. You're probably wondering whether there is a better way to do this instead of with the AWS CLI. Yes, there is; there are various tools to automate the creation of the infrastructure. Those tools are generally called **Infrastructure as Code (IaC)** and we will talk about them in *Chapter 12*. There are various options we could have used in this example, from CloudFormation, which is the go-to IaC tool for AWS, to Terraform, from HashiCorp to the Pulumi project, which is gaining traction.

Now that we have an EC2 instance, we can connect to it and install the **CloudWatch agent**. It's needed because AWS by default monitors only two metrics: CPU and memory usage. If you want to monitor disk space and send additional data to CloudWatch (such as logs or custom metrics), the agent is a must.

8. After getting into the SSH console, we will need to download the CloudWatch agent deb package and install it using the dpkg tool:

```
admin@myhome:~$ ssh -i ~/.ssh/admin-key ubuntu@3.121.74.46
ubuntu@ip-172-31-17-127:~$ wget https://s3.amazonaws.com/
amazoncloudwatch-agent/ubuntu/amd64/latest/amazon-cloudwatch-agent.deb
ubuntu@ip-172-31-17-127:~$ sudo dpkg -i -E ./amazon-cloudwatch-agent.
deb
```

Let's become the root user so we can omit sudo from every command:

```
ubuntu@ip-172-31-17-127:~$ sudo -i
root@ip-172-31-17-127:~# /opt/aws/amazon-cloudwatch-agent/bin/amazon-
cloudwatch-agent-config-wizard
```

```
====================================================================
= Welcome to the Amazon CloudWatch Agent Configuration Manager =
=                                                              =
= CloudWatch Agent allows you to collect metrics and logs from =
= your host and send them to CloudWatch. Additional CloudWatch =
= charges may apply.                                           =
====================================================================
On which OS are you planning to use the agent?
1. linux
2. windows
3. darwin
default choice: [1]:
```

It will ask a lot of questions, but it's safe to leave most of them as their default and just hit *Enter*. There are some questions, however, that will require additional attention from us:

```
Do you want to monitor metrics from CollectD? WARNING: CollectD must
be installed or the Agent will fail to start
1. yes
2. no
default choice: [1]:
```

9. If you answered yes (1) to this question, you will need to install collectd by invoking the following command:

```
root@ip-172-31-17-127:~# apt install -y collectd
```

10. To the following question, answer no (2) unless you want some particular log file to be uploaded to CloudWatch Logs:

```
Do you want to monitor any log files?
1. yes
2. no
default choice: [1]:
2
```

11. The final question is whether to save the agent configuration in AWS SSM, to which you can safely answer no (2) as well:

```
Do you want to store the config in the SSM parameter store?
1. yes
2. no
default choice: [1]:
2
Program exits now.
```

The wizard will save the configuration in /opt/aws/amazon-cloudwatch-agent/ bin/config.json. You will be able to alter it later or launch the wizard again if needed.

12. Before we start the agent, we will need to convert the output JSON file into new **Tom's Obvious, Minimal Language** (**TOML**) format, which is what the agent is using. Fortunately, there's also a command for this job, too. We will use the agent control script to load the existing schema, save the TOML file, and optionally, start the agent if everything is in order:

```
root@ip-172-31-17-127:~# /opt/aws/amazon-cloudwatch-agent/bin/amazon-
cloudwatch-agent-ctl -a fetch-config -m ec2 -s -c file:/opt/aws/
amazon-cloudwatch-agent/bin/config.json
root@ip-172-31-17-127:~# systemctl status amazon-cloudwatch-agent

amazon-cloudwatch-agent.service - Amazon CloudWatch Agent
    Loaded: loaded (/etc/systemd/system/amazon-cloudwatch-agent.
service; enabled; vendor preset: enabled)
    Active: active (running) since Wed 2023-03-08 15:00:30 UTC; 4min
54s ago
  Main PID: 20130 (amazon-cloudwat)
     Tasks: 6 (limit: 1141)
    Memory: 14.3M
    CGroup: /system.slice/amazon-cloudwatch-agent.service
            └─20130 /opt/aws/amazon-cloudwatch-agent/bin/amazon-
cloudwatch-agent -config /opt/aws/amazon-cloudwatch-agent/etc/amazon-
cloudwatch-agent.toml -envconfig /opt/aws/amazon-cloudwatch-agent/e>

Mar 08 15:00:30 ip-172-31-17-127 systemd[1]: Started Amazon CloudWatch
Agent.
Mar 08 15:00:30 ip-172-31-17-127 start-amazon-cloudwatch-agent[20130]:
/opt/aws/amazon-cloudwatch-agent/etc/amazon-cloudwatch-agent.json does
not exist or cannot read. Skipping it.
Mar 08 15:00:30 ip-172-31-17-127 start-amazon-cloudwatch-agent[20130]:
I! Detecting run_as_user...
```

Now, we can go to the AWS web console and navigate to CloudWatch to see whether we can see the metrics coming in. It may take several minutes until they're shown.

Before starting the CloudWatch agent, we will get about 17 different metrics for our EC2 instance, as seen in the following screenshot:

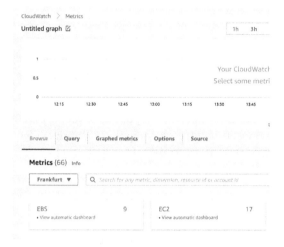

Figure 10.1 – Basic EC2 and EBS metrics in CloudWatch without the CloudWatch agent installed

After we've started the CloudWatch agent, we will start receiving a lot more metrics and we will see an additional namespace in the CloudWatch **Metrics** panel. See the following screenshot:

Figure 10.2 – CloudWatch metrics after successfully enabling the CloudWatch agent on the EC2 instance

All metrics we're receiving can be used to create dashboards and alerts (including anomaly detection) in the CloudWatch service.

AWS X-Ray

AWS X-Ray is a service that enables you to trace requests through distributed systems and microservice applications. It provides an end-to-end view of requests as they travel through an application, allowing developers to identify performance bottlenecks, diagnose errors, and improve overall application efficiency.

With X-Ray, it's possible to visualize the different components of your application and see how requests are being processed as they travel through each component. This includes details such as the time taken to complete each component, any errors that occur, and the cause of those errors.

X-Ray also provides a range of analysis tools, including statistical analysis and heat maps, to help developers identify trends and patterns in request processing. These insights can be used to optimize performance and ensure that the application is running as efficiently as possible.

AWS X-Ray supports a wide range of programming languages, including the following:

- Node.js
- Java
- .NET
- Go
- Python
- Ruby
- PHP

To instrument your application with the diagnostic tools provided by AWS X-Ray, you can use the AWS SDK. Consider the following code (found in the GitHub repository at `https://github.com/Sysnove/flask-hello-world`):

```python
from flask import Flask
app = Flask(__name__)

@app.route('/')
def hello_world():
    return 'Hello World!'

if __name__ == '__main__':
    app.run()
```

To gather tracing data about this service, you'll need to install the `aws_xray_sdk` package using the `pip` package manager. Then, import the `xray_recorder` subpackage into our code. In this case, we will also use this SDK's integration with the Flask framework. The modified code will look like this:

```
from aws_xray_sdk.core import xray_recorder
from aws_xray_sdk.ext.flask.middleware import XRayMiddleware

xray_recorder.configure(service='FlaskHelloWorldApp')
app = Flask(__name__)
XRayMiddleware(app, xray_recorder)
```

The rest of the code can remain unchanged. Here, we are configuring the X-Ray recorder to use the service name `FlaskHelloWorldApp`, which will show up in the X-Ray console as the name of our service. When the service starts running, you can go to the X-Ray console and see the service name `FlaskHelloWorldApp` with a list of traces.

The full documentation for the AWS X-Ray SDK can be found on this website: `https://docs.aws.amazon.com/xray-sdk-for-python/latest/reference/index.html`.

When running the preceding application on the EC2 instance we created in a previous section, you will see a complete picture of the running environment of your application including the internals of the running Flask processes.

There are multiple projects that deal with application monitoring, tracing, and gathering logs. Apart from cloud-hosted solutions that are available in the cloud environment, there are commercial and open source solutions worth knowing about. This awareness might prove useful when dealing with more and more common hybrid solutions.

Open source solutions for self-hosting

One of the most popular projects built around monitoring that is also adopted by commercial solutions is **OpenTelemetry**. It's an open source project for application monitoring and observability. It provides a set of APIs, libraries, agents, and integrations for collecting, processing, and exporting telemetry data such as traces, metrics, and logs from different sources in distributed systems. OpenTelemetry is designed to be vendor-agnostic and cloud-native, meaning it can work with various cloud providers, programming languages, frameworks, and architectures.

The main goal of OpenTelemetry is to provide developers and operators with a unified and standardized way to instrument, collect, and analyze telemetry data across the entire stack of their applications and services, regardless of the underlying infrastructure. OpenTelemetry supports different data formats, protocols, and export destinations, including popular observability platforms such as **Prometheus**, **Jaeger**, **Zipkin**, **Grafana**, and **SigNoz**. This allows users to mix and match their preferred tools and services to build a comprehensive observability pipeline that meets their specific needs.

Some examples of commercial software that adopts OpenTelemetry are **Datadog**, AWS, and **New Relic**. AWS provides OpenTelemetry Collector as a managed service for collecting and exporting telemetry data to AWS services such as Amazon CloudWatch, AWS X-Ray, and AWS App Runner.

Prometheus

Prometheus is an open source monitoring solution that is widely used for collecting and querying metrics from distributed systems. It was created by the developers at SoundCloud and is now maintained by the **Cloud Native Computing Foundation** (**CNCF**). Prometheus is designed to be highly scalable and adaptable, with support for a wide range of data sources and integration options. It allows users to define and collect custom metrics, visualize data through a built-in dashboard, and set alerts based on predefined thresholds or anomalies. Prometheus is often used in conjunction with Kubernetes and other cloud-native technologies, but it can also be used to monitor traditional infrastructure and applications.

One common use case is to track request latencies and error rates, which can help identify performance bottlenecks and potential issues in the application. To get started with monitoring a Flask application using Prometheus, you can use the Prometheus client library for Python. This library provides decorators that can be added to Flask routes to automatically generate metrics such as request count, request duration, and HTTP response codes. These metrics can then be collected by a Prometheus server and displayed on a Grafana dashboard for visualization and analysis.

Here's an example of how you can instrument the "*Hello World*" Flask application with Prometheus to send metrics. We used the same application with AWS X-Ray in a previous section.

First, you'll need to install the `prometheus_client` library using `pip`:

```
$ pip install prometheus_client
```

Next, you can modify the `app.py` file in the `flask-hello-world` repository to add the Prometheus client library and instrument the routes with metrics. Here's an example:

```
from flask import Flask
from prometheus_client import Counter, Histogram, start_http_server

app = Flask(__name__)

# Define Prometheus metrics
REQUEST_COUNT = Counter('hello_world_request_count', 'Hello World
Request Count')
REQUEST_LATENCY = Histogram('hello_world_request_latency_seconds',
'Hello World Request Latency',
                            bins=[0.1, 0.2, 0.5, 1.0, 5.0, 10.0, 30.0,
60.0])
```

```
# Instrument Flask routes with Prometheus metrics
@app.route('/')
@REQUEST_LATENCY.time()
def hello():
    REQUEST_COUNT.inc()
    return "Hello World!"

# Start the Prometheus server on port 8000
if __name__ == '__main__':
    start_http_server(8000)
    app.run(debug=True)
```

In this example, we've defined two Prometheus metrics: `hello_world_request_count` and `hello_world_request_latency_seconds`. The `hello()` route is instrumented with these metrics using decorators. The `REQUEST_LATENCY` histogram measures the request latency for each request, while the `REQUEST_COUNT` counter increments on each request.

We've started the Prometheus server on port `8000` using `start_http_server()`. This will make the metrics available for collection by a Prometheus server.

To view the metrics, you can navigate to `http://localhost:8000/metrics` in your web browser. This will display the raw metrics data in Prometheus format. You can also use a tool such as Grafana to visualize the metrics on a dashboard.

Grafana

Grafana is a popular open source dashboard and data visualization platform that enables users to create interactive and customizable dashboards for monitoring and analyzing metrics from various data sources. It is usually used alongside Prometheus.

With Grafana, users can create visualizations, alerting rules, and dashboards that provide insight into the performance and behavior of their applications and infrastructure. Grafana supports a wide range of data sources, including popular time-series databases such as Prometheus, InfluxDB, and Graphite, making it a versatile tool for monitoring and visualization. Once you have connected your data source, you can start creating dashboards by adding panels to visualize the data. These panels can include various types of visualizations, including line graphs, bar charts, and gauges. You can also customize the dashboard layout, add annotations, and set up alerts to notify you of anomalies or issues in your metrics. With its powerful features and flexibility, Grafana is a go-to tool for visualizing and analyzing application and infrastructure metrics.

Grafana Labs also created the Grafana Loki project, which can be used to extend your monitoring with logs visualization. **Grafana Loki** is a horizontally scalable log aggregation system that provides a way to centralize logs from various sources and quickly search and analyze them. It's being seen as an alternative to Prometheus, but both tools have different use cases and could be complementary to each other.

Loki, unlike traditional log management solutions, does not index or parse logs upfront. Instead, it uses a streaming pipeline that extracts log labels and stores them in a compact and efficient format, making it ideal for ingesting and querying large volumes of logs in real time. Grafana Loki integrates seamlessly with Grafana, allowing users to correlate logs with metrics and create powerful dashboards that provide insight into the behavior of their applications and infrastructure.

To use Grafana Loki, you need to set up a Loki server and configure it to receive log data from your applications and infrastructure. Once Loki is set up, you can use the Grafana Explore feature to search and visualize logs in real time. Explore provides a user-friendly interface that enables you to search logs using various filters, such as labels, time range, and query expressions

SigNoz

SigNoz is an observability platform that enables users to collect, store, and analyze application metrics' telemetry data and provides log management under a single web panel. It is built on top of the OpenTelemetry specification, which is an industry-standard framework for distributed tracing and metric collection. SigNoz provides a simple, intuitive interface for users to view real-time and historical data about their applications' performance and health.

SigNoz has its own agent that you can install on your servers, but it also supports Prometheus as a data source. So, if you're already using Prometheus, you can use SigNoz without any significant changes to your monitoring infrastructure.

To install SigNoz on your server, you can follow a comprehensive guide on the official project website: `https://signoz.io/docs/install/`.

New Relic Pixie

New Relic is a well-known monitoring SaaS solution; we will get back to it later in this chapter in the *SaaS solutions* section. Pixie is an open source project started by New Relic and was contributed to CNCF.

CNCF is an open source software foundation that was established in 2015 to advance the development and adoption of cloud-native technologies. CNCF is the home of many popular projects, such as Kubernetes, Prometheus, and Envoy, which are widely used in modern cloud-native applications. The foundation aims to create a vendor-neutral ecosystem for cloud-native computing, promoting interoperability and standardization among different cloud platforms and technologies. CNCF also hosts several certification programs that help developers and organizations validate their proficiency in cloud-native technologies. CNCF plays a critical role in driving innovation and standardization in the rapidly evolving cloud-native landscape.

New Relic Pixie is an open source, Kubernetes-native observability solution that provides real-time monitoring and tracing capabilities for modern applications. It can help developers and operations teams to quickly identify and troubleshoot performance issues in microservices-based applications running on Kubernetes clusters. Pixie can be easily deployed on any Kubernetes cluster and provides out-of-the-box support for popular open source tools such as Prometheus, Jaeger, and OpenTelemetry.

One of the key benefits of using New Relic Pixie is that it provides end-to-end visibility into the performance of applications and infrastructure, from the application code to the underlying Kubernetes resources. By collecting and analyzing data from various sources, including logs, metrics, and traces, Pixie can help pinpoint the root cause of performance bottlenecks and issues. This can significantly reduce the **Mean Time to Resolution** (**MTTR**) and improve application reliability and uptime.

Another advantage of New Relic Pixie is that it uses a unique instrumentation approach that does not require any code changes or configuration. Pixie uses **extended Berkeley Packet Filter** (**eBPF**) technology to collect performance data at the kernel level, allowing for low-overhead monitoring without adding any additional load to applications or infrastructure. This makes it an ideal solution for monitoring and tracing modern, cloud-native applications that are highly dynamic and scalable. Overall, New Relic Pixie provides a powerful and easy-to-use observability solution that can help teams to optimize the performance and reliability of their Kubernetes-based applications.

Graylog

Graylog is an open source log management platform that allows users to collect, index, and analyze log data from various sources. The platform provides a centralized location for monitoring and troubleshooting applications, systems, and network infrastructure. It is built on top of Elasticsearch, MongoDB, and Apache Kafka, which ensures high scalability and availability.

Graylog has the ability to scale horizontally, which means that you can add additional Graylog nodes to handle increased log data volume and query load. The system can also distribute the workload across multiple nodes, which allows for efficient use of resources and faster processing of data. This scalability makes Graylog suitable for organizations of any size, from small start-ups to large enterprises.

Graylog uses Elasticsearch as the primary data store for indexing and searching log data. Elasticsearch is a powerful search and analytics engine that enables fast and efficient querying of large datasets. MongoDB in Graylog is used to store metadata about log data and manage the configuration and state of the system.

Graylog also has a web-based user interface that allows users to search and visualize log data, as well as manage system configuration and settings:

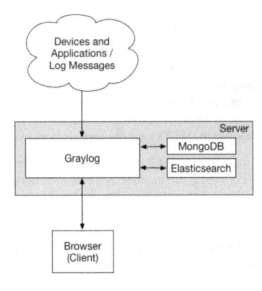

Figure 10.3 – Graylog logging system architecture

The architecture of this solution is pretty simple, as you can notice in the preceding diagram.

Sentry

Sentry is an open source error tracking tool that helps developers monitor and fix errors in their applications. It allows developers to track errors and exceptions in real time, enabling them to quickly diagnose and fix issues before they become critical. Sentry supports multiple programming languages, including Python, Java, JavaScript, and Ruby, among others.

One of the key benefits of using Sentry is its ease of setup and integration. Sentry can be easily integrated with popular frameworks and platforms, such as Django, Flask, and Rails, among others. It also provides a range of plugins and integrations with third-party tools, such as Slack and GitHub, to help developers streamline their workflows and collaborate more effectively.

Sentry provides developers with detailed error reports that include information about the error, such as the stack trace, environment variables, and request parameters. This allows developers to quickly identify the cause of the error and take corrective action. Sentry also provides real-time notifications when errors occur, so developers can respond immediately.

Another benefit of using Sentry is its ability to analyze errors over time. Sentry allows developers to track error rates and identify patterns in error occurrence, making it easier to identify and address systemic issues in the application. This data can also be used to improve the overall performance and reliability of the application.

Sentry provides integration with Jira, which is a popular ticketing and issue-tracking system. The integration allows developers to create Jira issues directly from within Sentry, making it easier to manage and track issues that are discovered through Sentry.

To set up the integration, you will first need to create a Jira API token and configure the integration settings in Sentry. Once the integration is set up, you can create Jira issues directly from Sentry by clicking the **Create JIRA issue** button on the **Error details** page. This will automatically populate the Jira issue with relevant information about the error, such as the error message, stack trace, and request parameters. You can find detailed instructions on how to do it on the official documentation page here: `https://docs.sentry.io/product/integrations/issue-tracking/jira/`.

Sentry provides integrations with several other popular ticketing and issue-tracking systems, such as GitHub, Trello, Asana, Clubhouse, and PagerDuty, which allows you to trigger PagerDuty incidents directly from Sentry.

In this section, we have shown you several leading solutions that are both open source and suitable for self-hosting. Self-hosting may, however, not be what you are looking for, if you wish to lower the complexity of both deployment and maintenance. The next section will cover monitoring and logging software hosted for you by third-party companies.

SaaS solutions

SaaS monitoring solutions are the easiest (and most expensive) to use. In most cases, what you'll need to do is install and configure a small daemon (agent) on your servers or inside a cluster. And there you go, all your monitoring data is visible within minutes. SaaS is great if your team doesn't have the capacity to implement other solutions but your budget allows you to use one. Here are some more popular applications for handling your monitoring, tracing, and logging needs.

Datadog

Datadog is a monitoring and analytics platform that provides visibility into the performance and health of applications, infrastructure, and networks. It was founded in 2010 by Olivier Pomel and Alexis Lê-Quôc and is headquartered in New York City, with offices around the world. According to Datadog's financial report for the fiscal year 2021 (ending December 31, 2021), their total revenue was $2.065 billion, which represents a 60% increase from the previous year (fiscal year 2020).

Datadog's platform integrates with more than 450 technologies, including cloud providers, databases, and containers, allowing users to collect and correlate data from across their entire technology stack. It provides real-time monitoring, alerting, and collaboration tools that enable teams to troubleshoot issues, optimize performance, and improve the user experience.

Datadog allows users to monitor the health and performance of their servers, containers, and cloud services, providing insights into CPU usage, memory utilization, network traffic, and more.

Datadog's APM tools provide detailed insights into the performance of web applications, microservices, and other distributed systems, allowing users to identify and diagnose bottlenecks and issues.

Log management tools in Datadog enable users to collect, process, and analyze logs from across their entire infrastructure, helping to troubleshoot issues and identify trends.

And finally, Datadog security monitoring helps detect and respond to threats by analyzing network traffic, identifying anomalies, and integrating with security solutions.

Dashboarding in Datadog allows users to visualize and analyze data from their applications, infrastructure, and network in a centralized location. Users can create a dashboard in Datadog by clicking on the **Create Dashboard** button and selecting the type of dashboard they want to create (e.g., **Infrastructure**, **APM**, **Log**, or **Custom**). They can then add widgets to the dashboard and configure their settings. There are multiple automated dashboards available; for instance, if you start sending data from a Kubernetes cluster, Datadog will show a dashboard for that. You can find more detailed information about using dashboards on the Datadog documentation website: `https://docs.datadoghq.com/getting_started/dashboards/`.

Widgets are the building blocks of a dashboard in Datadog. They can display metrics, logs, traces, events, or custom data. To add a widget, users can click on the + button and select the type of widget they want to add. They can then configure the widget's settings, such as selecting the data source, applying filters, and setting the time range. For instance, you can view an example dashboard for the nginx web server on the Datadog web page: `https://www.datadoghq.com/dashboards/nginx-dashboard/`.

In addition to displaying data on a dashboard, Datadog provides various tools for exploring and analyzing data, such as the query builder, Live Tail, and tracing. Users can use these tools to dive deeper into the data and troubleshoot issues.

New Relic

New Relic is a cloud-based software analytics company that provides real-time insights into the performance of web and mobile applications. Founded in 2008 by Lew Cirne (a software engineer and entrepreneur with experience at companies such as Apple and Wily Technology), New Relic has become a leading player in the **Application Performance Management** (**APM**) market. The company is headquartered in San Francisco and has offices in a number of other cities around the world. New Relic went public in 2014 and is traded on the New York Stock Exchange under the symbol *NEWR*.

New Relic reported its 2021 fiscal year financial results in May 2021. According to the report, New Relic's revenue for the full fiscal year 2021 was $600.8 million, which represents a 3% increase compared to the previous fiscal year.

It's worth noting that New Relic experienced some challenges in fiscal year 2021, including the impact of the COVID-19 pandemic and a strategic shift to a new pricing model.

New Relic's main purpose is to help companies optimize their application performance and identify issues before they become major problems. The platform provides real-time visibility into the entire application stack, from the frontend user interface to the backend infrastructure, allowing developers and operations teams to quickly identify bottlenecks and optimize performance.

New Relic's APM solution offers a variety of features, including code-level visibility, transaction tracing, real-time monitoring, and alerting. The platform also provides insights into application dependencies, database performance, and user behavior.

In addition to APM, New Relic also offers a range of other products and services, including infrastructure monitoring, mobile APM, and browser monitoring.

Ruxit

Ruxit is a comprehensive APM solution that helps businesses identify and troubleshoot performance issues across complex distributed applications, microservices, and cloud-native environments. It was initially founded in 2012 as an independent company and was later acquired by Dynatrace in 2015, expanding Dynatrace's APM capabilities.

One of the key features of Ruxit is its ability to provide end-to-end visibility into the performance of applications, including code-level diagnostics, user experience monitoring, and infrastructure monitoring. This means that it can help businesses quickly pinpoint the root cause of performance problems and identify opportunities for optimization.

Ruxit also has a range of other features designed to make monitoring and troubleshooting easier and more efficient. For example, it uses artificial intelligence and machine learning to automatically detect anomalies and performance degradations, alerting users in real time. It also provides a range of analytics and visualization tools to help users understand application performance trends and identify patterns over time.

In addition to its monitoring capabilities, Ruxit also provides a range of integrations with other tools and services commonly used in modern application environments. This includes integration with container orchestration platforms such as Kubernetes, as well as with popular application development frameworks and tools.

Splunk

Splunk was founded in 2003 by Erik Swan, Rob Das, and Michael Baum in San Francisco, California. Since then, the company has grown significantly and is now a publicly traded company with a global presence. Splunk's software solutions are used by organizations in various industries, including financial services, healthcare, government, and retail, to name a few.

Splunk, you guessed it, is a data analysis and monitoring software solution used to monitor, search, analyze, and visualize machine-generated data in real time. The software can gather and analyze data from various sources, including servers, applications, networks, and mobile devices, and provide insights into the performance and behavior of an organization's IT infrastructure.

The main uses of Splunk include security monitoring, application monitoring, log management, and business analytics. With Splunk, users can identify security threats, troubleshoot application performance issues, monitor network activity, and gain insights into business operations.

One of the key features of Splunk is its ability to collect and analyze data from a wide range of sources, including structured and unstructured data. The software can also scale to handle large volumes of data, making it a powerful tool.

In this section, we have presented you with a few leading solutions hosted by third-party companies that are ready to use; they just require integration with your systems. In the next section, we are going to describe and explain retention policies for both logs and metrics.

Log and metrics retention

Data retention refers to the practice of retaining data, or keeping data stored for a certain period of time. This can involve storing data on servers, hard drives, or other storage devices. The purpose of data retention is to ensure that data is available for future use or analysis.

Data retention policies are often developed by organizations to determine how long specific types of data should be retained. These policies may be driven by regulatory requirements, legal obligations, or business needs. For example, some regulations may require financial institutions to retain transaction data for a certain number of years, while businesses may choose to retain customer data for marketing or analytics purposes.

Data retention policies typically include guidelines for how data should be stored, how long it should be retained, and when it should be deleted. Effective data retention policies can help organizations to manage their data more efficiently, reduce storage costs, and ensure compliance with applicable regulations and laws.

When it comes to data retention strategies, organizations have a number of options to consider. Depending on the specific needs of the organization, different strategies may be more or less suitable.

Full retention

In this strategy, all data is kept indefinitely. This is often used for compliance purposes, such as for regulatory requirements that mandate data retention for a specific period of time. This strategy can be expensive as it requires a large amount of storage, but it can also provide significant benefits in terms of historical analysis and trend spotting.

Time-based retention

Time-based retention is a strategy where data is kept for a specific period of time before it is deleted. This strategy is often used to balance the need for data with storage costs. The retention period can be set based on regulatory requirements, business needs, or other factors.

Event-based retention

Event-based retention is a strategy where data is kept based on specific events or triggers. For example, data may be retained for a specific customer or transaction, or based on the severity of an event. This strategy can help to reduce storage costs while still maintaining access to important data.

Selective retention

Selective retention is a strategy where only certain types of data are retained. This strategy can be used to prioritize the retention of the most important data while reducing storage costs. For example, an organization may choose to retain only data related to financial transactions or customer interactions.

Tiered retention

Tiered retention is a strategy where data is stored in different tiers based on its age or importance. For example, recent data may be stored on fast, expensive storage, while older data is moved to slower, less expensive storage. This strategy can help to balance the need for fast access to recent data with the need to reduce storage costs over time.

Each of these data retention strategies has its own benefits and drawbacks, and the best strategy for an organization will depend on its specific needs and goals. It's important to carefully consider the trade-offs between cost, storage capacity, and the value of the data being retained when choosing a data retention strategy.

The most common mistake in organizations is to use full retention strategies *just in case*, which often leads to exhausted disk space and increased cloud costs. Sometimes this strategy is justified, but not in most cases.

Summary

In this chapter, we covered the differences between monitoring, tracing, and logging. Monitoring is the process of observing and collecting data on a system to ensure it's running correctly. Tracing is the process of tracking requests as they flow through a system to identify performance issues. Logging is the process of recording events and errors in a system for later analysis.

We also discussed cloud solutions for monitoring, logging, and tracing in Azure, GCP, and AWS. For Azure, we mentioned Azure Monitor for monitoring and Azure Application Insights for tracing. For AWS, we mentioned CloudWatch for monitoring and logging, and X-Ray for tracing.

We then went on to explain and provide an example of configuring the AWS CloudWatch agent on an EC2 instance. We also introduced AWS X-Ray with a code example to show how it can be used to trace requests in a distributed system.

Finally, we named some open source and SaaS solutions for monitoring, logging, and tracing, including Grafana, Prometheus, Datadog, New Relic, and Splunk. These solutions provide various features and capabilities for monitoring and troubleshooting systems, depending on the user's requirements and preferences.

In the next chapter, we will get hands-on with automating server configuration with the use of a configuration as code solution: Ansible.

11
Using Ansible for Configuration as Code

In this chapter, we are going to cover **configuration management (CM)**, **Configuration as Code (CaC)**, and our tool of choice for it: Ansible.

We will cover the following topics:

- CM systems and CaC
- Ansible
- Ansible Galaxy
- Handling secrets
- Ansible Tower and alternatives
- Advanced topics

Technical requirements

For this chapter, you will need a Linux system that you can access through `ssh`. If your main operating system is Windows, you will need another Linux system to play the role of the control node. As of now, the Ansible project does not support Windows as a control node.

CM systems and CaC

Setting up and maintaining a system other than a hobbyist server (and even those, maybe, too) poses a serious challenge: how do you ensure that the system is installed and configured correctly and according to expectations? When you have to install a new server that is identical in configuration, how do you ensure that? In the past, a way of doing it was documenting the current configuration after the installation process was done. This would be a document describing the hardware, operating system, installed software, created users, and configuration applied. Any person who wanted to recreate it would have to follow steps to achieve the configuration described in the document.

The very next logical step is to write shell scripts that achieve the same goal with one additional improvement over the manual process: the scripts—properly written, tested, and maintained—do not require manual work, except, maybe, the initial system installation. But a properly set up environment would take care even of this.

The scripts, however, also have some defects or deficiencies. One of them is the fact that you need to account in your scripts for unfinished execution. This could happen for various reasons and would leave the system in a partially configured state. Executing the script again would perform all configuration actions from the start, sometimes leading to unexpected results. One way to account for incomplete runs would be to wrap every configuration action in a check, to see whether it had been performed previously. That would lead to the configuration script becoming larger and, eventually, evolving into a library of configuration functions and check functions.

The task of developing and maintaining such a tool can be daunting and will probably require a whole team. Still, the results are probably worth the effort.

Writing and maintaining documentation that describes the desired state of the system may, at first glance, be simpler and more desirable than the previously mentioned method of automating. The script cannot recover from an incomplete execution. The best it can do is inform the sysop about failure, log the error, and stop gracefully. Manually performed configuration allows the sysop to work around any obstacles and inadequacies in the procedure and edit the document to reflect the current state on the go.

Still, a properly developed and tested script turns out to be better. Let us enumerate the reasons:

- If the script executes without an error, it is guaranteed to perform actions without a mistake. Time and again, it has been proven that a human is the element most prone to errors in IT.

- If the script exits prematurely, the action of updating it to account for the new requirements is a perfect equivalent to updating the documentation.

- People are known to be pretty bad at maintaining documentation. The Holy Grail of programming is self-documenting code, rendering comments unnecessary, thus eliminating the risk of comments being out of sync with the code.

- The script can be executed on multiple systems at once, scaling very well, if not infinitely. Humans can perform the configuration of one system at a time with minimal risk of making a mistake.

- Configuration kept in the form of a script or program benefits from typical programming techniques, such as automated testing, dry runs, and static analysis. More so, keeping the code in a repository allows us to easily track a history of changes and integrate it with ticket-tracking tools.

- Code is unequivocal, which cannot be said about written language. A document may leave space for interpretation; a script won't.

- Automating configuration lets you move to other, more interesting tasks, leaving the computers to do what they do best—performing repetitive, boring tasks well.

The world of programming and system administration has a tendency to turn small projects into larger ones with a vibrant community of developers and users. It was only a matter of time before CM systems were born. They take the burden of developing and managing portions of the code responsible for configuration actions off your shoulders. The CM system developers write the code, test it, and deem it stable. What you are left with is an action of writing configuration files or directives that tell the system what to do. Most of these systems will be able to cover the most popular platforms, allowing you to describe configuration once and run it with the same expected results on commercial Unix systems, such as AIX or Solaris, as on Linux or Windows.

Configuration files for these systems are easily stored in a version control system such as Git. They are easily understandable by a human, which allows for simple review by your colleagues. They can be checked for syntax errors by automated tools and allow you to concentrate on the most important part of the whole endeavor: the configuration.

This approach of keeping your configuration as a set of scripts or other data instead of a procedure to be followed manually is known as CaC.

The CaC approach is becoming more important as the number of systems to be managed grows and the demand for fast and efficient configuration scales up. In the world of DevOps, it is usual practice to set up tens and hundreds of systems a day: systems for developers, testers, and production systems to manage new levels of demand for the service. Managing it manually would be an impossible task. Well-implemented CaC allows to run this task with a click of a button. Thus, developers and testers can deploy their own systems without bothering sysops. Your task will be to develop, maintain, and test the configuration data.

If there is one thing sure in the world of programming, it is that there's never going to be only one solution. The same goes for CM tools.

Alternatives for Ansible include **SaltStack**, **Chef**, **Puppet**, and **CFEngine**, which is the oldest one; its initial release date was 1993, so it's 30 years old as of the time of writing this book. In general, those solutions differentiate between each other with a method of enforcing configuration (pull or push) and an approach of describing the system's state (imperative or declarative).

Imperative means that we describe the state of the server with commands for the tool to perform. Imperative programming focuses on describing how a given tool operates step by step.

Declarative, on the other hand, means we focus on what the CaC tool should accomplish without specifying all the details of how it should achieve the result.

SaltStack

SaltStack is an open source CM tool that allows for the management of complex IT infrastructure at scale. It enables the automation of routine tasks such as package installation, user management, and software configuration, and is designed to work across a wide range of operating systems and platforms. SaltStack was founded by Thomas Hatch in 2011. The first release of SaltStack, version 0.8.0, was also made in 2011.

SaltStack works by utilizing a master-slave architecture, where a central salt-master communicates with salt-minions running on remote machines to execute commands and manage configurations. It operates in a pull method of enforcing configuration: minions pull the latest manifest from the master server.

Once the minion is installed and configured, we can use SaltStack to manage the server's configuration. Here's an example `nginx.sls` file that would install and configure `nginx`:

```
nginx:
  pkg.installed

/etc/nginx/sites-available/yourdomain.tld.conf:
  file.managed:
    - source: salt://nginx/yourdomain.tld.conf
    - user: root
    - group: root
    - mode: 644
```

In this example, the first line specifies that the `nginx` package should be installed on the target server. The next two lines define the configuration file for a hypothetical website, `example.com`, which is copied to `/etc/nginx/sites-available/yourdomain.tld.conf`.

To apply this state file to a server, we would use the `state.apply` command in the SaltStack command-line interface, specifying the name of the state file as the argument:

```
admin@myhome:~$ salt 'webserver' state.apply nginx
```

This would send the instructions in the `nginx.sls` file to the salt-minion running on the web server machine, which would execute the necessary steps to ensure that `nginx` is installed and configured correctly.

Chef

Chef is a powerful open source CM tool that allows users to automate the deployment and management of infrastructure, applications, and services. It was first released in 2009 by Opscode, which was later acquired by Chef Software Inc. Since then, Chef has been widely adopted by IT professionals and DevOps teams to streamline their workflows and reduce the time and effort required for managing complex systems.

Chef works by defining the desired state of an infrastructure in a set of code files, called cookbooks. A **cookbook** is a collection of recipes that describe how to install, configure, and manage a specific piece of software or service. Each recipe contains a series of resources, which are pre-built modules that can perform specific tasks, such as installing a package or configuring a file. Chef uses a declarative approach to CM, meaning that users define what they want the system to look like, and Chef takes care of the details of how to get there.

To install nginx using Chef, you would first need to create a cookbook that includes a recipe for installing nginx. This recipe would use the package resource to install the nginx package and the service resource to ensure that the nginx service is running. You could also use other resources, such as file, directory, or template, to configure nginx's settings, depending on your requirements.

Once you had created the cookbook, you would upload it to a Chef server, which acts as a central repository for cookbooks and their associated metadata. You would then use Chef's command-line tool, called knife, to configure the target system to use the cookbook. This involves associating the system with a Chef environment, which defines the set of cookbooks and their versions that should be applied to the system. You would then use the chef-client command to run the Chef client on the target system, which will download and apply the necessary cookbooks and recipes to bring the system into the desired state.

Here's an example of installing and configuring nginx:

```
# Install Nginx package
package 'nginx'

# Configure Nginx service
service 'nginx' do
  action [:enable, :start]
end

# Configure Nginx site
template '/etc/nginx/sites-available/yourdomain.tld.conf' do
  source 'nginx-site.erb'
  owner 'root'
  group 'root'
  mode '0644'
  notifies :restart, 'service[nginx]'
end
```

This recipe uses three resources, as follows:

- package: This installs the nginx package using the default package manager on the system.
- service: This starts and enables the nginx service so that it will automatically start on boot and stay running.
- template: This creates a configuration file for nginx by generating it from a template file. The template file (nginx-site.erb) is written in **Embedded Ruby** (**ERB**) format and is located in the templates directory of the cookbook. The notifies attribute tells Chef to restart the nginx service if the configuration file changes.

Once you have created this recipe in a cookbook, you can use the `knife` command to upload the cookbook to a Chef server. You can then use the `chef-client` command to apply the recipe to a target system, which will install and configure `nginx` according to the recipe.

Puppet

Puppet is an open source CM tool that allows system administrators to automate the deployment, configuration, and management of infrastructure. It was created by Luke Kanies in 2005 and released under the Apache License `2.0`.

Puppet works by defining the desired state of infrastructure resources in a declarative language, known as the Puppet language. Administrators can define the configuration of servers, applications, and other infrastructure components in Puppet code, which can then be applied consistently across multiple systems.

Puppet consists of a master server and multiple agent nodes. The master server acts as a central repository for Puppet code and configuration data, while the agent nodes execute the Puppet code and apply the desired state to the system.

Puppet has a robust ecosystem of modules, which are pre-written Puppet code that can be used to configure common infrastructure resources. These modules are available in **Puppet Forge**, a public repository of Puppet code.

Here's an example Puppet manifest that installs `nginx` and creates a configuration file similar to what we did with SaltStack and Chef:

```
# Install Nginx
package { 'nginx':
  ensure => installed,
}

# Define the configuration template for the domain
file { '/etc/nginx/sites-available/yourdomain.tld.conf':
  content => template('nginx/yourdomain.tld.conf.erb'),
  owner   => 'root',
  group   => 'root',
  mode    => '0644',
  notify  => Service['nginx'],
}

# Enable the site by creating a symbolic link from sites-available to
sites-enabled
file { '/etc/nginx/sites-enabled/yourdomain.tld.conf':
  ensure => 'link',
  target => '/etc/nginx/sites-available/yourdomain.tld.conf',
```

```
  require => File['/etc/nginx/sites-available/yourdomain.tld.conf'],
}


# Restart Nginx when the configuration changes
service { 'nginx':
  ensure      => running,
  enable      => true,
  subscribe   => File['/etc/nginx/sites-enabled/yourdomain.tld.conf'],
}
```

Once you've created a manifest and put it on the Puppet server, it will be picked up by the Puppet agent installed on your server and executed. Communication, the same as in SaltStack, is being secured by the TLS protocol using the same mechanism as the HTTPS servers on the internet.

The agent nodes run a Puppet agent process, which connects to the master server over TCP port 8140. The agent sends a **certificate signing request** (**CSR**) to the server, which the administrator must approve. Once the CSR is approved, the agent is granted access to the server's Puppet configuration.

When the agent runs, it sends a request to the master server for its configuration. The server responds with a catalog of resources that should be applied to the node. The catalog is generated based on the Puppet code and manifests stored on the server, as well as any external data sources or hierarchies that are configured.

The agent then applies the catalog to the node, which involves making any necessary changes to the node's configuration to ensure it matches the desired state defined in the catalog. This may involve installing packages, updating configuration files, or starting or stopping services.

The agent sends reports back to the server after applying the catalog, which can be used for monitoring and auditing purposes. The server can also use this information to detect changes to the node's configuration that were not made through Puppet and to take corrective action if necessary.

CFEngine

CFEngine is an open source CM system that allows users to automate the deployment, configuration, and maintenance of IT systems. It was founded by Mark Burgess in 1993 and has since become a popular tool for managing large-scale IT infrastructures. CFEngine is known for its powerful and flexible language for describing system configurations and enforcing policies, making it a great choice for complex IT environments.

CFEngine's first release was in 1994, making it one of the oldest CM tools in existence. Since then, CFEngine has undergone numerous updates and improvements to keep up with changing IT environments and emerging technologies. The latest release of CFEngine, version 3.18, includes features such as improved encryption, enhanced monitoring capabilities, and better support for cloud infrastructure.

CFEngine has gained popularity over the years due to its robust functionality, ease of use, and strong community support. It's still being used today by many organizations and is actively developed, so it is a safe option to manage the configuration of your servers using this tool.

An example CFengine configuration will be presented here. It is, out of necessity, only a snipped and not a complete configuration:

```
###############################################################
# cf.main - for master infrastructure server
###############################################################


###
# BEGIN cf.main
###

control:
    access    = ( root )          # Only root should run this

    site      = ( main )
    domain    = ( example.com )
    sysadm    = ( admin@example.com )

    repository = ( /var/spool/cfengine )

    netmask   = ( 255.255.255.0 )
    timezone  = ( CET )


###############################################################
files:
  Prepare::
      /etc/motd                m=0644 r=0 o=root act=touch
```

In this section, we have explained what CaC is and why it is an important tool in the toolbelt of system administrators. We have briefly described the most popular tools available to you. In the next section, we will introduce our tool of choice—Ansible.

Ansible

In this section, we are going to introduce you to **Ansible**, our tool of choice when it comes to CaC.

Ansible is a tool written for managing the configuration of systems and devices. It is written in Python and its source code is freely available to anyone for downloading and modification (within the limits of its license, which is Apache License 2.0). The name "Ansible" comes from Ursula K. Le Guin's book *Rocannon's World* and denotes a device that allows instantaneous communication no matter the distance.

Some interesting characteristics of Ansible are set out here:

- **Modularity**: Ansible is not a monolithic tool. Rather, it's a core program with each task it knows how to perform written as a separate module—a library, if you will. Since this was the design from the start, it produced a clean API that anyone can use to write their own modules.

- **Idempotence**: No matter how many times you perform a configuration, the result is always the same. This is one of the most important and fundamental characteristics of Ansible. You don't have to know which actions have been performed. When you extend the configuration and run the tool again, it is its job to find out the state of the system and only apply new actions.

- **Agentlessness**: Ansible doesn't install its agent on the configured system. That is not to say it doesn't need anything at all. To execute the Ansible scripts, the target system will need some means of connecting to it (most often, the SSH server running) and the Python language installed. There are several advantages born from this paradigm, including the following:

  - Ansible is not concerned with communication protocol. It uses SSH, but it doesn't implement it, leaving the details to the operating system, SSH server, and client. An advantage is that you can freely swap one SSH solution with another for whatever reasons, and your Ansible playbooks should work as intended. Also, Ansible doesn't concern itself with securing the SSH configuration. This leaves developers to concentrate on what the system is really about: configuring your systems.

  - The Ansible project does not need to develop and maintain separate programs for managed nodes. This not only frees developers from unneeded burdens but also limits the possibility of security exploits being discovered and used against target machines.

  - In an agent-utilizing solution, if, for any reason, the agent program stops working, there is no way to deliver new configurations to the system. SSH servers are usually very widely used, and the probability of failure is negligible.

  - Using SSH as the communication protocol lowers the risk of a firewall blocking the communication port for the CM system.

- **Declarativeness**: A person writing Ansible playbooks shouldn't trouble themselves much with how things should be done, just what the desired state of the system is after Ansible is done. For example, if the sysop wants Ansible to ensure that `nginx` is installed, the proper configuration entry would look like this:

```
- name: install nginx
  package:
    name: nginx
    state: present
```

The main way of working with Ansible is through writing configuration files in a special syntax called YAML. **YAML** is a syntax created specifically for configuration files and is loosely based on Python formatting. Indentations play a significant role in YAML files. YAML's home page presents a full cheat sheet card (https://yaml.org/refcard.html) for the syntax. However, the most important parts are presented here, as we will be working mostly with those files in this chapter:

- They are clear text files. This means they can be viewed and edited using the simplest editors such as Notepad, Vim, Emacs, or whatever is your favorite tool for working with text files.

- Indentations are used to denote scope. Tabulators are not permitted for indentations and it is customary to use spaces for this purpose.

- A new document opens with three hyphens (-). One file can have more than one document.

- Comments, as in Python, start with a hash (#) and continue until the end of the line. The comment must be surrounded by whitespace characters; otherwise, it will be treated as a literal hash (#) within a text.

- Text (strings) can be unquoted, single-quoted ('), or double-quoted (").

- When a list is specified, each member is denoted by a hyphen (-) character. Each item will be on a separate line. If single-line representation is required, the list items can be enclosed in square brackets ([ ]) with entries separated by commas (,).

- Associative arrays are represented by a key-value pair with each key separated from the value by a colon and space. If they have to be presented in one line, the array is enclosed in curly brackets ({ }) and pairs are separated by commas (,).

If the preceding rules are not very transparent now, don't panic. We are going to write proper Ansible configuration files, and things will become clear as we go.

Ansible divides machines into two groups: the **control node** is the computer that stores configuration directives and will connect to the target machines and configure them. There can be more than one control node. The target machines are called **inventories**. Ansible runs actions against computers in the inventory listed. Inventories are often written in an **initialization** (INI) format, which is simple and easy to follow, as explained here:

- Comments start with a semicolon (;)

- Sections are named and the names are enclosed in square brackets ([ ])

- Configuration directives are stored in pairs, each pair on its own line, with the key and value separated by an equals sign (=)

We will see an example of an inventory file shortly. There is, however, the possibility of having so-called dynamic inventories, which are generated automatically by means of a script or a system with each run of Ansible.

The main configuration file that we will be interacting with is called a playbook. A **playbook** is an entry point for Ansible. This is where the tool will start the execution. Playbooks can include other files (and this is often done).

Target hosts can be broken down into groups based on custom criteria: operating system, role within organization, physical location—whatever is required. The groups are called **roles**.

A single action that is to be performed is called a **task**.

Basics of using Ansible

The first thing to do is to install Ansible. This is pretty straightforward on all major Linux distributions and can be equally so on macOS. We encourage you to figure out the solution for your chosen operating system.

For Debian-based distributions, the following command should suffice:

```
$ sudo apt-get install ansible
```

For Fedora Linux distributions, you'd run the following command:

```
$ sudo dnf install ansible
```

To install the latest version of Ansible, however, we recommend using the Python virtual environment and its `pip` tool, like so:

```
$ python3 -m venv venv
```

In the preceding code, we have activated the virtual environment using the `venv` `python3` module. It will create a special `venv` directory that contains all important files and libraries that allow us to set up a Python virtual environment. Next, we have the following:

```
$ source venv/bin/activate
```

In the preceding code line, we have read a special file that sets the environment by configuring the shell. Next, we have this:

```
$ pip install -U pip
Requirement already satisfied: pip in ./venv/lib/python3.11/site-
packages (22.3.1)
Collecting pip
  Using cached pip-23.0.1-py3-none-any.whl (2.1 MB)
Installing collected packages: pip
  Attempting uninstall: pip
    Found existing installation: pip 22.3.1
    Uninstalling pip-22.3.1:
      Successfully uninstalled pip-22.3.1
Successfully installed pip-23.0.1
```

In the preceding code, we have upgraded `pip`, a Python package installer. In the next step, we are going to actually install Ansible, also using `pip`. The output will be shortened for brevity:

```
$ pip install ansible
Collecting ansible
  Using cached ansible-7.3.0-py3-none-any.whl (43.1 MB)
Collecting ansible-core~=2.14.3
Installing collected packages: resolvelib, PyYAML, pycparser,
packaging, MarkupSafe, jinja2, cffi, cryptography, ansible-core,
ansible
Successfully installed MarkupSafe-2.1.2 PyYAML-6.0 ansible-7.3.0
ansible-core-2.14.3 cffi-1.15.1 cryptography-39.0.2 jinja2-3.1.2
packaging-23.0 pycparser-2.21 resolvelib-0.8.1
```

As you can see in the preceding code, `pip` informs us that Ansible and its dependencies were installed successfully.

Either of those ways of installing Ansible will download and install all the packages required to run Ansible on your computer.

The simplest command you can run is called ad hoc `ping`. It's so basic that it is one of the most prevalent first uses of Ansible in tutorials and books. We are not going to deviate from it. The following command tries to connect to a specified host and then prints the result of the trial:

```
$ ansible all -m ping -i inventory
hostone | SUCCESS => {
    "ansible_facts": {
        "discovered_interpreter_python": "/usr/bin/python3"
    },
    "changed": false,
    "ping": "pong"
}
```

In the preceding command, we have told Ansible to run against all hosts in inventory, use the `ping` module (`-m ping`), and use an inventory file named `inventory` (`-i inventory`). If you don't specify the inventory file, Ansible will try to use `/etc/ansible/hosts`, which is generally a bad idea.

We will delve into the inventory file in a moment, a few paragraphs ahead.

A **module** is a small Python library written for Ansible and should ideally map 1:1 with a command. In the preceding example, we run the `ping` module (which we can also understand as a `ping` command). It is not, however, the same as the OS `ping` command, which sends a specially crafted network packet to determine whether a host is up. The Ansible `ping` command will also try to log in, to determine whether the credentials are correct.

By default, Ansible uses the SSH protocol with a private-public key pair. In a normal operation, you don't want to use password-based authentication, and Ansible will choose a key-based one.

Ansible is very good at providing self-explanatory information as a result of the execution. The preceding output tells us that Ansible was able to connect to all nodes in the inventory (there's only one), `python3` is installed there, and nothing was changed.

The inventory file is a simple but powerful tool to list, group, and provide variables for managed nodes. Our example inventory file is pasted as follows:

```
[www]
hostone ansible_host=192.168.1.2  ansible_ssh_private_key_file=~/.ssh/
hostone.pem  ansible_user=admin
```

The preceding code is two lines. The first one declares a group of nodes called www. The second declares a node called `hostone`. Since this is not a name resolvable by DNS, we have declared its IP address using the `ansible_host` variable. Then, we point to the proper `ssh` key file and declare which username should be used while logging in (`admin`). There are more things we can define in this file. Very detailed information about writing your inventory file can be found in the Ansible project documentation (`https://docs.ansible.com/ansible/latest/inventory_guide/intro_inventory.html`).

We are unable to cover all aspects of Ansible and dig deeper into most of the features that we are going to use here. Packt Publishing, however, has a very good selection of books on Ansible that you may wish to choose if you want to deepen your knowledge—for example, *Mastering Ansible, Fourth Edition,* by James Freeman and Jesse Keating.

Tasks

Tasks lie at the heart of the Ansible configuration. They are exactly that: tasks to be run against managed nodes. Tasks are contained in plays. They can be placed there directly (put into a playbook) or indirectly (included via a role).

There's a special type of task called **handles**. This is a task that will be executed only when notified by another task.

Roles

Ansible's documentation defines **roles** as "*A limited distribution of reusable Ansible content (tasks, handlers, variables, plugins, templates and files) for use inside of a play.*" For our purpose, we can think of them as a mechanism to group parts of the play. A role can be a type of host we are going to run the play against: web server, database server, Kubernetes node, and so on. By breaking down playbooks into roles, we can manage separate required tasks more easily and efficiently. Not to mention the files containing those become more readable, since we limit the number of tasks within and group them by their function.

Plays and playbooks

Plays are the context in which tasks are performed. While plays are somewhat of an ephemeral concept, **playbooks** are their physical representations: YAML files in which plays are defined.

Let's look at an example of a playbook. The following playbook will install `nginx` and `php` packages on a managed node:

```yaml
---
- name: Install nginx  and php
  hosts: www
  become: yes
  tasks:
  - name: Install nginx
    package:
      name: nginx
      state: present
  - name: Install php
    package:
      name: php8
      state: present
  - name: Start nginx
    service:
      name: nginx
      state: started
```

The first line (three dashes) marks the beginning of a new YAML document. The next line names the whole playbook. Playbook names should be short but descriptive. They are going to end up in logs and debugging information.

Next, we inform Ansible that this play should be run against nodes in the inventory placed in the www group. We also tell Ansible to use `sudo` when executing the commands. This is required as all distributions that we cover in our guide require root privileges to install and remove packages.

Then, we start the `tasks` section. Each task is given a name, the name of the module (command) we are going to use, and the command is given options and arguments. As you can see, the indentation declares the scope. If you are familiar with the Python programming language, this should be intuitive for you.

Before we run this playbook, let's use a very useful tool, `ansible-lint`:

```
$ ansible-lint install.yaml
WARNING: PATH altered to expand ~ in it. Read https://stackoverflow.
com/a/44704799/99834 and correct your system configuration.
WARNING  Listing 9 violation(s) that are fatal
yaml[trailing-spaces]: Trailing spaces
```

```
install.yaml:1


yaml[truthy]: Truthy value should be one of [false, true]
install.yaml:4


fqcn[action-core]: Use FQCN for builtin module actions (package).
install.yaml:6 Use `ansible.builtin.package` or `ansible.legacy.
package` instead.
[...]
yaml[empty-lines]: Too many blank lines (1 > 0)
install.yaml:18


Read documentation for instructions on how to ignore specific rule
violations.

                    Rule Violation Summary
 count tag                    profile     rule associated tags
     1 yaml[empty-lines]      basic       formatting, yaml
     1 yaml[indentation]      basic       formatting, yaml
     3 yaml[trailing-spaces]  basic       formatting, yaml
     1 yaml[truthy]           basic       formatting, yaml
     3 fqcn[action-core]      production  formatting


Failed after min profile: 9 failure(s), 0 warning(s) on 1 files.
```

I have cut part of the output for brevity, but you can see that the tool printed information about violations of YAML syntax and Ansible best practices. Failures are types of errors that will stop the execution of the playbook. Warnings are just that: the playbook will be executed, but there are some errors that go against best practices. Let's correct our playbook, as follows:

```
---
- name: Install nginx  and php
  hosts: www
  become: true
  tasks:
    - name: Install nginx
      ansible.builtin.package:
        name: nginx
        state: present
    - name: Install php
      ansible.builtin.package:
        name: php8
        state: present
    - name: Start nginx
```

```
        ansible.builtin.service:
          name: nginx
          state: started
```

We can now run the playbook using the `ansible-playbook` command:

```
$ ansible-playbook -i inventory install.yaml

PLAY [Install nginx and php] ****************************************
****************************

TASK [Gathering Facts] *********************************************
********************
ok: [hostone]

TASK [Install nginx] ***********************************************
******************
changed: [hostone]

TASK [Install php] *************************************************
****************
fatal: [hostone]: FAILED! => {"changed": false, "msg": "No package
matching 'php8' is available"}

PLAY RECAP *********************************************************
**********
hostone      : ok=2    changed=1   reachable=0    failed=1
skipped=0    rescued=0   ignored=0
```

In the preceding output, we have instructed the `ansible-playbook` command to use an inventory file called `inventory` and run a playbook named `install.yaml`. The output should be self-explanatory: we are informed of the name of the play we run. Then, we see a list of managed nodes that Ansible will try to execute action against. Then, we see tasks and a list of nodes that the tasks succeeded or failed at. The `nginx` task was a success on `hostone`. However, installing php failed. Ansible gives us the exact reason for it: there is no `php8` package available for our managed node. Sometimes, a resolution is pretty obvious, but sometimes it requires a bit of digging around. After checking with our distribution, we find out that the actual php package available there is `php7.4`. After quickly correcting the offending line, we run the playbook again, as follows:

```
$ ansible-playbook -i inventory install.yaml

PLAY [Install nginx and php] ****************************************
****************************

TASK [Gathering Facts] *********************************************
********************
```

```
ok: [hostone]

TASK [Install nginx] **********************************************
*******************
ok: [hostone]

TASK [Install php] ************************************************
*****************
changed: [hostone]

TASK [Start nginx] ************************************************
*****************
ok: [hostone]

PLAY RECAP *******************************************************
**********
hostone                    : ok=4     changed=1     unreachable=0
failed=0      skipped=0     rescued=0     ignored=0
```

Notice the change in the output. First, Ansible tells us that `nginx` on `hostone` is okay. This means that Ansible was able to determine that the package was already installed, and it took no action. Then, it told us that the `php7.4` installation was successful (`changed: [hostone]`) on the `hostone` server.

The preceding playbook is short, but we hope it demonstrates the usefulness of the tool. Playbooks are executed in a linear manner, from top to bottom.

There's one problem with our playbook. While it only installs two packages, you may be worried about maintainability and readability if you need to install tens or hundreds of packages. Having a separate task for each of them is troublesome. There's a solution. You can create a list of items that a given task will be executed for—something akin to a loop. Let's edit the `tasks` part of the playbook, as follows:

```
tasks:
  - name: Install nginx
    ansible.builtin.package:
      name: '{{ item }}'
      state: present
    with_items:
      - nginx
      - php7.4
      - gcc
      - g++
  - name: Start nginx
    ansible.builtin.service:
      name: nginx
```

```
        state: started
```

We have added two packages to demonstrate a more complete run. Notice the lack of a separate task for php7.4.

Before running this play, it is always a good idea to check it with ansible-lint. Here's how we can do that:

```
$ ansible-lint install.yaml
Passed with production profile: 0 failure(s), 0 warning(s) on 1 files.
```

Now, after ansible-lint gives us the green light, let's play this playbook:

```
$ ansible-playbook -i inventory install.yaml

PLAY [Install nginx  and php] ****************************************
****************************

TASK [Gathering Facts] **********************************************
*********************
ok: [hostone]

TASK [Install nginx] ************************************************
*******************
ok: [hostone] => (item=nginx)
ok: [hostone] => (item=php7.4)
changed: [hostone] => (item=gcc)
changed: [hostone] => (item=g++)

TASK [Start nginx] **************************************************
*****************
ok: [hostone]

PLAY RECAP **********************************************************
**********
hostone                 : ok=3    changed=1    unreachable=0    failed=0
skipped=0     rescued=0    ignored=0
```

Assume we want to install a page configuration in nginx. Ansible is able to copy files. We can set it to copy nginx virtual server configuration, but we only want to restart nginx once: at the service setup.

We can do it via notify and handlers. I'll paste the whole playbook ahead:

```
---
- name: Install nginx  and php
  hosts: www
  become: true
```

```
tasks:
  - name: Install nginx
    ansible.builtin.package:
      name: '{{ item }}'
      state: present
    with_items:
      - nginx
      - php7.4
      - gcc
      - g++
    notify:
      - Start nginx

  - name: Copy service configuration
    ansible.builtin.copy:
      src: "files/service.cfg"
      dest: "/etc/nginx/sites-available/service.cfg"
      owner: root
      group: root
      mode: '0640'

  - name: Enable site
    ansible.builtin.file:
      src: "/etc/nginx/sites-available/service.cfg"
      dest: "/etc/nginx/sites-enabled/default"
      state: link
    notify:
      - Restart nginx

handlers:
  - name: Start nginx
    ansible.builtin.service:
      name: nginx
      state: started

  - name: Restart nginx
    ansible.builtin.service:
      name: nginx
      state: restarted
```

Notice the whole new section related to copying files. We also create a **symlink** (an alias to a file) so that the site is enabled. The preceding example is incomplete. It takes more than that to actually set up a virtual server with nginx. We have shortened it for brevity's sake and only to demonstrate a principle.

The execution of this play yields the following output:

```
$ ansible-playbook -i inventory install.yaml

PLAY [Install nginx  and php] ****************************************
****************************

TASK [Gathering Facts] **********************************************
*********************
ok: [hostone]

TASK [Install nginx] ************************************************
******************
ok: [hostone] => (item=nginx)
ok: [hostone] => (item=php7.4)
ok: [hostone] => (item=gcc)
ok: [hostone] => (item=g++)

TASK [Copy service configuration] ***********************************
*******************************
ok: [hostone]

TASK [Enable site] **************************************************
****************
ok: [hostone]

PLAY RECAP **********************************************************
*********
hostone                        ok=4     changed=0     unreachable=0
failed=0     skipped=0     rescued=0     ignored=0
```

As soon as you start using Ansible in a real production environment, it becomes obvious that even with the *looping* technique we demonstrated previously, the playbook grows pretty fast and becomes unwieldy. Thankfully, there's a way to break the playbook into smaller files.

One of the ways to do it is to segregate tasks into roles. As mentioned earlier, a role is a bit of an abstract concept, so the easiest way to think about it is to consider it a group. You group tasks depending on the criteria that are relevant to you. While the criteria are totally up to you, it is pretty common for roles to refer to the type of function a given managed node performs. One managed node can perform more than one function, but it still is probably wise to separate the functions into their own roles. Thus, the HTTP server would be one role, the database server would be another, and the file server would be yet another.

Roles in Ansible have a predetermined directory structure. There are eight standard directories defined, although you only need to create one.

The following code is copied from the Ansible documentation (https://docs.ansible.com/ansible/latest/playbook_guide/playbooks_reuse_roles.html):

```
roles/
    common/                 # this hierarchy represents a "role"
        tasks/              #
            main.yml        #  <-- tasks file can include smaller files
if warranted
        handlers/           #
            main.yml        #  <-- handlers file
        templates/          #  <-- files for use with the template
resource
            ntp.conf.j2     #  <------- templates end in .j2
        files/              #
            bar.txt         #  <-- files for use with the copy resource
            foo.sh          #  <-- script files for use with the script
resource
        vars/               #
            main.yml        #  <-- variables associated with this role
        defaults/           #
            main.yml        #  <-- default lower priority variables for
this role
        meta/               #
            main.yml        #  <-- role dependencies
        library/            # roles can also include custom modules
        module_utils/       # roles can also include custom module_utils
        lookup_plugins/     # or other types of plugins, like lookup in
this case
```

The `roles` directory exists on the same level as your playbook. The most common subdirectories you will probably see are `tasks`, `templates`, `files`, `vars`, and `handlers`. Within each subdirectory, Ansible will look for a `main.yml`, `main.yaml`, or `main` file (all of them have to be valid YAML files). Their contents will be automatically available to the playbook. So, how would this work in practice? Within the same directory that our `install.yaml` playbook exists, we will create a `roles` directory. Within this directory, we will create another one: www. Inside that one, we will create directories: `tasks`, `files`, and `handlers`. A similar structure will be created with `development` as a role:

```
roles/
    www/                    # this hierarchy represents a "role"
        tasks/              #
```

```
            main.yml      #  <-- tasks file can include smaller files
if warranted
        handlers/         #
            main.yml      #  <-- handlers file
        files/            #
            service.cfg      #  <-- files for use with the copy
resource
    development/             # this hierarchy represents a "role"
        tasks/            #
            main.yml      #  <-- tasks file can include smaller files
if warranted
```

The `development` role has only tasks subdirectory because we don't require any additional bells and whistles. Why the development role, though? When you look back at our current playbook, you'll notice that we mix up installing things for the www server and development packages—namely, compilers. It is bad practice, even if they will end up on the same physical server.

Thus, we are going to edit our inventory file so that we have two separate roles:

```
[www]
hostone ansible_host=192.168.1.2  ansible_ssh_private_key_file=~/.ssh/
hostone.pem  ansible_user=admin

[development]
hostone ansible_host=192.168.1.3  ansible_ssh_private_key_file=~/.ssh/
hostone.pem  ansible_user=admin
```

Both groups contain only one node, which is a bad thing to do overall. We do it only for the purpose of this guide. Don't install compilers and development software on your HTTP server, especially on the production one.

Now, we have to move some stuff around within the playbook. The `install.yml` file will become much shorter, as we can see here:

```
---
- name: Install nginx and php
  hosts: www
  roles:
    - www
  become: true

- name: Install development packages
  hosts: development
  roles:
    - development
  become: true
```

We actually have two plays in this playbook. One of them is for www hosts, and the other is for development hosts. After we give the play a name and list host groups that we wish it to run against, we use keyword roles and list actual roles to be used. As you can see, Ansible is going to locate the proper roles automatically, as long as you adhere to the directory structure explained previously. There are ways to include roles directly by specifying their full path, but we're not going to cover them here.

Now, for the www role, we'll execute the following code:

```
---
- name: Install nginx
  ansible.builtin.package:
    name: '{{ item }}'
    state: present
  with_items:
    - nginx
    - php7.4
  notify:
    - Start nginx

- name: Copy service configuration
  ansible.builtin.copy:
    src: "files/service.cfg"
    dest: "/etc/nginx/sites-available/service.cfg"
    owner: root
    group: root
    mode: '0640'

- name: Enable site
  ansible.builtin.file:
    src: "/etc/nginx/sites-available/service.cfg"
    dest: "/etc/nginx/sites-enabled/default"
    state: link
  notify:
    - Restart nginx
```

You should notice several changes from the get-go:

- There is no tasks keyword in this document. This is implicit by the fact that this is the main.yaml file within the tasks subdirectory.

- Indentation has been moved left.

- There are no handlers here.

- We have removed the installation of the gcc and g++ packages.

Let's now see the handlers:

```
---
- name: Start nginx
  ansible.builtin.service:
    name: nginx
    state: started
  listen: "Start nginx"

- name: Restart nginx
  ansible.builtin.service:
    name: nginx
    state: restarted
  listen: "Restart nginx"
```

The overall changes are the same, as noted here:

- We have removed the `handlers` keyword
- We have moved the indentation to the left

Now, let's see `roles/development/tasks/main.yaml`:

```
---
- name: Install compilers
  ansible.builtin.package:
    name: '{{ item }}'
    state: present
  with_items:
    - gcc
    - g++
```

This is very simple. We may have increased the complexity of the directory structure, but we have gained the simplicity of tasks and plays. The gains are well worth the trade-off, especially when your playbooks grow in size and complexity.

There are many advantages to using a CaC tool, as outlined here:

- Configuration can be linted—that is, checked for syntax errors by automation tools
- It can be applied infinitely with the same outcomes
- It can be run on many systems in parallel
- It can be kept in a version control system, such as Git, where the history of changes and comments for them are stored and can be viewed at any time
- It can be run by automation tools, removing the need for a human factor besides writing playbooks

In this subsection, we have shown how to write simple Ansible playbooks. We have explained what Ansible is and what the building blocks of the configuration scripts are. We have introduced playbooks, roles, and the inventory. There are many more things Ansible can do for you. You can manage devices, filesystems, users, groups, permissions, networking, and so on. The list of all modules Ansible is shipped with out of the box is impressive. You can always check the list at `https://docs.ansible.com/ansible/latest/module_plugin_guide/index.html`. Remember to check the list for your version of Ansible.

So, now that we've covered installing, configuring, and using Ansible to manage your servers (installing software, creating configuration files, and managing services), we are ready to look into Ansible Galaxy: community-developed modules that increase Ansible's usefulness.

Ansible Galaxy

Ansible is a powerful automation tool that enables users to configure, deploy, and manage complex IT infrastructures with ease. However, creating and maintaining Ansible playbooks can be time-consuming, especially when working with large-scale environments. Fortunately, Ansible Galaxy exists to help streamline this process by providing a centralized repository of pre-built roles and playbooks that can be easily integrated into an existing Ansible project.

Ansible Galaxy is a community-driven platform that hosts an extensive collection of Ansible roles and playbooks. These roles and playbooks are submitted by users from around the world and are reviewed and curated by Ansible's maintainers. Ansible Galaxy provides a simple, efficient way to find and use pre-built automation content that can save users time and effort while ensuring quality and consistency.

Using Ansible Galaxy, users can quickly find, download, and use pre-built roles and playbooks for popular applications, services, and infrastructure components. These pre-built components can help speed up deployment times, ensure best practices are followed, and reduce the likelihood of errors or inconsistencies. Ansible Galaxy can also help users learn from others' experiences and gain insights into the best practices of their peers.

Let's use one of the Galaxy roles to install the `nginx` web server on our `webserver` role. In order to do that, we will need to install the role from Ansible Galaxy. First, ensure that Ansible is installed on your system by running the following command:

```
admin@myhome:~$ ansible-galaxy install nginxinc.nginx
```

This command will download and install the `nginx` role from Ansible Galaxy. By default, all roles installed are placed in the `~/.ansible/roles` directory. It's possible to change that by creating a global Ansible configuration file in your home directory: `~/.ansible.cfg`.

An example of a configuration file changing the `roles_path` directory looks like this:

```
[defaults]
roles_path = /home/admin/myansibleroles
```

A good practice is to pin role version numbers and put this version in a YAML file saved in the same Git repository where you will keep your Ansible playbooks. To achieve this, let's create an `ansible_requirements.yml` file:

```
---
- src: nginxinc.nginx
  version: 0.24.0
```

To install the role from Ansible Galaxy using that file, you would run the following command:

```
admin@myhome:~$ ansible-galaxy install -r ansible_requirements.yml
```

Once the role is installed, it can be used in an Ansible playbook by adding the following line to the playbook:

```
roles:
    - nginxinc.nginx
```

Here's an example playbook that uses the `nginx` role from Ansible Galaxy to install and configure `nginx` on a remote server:

```
---
- name: Install and configure Nginx
  hosts: webservers
  become: true
  roles:
    - nginxinc.nginx
  vars:
    nginx_sites:
      myapp:
        template: "{{ playbook_dir }}/templates/myapp.conf.j2"
```

In this playbook, we specify the `webservers` group as the target hosts and use the `nginxinc.nginx` role to install and configure `nginx`. We also define a variable called `nginx_sites` that specifies the configuration for a `nginx` server block that will be created using a Jinja2 template located in the playbook's `templates` directory.

By using Ansible Galaxy and pre-built roles such as `nginxinc.nginx`, users can automate complex tasks quickly and reliably, ensuring consistency and reducing the risk of errors.

Handling secrets

Protecting secrets such as passwords, tokens, and certificates is crucial in any IT infrastructure. These secrets are the keys to accessing sensitive information and services, and their exposure can lead to severe security breaches. Therefore, it is crucial to keep them safe and secure. Ansible provides several methods for managing secrets, such as Ansible Vault, which allows users to encrypt and decrypt sensitive data using a password or key file. This feature helps to protect secrets and ensures that only authorized users have access to them.

Saving secrets in a Git repository or any other public place is a significant security risk. Such repositories are often accessible to multiple users, some of whom may not have the necessary permissions to access sensitive data. Additionally, version control systems such as Git retain the history of changes made to files, making it possible for secrets to be exposed inadvertently. This could happen if a user inadvertently commits secrets to a repository or if a hacker gains access to a repository's commit history. Therefore, it is vital to discourage saving secrets in public places to reduce the risk of unauthorized access. Instead, Ansible provides secure ways to manage secrets, ensuring that they are encrypted and only accessible to authorized users. By doing so, users can be confident that their secrets are safe and secure.

Ansible Vault

Ansible Vault is a feature provided by Ansible that enables users to encrypt and decrypt sensitive data, such as passwords, keys, and certificates. The vault creates an encrypted file that can be decrypted only by authorized users, ensuring that sensitive data remains secure. The vault can be used to store secrets in files, variables, or other sources that are used by Ansible.

Ansible Vault uses a variety of encryption methods to secure the secrets stored within it. By default, Ansible Vault uses AES 256 encryption, a widely accepted and secure encryption algorithm. Additionally, Ansible Vault supports other encryption algorithms, such as AES 192 and AES 128, providing flexibility in the encryption strength used to secure secrets. When encrypting data with Ansible Vault, users can choose to encrypt it with a password or with a key file. This ensures that only authorized users who possess the password or key file can decrypt the secrets stored within the vault.

To create a new vault using Ansible Vault, you can use the following command:

```
admin@myhome:~$ ansible-vault create secrets.yml
New Vault password:
Confirm New Vault password:
```

This will create a new encrypted vault file called `secrets.yml`. You will be prompted to enter a password to encrypt the file. Once you've entered the password, the vault file will be created and opened in your default editor. Here's an example `secrets.yml` file:

```
somesecret: pleaseEncryptMe
secret_pgsql_password: veryPasswordyPassword
```

To edit this secret file, you'll need to use the `ansible-vault edit secrets.yml` command and type the encryption password afterward.

To write an Ansible task that reads the `pgsql_password` secret from the vault, you can use the `ansible.builtin.include_vars` module with the `vault_password_file` parameter. Here is an example task:

```
- name: Read pgsql_password from Ansible Vault
  include_vars:
    file: secrets.yml
    vault_password_file: /path/to/vault/password/file
  vars:
    pgsql_password: "{{ secret_pgsql_password }}"
```

In this task, we're using the `include_vars` module to read the variables from the `secrets.yml` vault file. The `vault_password_file` parameter specifies the location of the file containing the password to decrypt the vault. We then assign the value of `secret_pgsql_password` to the `pgsql_password` variable, which can be used elsewhere in the playbook.

Note that the `secret_pgsql_password` variable should be defined in the vault. The `secret_` prefix is to indicate that the password was retrieved from the vault. Ansible does not differentiate between regular variables or secret variables.

When you run this playbook with an increased Ansible debug level, you will notice that the PostgreSQL password is exposed in the debug output. To prevent that, every task dealing with sensitive information can be executed with the `no_log: True` option enabled.

SOPS

Secrets OPerationS (SOPS) is an open source tool developed by Mozilla that allows users to securely store and manage their secrets within various configuration files, including Ansible playbooks. SOPS uses a hybrid encryption approach, which means that it combines symmetric and asymmetric encryption to ensure maximum security.

SOPS encrypts secrets using a master key, which can be either symmetric or asymmetric. **Symmetric encryption** uses a password or passphrase to encrypt and decrypt secrets, while **asymmetric encryption** uses a pair of keys, one public and one private, to encrypt and decrypt secrets. SOPS encrypts the master key using key wrapping, a technique that encrypts the key with another key. In SOPS, the key that is used for key wrapping is often an AWS **Key Management Service (KMS)** key, but it can also be a **Pretty Good Privacy (PGP)** key or a Google Cloud KMS key.

SOPS integrates seamlessly with Ansible and supports various file formats, including YAML, JSON, and INI. It also supports various cloud providers, including AWS, Google Cloud, and Azure, making it a versatile tool for managing secrets across different environments.

Here's an example of how to create a SOPS-encrypted YAML file and load its contents into an Ansible playbook using `community.sops.load_vars`.

First, create a YAML file named `secrets.yaml` with the following content:

```
postgresql_password: So.VerySecret
```

Then, use SOPS to encrypt the file using the following command:

```
admin@myhome:~$ sops secrets.yaml > secrets.sops.yaml
```

This will create an encrypted version of the `secrets.yaml` file named `secrets.sops.yaml`. It's safe to remove the `secrets.yaml` plain text file now. The most common mistake is leaving these files behind and committing them to the Git repository.

Next, create a new Ansible playbook named `database.yml` with the following content:

```
---
- hosts: dbserver
  become: true
  vars:
    postgresql_password: "{{ lookup('community.sops.load_vars',
'secrets.sops.yaml')['postgresql_password'] }}"

  tasks:
    - name: Install PostgreSQL
      apt:
        name: postgresql
        state: present

    - name: Create PostgreSQL user and database
      postgresql_user:
        db: mydatabase
        login_user: postgres
        login_password: "{{ postgresql_password }}"
        name: myuser
        password: "{{ postgresql_password }}"
        state: present
```

In this example, we're using `community.sops.load_vars` to load the `postgresql_password` variable from the encrypted `secrets.sops.yaml` file. The `postgresql_password` variable is then passed to the `postgresql_user` task using the `{{ postgresql_password }}` Jinja2 syntax.

When you run this playbook with Ansible, it will decrypt the `secrets.sops.yaml` file using SOPS and load the `postgresql_password` variable into the `postgresql_password` variable in the playbook. This ensures that the password is not stored in plain text in the playbook, providing an extra layer of security for your sensitive information.

More information about SOPS is available in its official GitHub repository: `https://github.com/mozilla/sops`.

Other solutions

There are several alternatives to Ansible Vault and SOPS that you can use to manage your sensitive data securely, as follows:

- **HashiCorp Vault**: An open source tool for securely storing and accessing secrets. It provides a service for secure and centralized storage of secrets, access control, and auditing.

- **Blackbox**: A command-line utility that encrypts and decrypts files using **GNU Privacy Guard (GPG)**. It works by creating a separate GPG key for each user or team that needs access to the encrypted data.

- **Keywhiz**: Another open source secrets management system that provides a simple way to store and distribute secrets securely. It includes a web interface for managing secrets and a command-line tool for accessing them.

- **Azure Key Vault, AWS Secrets Manager, or Google Cloud Secret Manager**: Solutions you might want to consider keeping secrets in when dealing with a cloud environment.

This is, of course, not a complete list of all available options.

In this section, we introduced Ansible Vault and SOPS as means to handle secrets such as passwords. In the next section, we will be introducing a graphical frontend (GUI) to Ansible.

Ansible Tower and alternatives

Ansible Tower provides a centralized platform for managing Ansible automation workflows, making it easier for IT teams to collaborate, share knowledge, and maintain their infrastructure. Some of its key features include a web-based interface for managing Ansible playbooks, inventories, and job runs, **role-based access control** (**RBAC**) for managing user permissions, a built-in dashboard for monitoring job status and results, and an API for integrating with other tools and platforms.

It was first released in 2013 by Ansible, Inc. (now part of Red Hat), and has since become one of the most popular tools for automating IT workflows.

Since its initial release, Ansible Tower has undergone numerous updates and enhancements, including support for more complex automation workflows, integration with cloud platforms such as AWS and Azure, and improved scalability and performance. Ansible Tower is a commercial product shipped by the Red Hat company. The closest alternative to Ansible Tower is **Ansible WorX** (**AWX**).

Ansible AWX is an open source alternative to Ansible Tower, offering many of the same features as Tower but with a greater degree of customization and flexibility. AWX was first released in 2017 and has since become a popular choice for organizations looking to implement Ansible automation at scale without the cost of a commercial license.

One of the primary differences between Ansible Tower and Ansible AWX is their licensing model. While Ansible Tower requires a commercial license and is sold as part of Red Hat Ansible Automation Platform, Ansible AWX is open source and freely available for download from the Ansible website. This means that organizations can deploy AWX on their own infrastructure and customize it to their specific needs, rather than relying on a pre-built commercial solution.

Feature-wise, Ansible Tower and Ansible AWX are quite similar, with both platforms offering a web-based interface for managing Ansible playbooks, inventories, and job runs, RBAC for managing user permissions, and a built-in dashboard for monitoring job status and results. However, Ansible Tower does offer some additional features not found in Ansible AWX, such as native integration with Red Hat Ansible Automation Platform, advanced analytics and reporting, and certified modules and collections.

Another open source alternative to Ansible Tower is **Ansible Semaphore**. Similar to Tower, it is a web-based application designed to simplify the management of Ansible playbooks and projects. It is an open source, free, and easy-to-use alternative to Ansible Tower that allows users to easily automate their infrastructure tasks without the need for extensive coding knowledge. The first release of Ansible Semaphore was in 2016, and since then it has become a popular choice for those who want a simple yet powerful web-based interface for managing their Ansible automation workflows.

You can read more about these alternatives on their respective websites:

- **Ansible Tower**: `https://access.redhat.com/products/ansible-tower-red-hat`
- **Ansible AWX**: `https://github.com/ansible/awx`
- **Ansible Semaphore**: `https://www.ansible-semaphore.com/`

Advanced topics

In this section, we will show you how to handle advanced Ansible features and techniques for debugging and automatically checking your playbooks for possible errors.

Debugging

In order to debug issues with your Ansible playbook runs, it is often useful to increase the verbosity level to get more detailed output about what Ansible is doing. Ansible has four verbosity levels: `-v`, `-vv`, `-vvv`, and `-vvvv`. The more vs you add, the more verbose the output becomes.

By default, Ansible runs with -v, which provides basic information about the tasks that are executed. However, if you are experiencing issues with your playbook, it may be helpful to increase the verbosity level to get more detailed output. For example, using -vv will provide additional information about the playbooks, roles, and tasks that are being executed, while using -vvv will also show the tasks that Ansible is skipping.

To increase the verbosity level of an Ansible playbook run, simply add one or more -v options to the ansible-playbook command. Here's an example:

```
admin@myhome:~$ ansible-playbook playbook.yml -vv
```

This will run the playbook with verbose output at the -vvlevel. If you need even more verbose output, you can add extra -v options. You can see an example here:

```
admin@myhome:~$ ansible-playbook playbook.yml -vvv
```

In addition to using the -v options, you can also set the verbosity level in your ansible.cfg file by adding the following lines:

```
[defaults]
verbosity = 2
```

This will set the verbosity level to -vv for all ansible-playbook commands. You can change the value to 3 or 4 to increase the verbosity level even further.

If you need to add some custom communication (such as printing a variable) in a verbose mode, you can do it by using the debug task.

Here's an example playbook that demonstrates how to print out a variable in -vv debug mode:

```
---
- hosts: all
  gather_facts: true
  vars:
    my_variable: "Hello, World!"

  tasks:
    - name: Print variable in debug mode
      debug:
        msg: "{{ my_variable }}"
      verbosity: 2
```

In this playbook, we define a my_variable variable that holds the "Hello, World!" string. Then, we use the debug module to print out the value of this variable using the msg parameter.

The `verbosity: 2` line is what enables `debug` mode. This tells Ansible to increase the verbosity level to `-vv`, which will show us the output of the `debug` module.

Linting Ansible playbooks

Ansible code linting is the process of analyzing and verifying the syntax and style of Ansible code to ensure that it conforms to best practices and standards. The purpose of linting is to catch potential errors or issues before the code is executed, which can save time and effort in the long run.

The most common tool used for Ansible code linting is `ansible-lint`. This is an open source command-line tool that analyzes Ansible playbooks and roles for potential problems and provides suggestions for improvement.

To run `ansible-lint` against a sample playbook, you can execute the following command in the terminal:

```
admin@myhome:~$ ansible-lint sample-playbook.yml
```

Assuming the sample playbook is saved as `sample-playbook.yml` in the current working directory, its content looks like this:

```
---
- name: (Debian/Ubuntu) {{ (nginx_setup == 'uninstall') |
ternary('Remove', 'Configure') }} NGINX repository
  ansible.builtin.apt_repository:
    filename: nginx
    repo: "{{ item }}"
    update_cache: true
    mode: "0644"
    state: "{{ (nginx_state == 'uninstall') | ternary('absent',
'present') }}"
  loop: "{{ nginx_repository | default(nginx_default_repository_
debian) }}"
  when: nginx_manage_repo | bool

- name: (Debian/Ubuntu) {{ (nginx_setup == 'uninstall') |
ternary('Unpin', 'Pin') }} NGINX repository
  ansible.builtin.blockinfile:
    path: /etc/apt/preferences.d/99nginx
    create: true
    block: |
      Package: *
      Pin: origin nginx.org
      Pin: release o=nginx
      Pin-Priority: 900
    mode: "0644"
```

```
      state: "{{ (nginx_state == 'uninstall') | ternary('absent',
'present') }}"
    when: nginx_repository is not defined

  - name: (Debian/Ubuntu) {{ nginx_setup | capitalize }} NGINX
    ansible.builtin.apt:
      name: nginx{{ nginx_version | default('') }}
      state: "{{ nginx_state }}"
      update_cache: true
      allow_downgrade: "{{ omit if ansible_version['full'] is
version('2.12', '<') else true }}"
    ignore_errors: "{{ ansible_check_mode }}"
    notify: (Handler) Run NGINX
```

Note that the output of `ansible-lint` may vary depending on the version of Ansible and `ansible-lint` being used, as well as the specific rules enabled.

As an example, the following is the output of running `ansible-lint` against the sample playbook provided:

```
[WARNING]: empty path for ansible.builtin.blockinfile, path set to ''
[WARNING]: error loading version info from /usr/lib/python3.10/site-
packages/ansible/modules/system/setup.py: __version__ = '2.10.7'
[WARNING]: 3.0.0 includes an experimental document syntax parser
that could result in parsing errors for documents that used the
previous parser. Use `--syntax-check` to verify new documents before
use or consider setting `document_start_marker` to avoid using the
experimental parser.

sample-playbook.yml:1:1: ELL0011: Trailing whitespace
sample-playbook.yml:7:1: ELL0011: Trailing whitespace
sample-playbook.yml:9:1: ELL0011: Trailing whitespace
sample-playbook.yml:17:1: ELL0011: Trailing whitespace
sample-playbook.yml:22:1: ELL0011: Trailing whitespace
sample-playbook.yml:26:1: ELL0011: Trailing whitespace
sample-playbook.yml:29:1: ELL0011: Trailing whitespace
sample-playbook.yml:35:1: ELL0011: Trailing whitespace
sample-playbook.yml:38:1: ELL0011: Trailing whitespace
sample-playbook.yml:41:1: ELL0011: Trailing whitespace
sample-playbook.yml:44:1: ELL0011: Trailing whitespace
sample-playbook.yml:47:1: ELL0011: Trailing whitespace
sample-playbook.yml:50:1: ELL0011: Trailing whitespace
sample-playbook.yml:53:1: ELL0011: Trailing whitespace
```

The output indicates that the playbook contains trailing whitespace on several lines, which violates the ELL0011 rule of ansible-lint. The warning messages regarding the empty path and version info loading are not critical issues and can be safely ignored.

To fix the trailing whitespace issue, simply remove the extra spaces at the end of each affected line. Once the issues are fixed, you can rerun ansible-lint to ensure that there are no further problems with the playbook.

Speeding up SSH connections

Ansible is an open source automation tool that is widely used to deploy and manage IT infrastructure. One of the key features of Ansible is its ability to use SSH for secure communication with remote servers. However, using SSH for every single task can be time-consuming and can impact performance, especially if you're dealing with a lot of servers. To address this issue, Ansible supports **SSH multiplexing**, which allows multiple SSH connections to share a single TCP connection.

SSH multiplexing works by reusing the existing SSH connection rather than creating a new one for every task. When Ansible establishes an SSH connection to a remote server, it opens a TCP socket and creates a control socket for that connection. A **control socket** is a special socket that is used to manage the SSH connection. When another SSH connection is requested to the same host, Ansible checks whether a control socket already exists for that connection. If it does, Ansible reuses the existing control socket and creates a new channel within the same SSH connection for the new task.

The benefits of SSH multiplexing are that it saves time and resources by reducing the number of SSH connections that Ansible has to establish. Additionally, it can improve the performance of Ansible by reducing the overhead of creating and tearing down SSH connections for every task.

To enable SSH multiplexing in Ansible, you need to configure the ControlMaster and ControlPath options in the SSH client configuration file. The ControlMaster option enables the use of SSH multiplexing, while the ControlPath option specifies the location of the control socket. By default, Ansible uses the ~/.ansible/cp directory to store control sockets. You can also configure the maximum number of SSH connections that can be multiplexed simultaneously by setting the ControlPersist option.

To customize the SSH multiplexing configuration, you can put SSH options into your default Ansible configuration file placed in ~/.ansible.cfg by adding the [ssh_connection] section, as follows:

```
[ssh_connection]
ssh_args = -o ControlMaster=auto -o ControlPersist=3600
control_path = ~/.ssh/multiplexing/ansible-ssh-%%r@%%h:%%p
```

Make sure to also create a ~/.ssh/multiplexing directory after adding this configuration.

The `ControlMaster=auto` option creates a master session automatically, and if there is a master session already available, subsequent sessions are automatically multiplexed. Setting `ControlPersist=3600` will leave the master connection open in the background to accept new connections for 3600 seconds (1 hour).

Dynamic inventory

In cloud environments such as AWS, servers can be created and terminated dynamically based on demand. Managing these servers manually can be a daunting task, which is why automation tools such as Ansible are essential. One of the critical features of Ansible that makes it well suited for cloud environments is dynamic inventory.

Dynamic inventory is an Ansible feature that enables the automatic discovery of hosts (servers) and groups (tags) in a cloud environment. In AWS, Ansible can use the **Elastic Compute Cloud** (**EC2**) inventory plugin to query the AWS API to retrieve information about EC2 instances and groups.

To use dynamic inventory in AWS, you need to configure the EC2 inventory plugin in your Ansible configuration.

The `amazon.aws.aws_ec2` inventory plugin is an official Ansible plugin that enables dynamic inventory for Amazon EC2 instances. To use this plugin, you need to follow the steps set out next.

Depending on which version of Ansible you're using and whether you've installed full Ansible (not only Ansible Core), you might need to use Ansible Galaxy to install an AWS collection plugin, as follows:

```
admin@myhome:~$ ansible-galaxy collection install amazon.aws
```

Install the `boto3` and `botocore` libraries on your Ansible control node. You can install them using the `pip` package manager, like so:

```
admin@myhome:~$ pip install boto3 botocore
```

Create an **Identity and Access Management** (**IAM**) user with the necessary permissions to access EC2 instances and groups. You can create an IAM user using the AWS Management Console or the AWS CLI. Make sure to save the IAM user's access key and secret access key.

Create an AWS credentials file (`~/.aws/credentials`) on your Ansible control node and add the IAM user's access key and secret access key, as follows:

```
[default]
aws_access_key_id = YOUR_ACCESS_KEY
aws_secret_access_key = YOUR_SECRET_KEY
```

Create an Ansible inventory file (`inventory.yml`) in your project directory and configure it to use the `amazon.aws.aws_ec2` plugin, like so:

```
plugin: amazon.aws.aws_ec2
regions:
  - eu-central-1
filters:
  tag:Environment:
    - webserver
    - frontend
```

Here's a brief explanation of the configuration options:

- `plugin`: Specifies the inventory plugin to use

- `regions`: Specifies the AWS regions to search for instances in

- `filters`: Allows you to filter EC2 instances by tags

Test the inventory by running the following command:

```
admin@myhome:~$ ansible-inventory -i inventory.yml --list
```

This command should output a JSON object that lists all the EC2 instances in the specified regions, grouped by their Ansible tags.

This way, you won't have to update your inventory file on every Ansible playbook run to make sure you have an up-to-date servers list.

Summary

In this chapter, we have presented you with the Ansible CaC tool. We have explained and demonstrated how moving configuration, from tribal knowledge and documents (as well as describing steps required to get your system to a desired state) to tools that can implement said configuration based on a well-defined syntax brings benefits to your organization, such as repeatability, ability to run many configurations in parallel, automated tests, and execution.

In the next chapter, we are going to introduce you to **Infrastructure as Code (IaC)**.

Further reading

- *Mastering Ansible, Fourth Edition* by James Freeman and Jesse Keating

- *Ansible Playbook Essentials* by Gourav Shah

- *Ansible for Real-Life Automation* by Gineesh Madapparambath

12

Leveraging Infrastructure as Code

In today's digital landscape, managing and deploying infrastructure is a complex and time-consuming process. Traditionally, infrastructure deployment involves manually configuring each server, network, and storage device. This process is not only time-consuming but also prone to errors and inconsistencies. **Infrastructure as Code** (**IaC**) solutions provide an automated way to manage and deploy infrastructure. IaC solutions allow developers to treat infrastructure as code, enabling them to define, manage, and provision infrastructure in the same way they do with code.

In this chapter, we will explore IaC solutions, with a focus on Terraform. **Terraform** is an open source IaC tool that enables developers to define, manage, and provision infrastructure across multiple cloud providers and on-premises data centers. HashiCorp, the owner of Terraform and many other automation tools, changed the license from **Mozilla Public License** (**MPL**) version 2.0 to **Business Source License** (**BSL**) version 1.1. BSL allows you to use Terraform freely and you can access its source code, so there is no change for the end users.

With Terraform, developers can write code to define their infrastructure requirements, and Terraform will handle the provisioning and configuration of the required resources. Terraform has gained popularity in recent years due to its simplicity, flexibility, and support for multiple cloud providers. In the following sections, we will discuss the key features and benefits of Terraform, as well as how to use it to provision infrastructure on popular cloud providers.

We will learn about the following topics in this chapter:

- What is IaC?
- IaC versus Configuration as Code
- IaC projects worth knowing
- Terraform
- HCL in depth
- Terraform examples with AWS

Technical requirements

For this chapter, you will need a system capable of running Terraform. Terraform is a single binary program written in the Go programming language. Its installation is straightforward and is explained on the HashiCorp Terraform project page (`https://developer.hashicorp.com/terraform/downloads`). HashiCorp is the company behind Terraform and other cloud management tools that have become de facto standards in the DevOps world. You will also need an AWS account. AWS provides a Free Tier of services for a limited time. We are using services that have free tiers at the time of writing this book. Before you run the examples, please consult the AWS Free Tier listing to avoid unnecessary costs.

What is IaC?

IaC is a software development practice that involves defining and managing infrastructure through code. In essence, it means that infrastructure is treated as if it were a piece of software, and is managed through the same processes and tools. IaC solutions enable developers to define, provision, and manage infrastructure using code, instead of manually configuring servers, networks, and storage devices. This approach to infrastructure management is highly automated, scalable, and efficient, allowing organizations to reduce deployment times and improve consistency and reliability.

IaC solutions come in different forms, including configuration management tools, provisioning tools, and cloud orchestration tools. Configuration management tools, such as **Ansible** and **Chef**, are used to manage the configuration of individual servers or groups of servers. Provisioning tools, such as Terraform and **CloudFormation**, are used to provision and configure infrastructure resources. Cloud orchestration tools, such as **Kubernetes** and **OpenShift**, are used to manage containerized applications and their associated infrastructure. Regardless of the specific tool used, IaC solutions offer several benefits, including repeatability and consistency.

IaC versus Configuration as Code

You might be wondering, didn't we just cover this in *Chapter 11*, when we spoke about Ansible? The answer is no, we didn't. There's a very distinctive difference between IaC and **Configuration as Code** (**CaC**). IaC tools are concerned with exactly that: infrastructure. This means networking, DNS names, routes, and servers (VM or physical) up to the installation of the operating system. CaC is concerned with what lives inside the operating system. People try to use one tool for everything, so you'll see modules for Ansible that can configure switches and routers, but the tool shines best where it is intended to be used. Nobody is going to die if you mix these two, but your life will become more difficult.

IaC projects worth knowing

Since the rise of the public cloud, especially AWS, the need for a repeatable and reliable way of setting up an infrastructure and configuring cloud services started to grow as well. Since then, a lot of tools have come to be and more of them are being developed. In this section, we will review the most popular and innovative tools out there.

AWS CloudFormation

AWS CloudFormation is a popular IaC tool offered by **Amazon Web Services** (**AWS**) to automate the provisioning of AWS resources. It was first released in 2011 and has since become a widely used tool for managing infrastructure in the cloud.

CloudFormation allows you to define the infrastructure in a declarative language, such as YAML or JSON, and then create, update, or delete stacks of resources based on those definitions. This allows for consistent and reproducible infrastructure deployments, as well as easy rollback and version control. It's not all sparkles and rainbows, though – at times, you might find yourself stuck in a rollback loop after making an untested change. For example, let's say you're changing your AWS Lambda environment version. Unfortunately, it failed as the version you're currently using is no longer supported. So, it is now stuck at rolling back and it's in an UPDATE_ROLLBACK_FAILED state. You will need to resolve the issue by hand as there is no automated way of doing that.

CloudFormation integrates with other AWS services, such as AWS **Identity and Access Management** (**IAM**), AWS **Elastic Load Balancing** (**ELB**), and AWS Auto Scaling, to enable the creation of complex architectures with ease.

Here's an example CloudFormation stack written in YAML that creates an EC2 instance called t4g. small in a public subnet in the default VPC:

```
AWSTemplateFormatVersion: '2010-09-09'
Parameters:
  KeyName:
    Type: AWS::EC2::KeyPair::KeyName
    Default: admin-key
  InstanceType:
    Type: String
    Default: t4g.small
  SSHCIDR:
    Type: String
    MinLength: 9
    MaxLength: 18
    Default: 0.0.0.0/0
    AllowedPattern: (\d{1,3})\.(\d{1,3})\.(\d{1,3})\.(\d{1,3})/
(\d{1,2})
  LatestAmiId:
    Type:  'AWS::SSM::Parameter::Value<AWS::EC2::Image::Id>'
    Default: '/aws/service/canonical/ubuntu/server/jammy/stable/
current/amd6/hvm/ebs-gp2/ami-id'
Resources:
```

```
  EC2Instance:
    Type: AWS::EC2::Instance
    Properties:
      InstanceType: !Ref 'InstanceType'
      SecurityGroups: [!Ref 'InstanceSecurityGroup']
      KeyName: !Ref 'KeyName'
      ImageId: !Ref 'LatestAmiId'
  InstanceSecurityGroup:
    Type: AWS::EC2::SecurityGroup
    Properties:
      GroupDescription: Enable SSH access
      SecurityGroupIngress:
      - IpProtocol: tcp
        FromPort: 22
        ToPort: 22
        CidrIp: !Ref 'SSHCIDR'
Outputs:
  InstanceId:
    Description: InstanceId of the newly created EC2 instance
    Value: !Ref 'EC2Instance'
  PublicDNS:
    Description: Public DNSName of the newly created EC2 instance
    Value: !GetAtt [EC2Instance, PublicDnsName]
  PublicIP:
    Description: Public IP address of the newly created EC2 instance
    Value: !GetAtt [EC2Instance, PublicIp]
```

In this stack, we're creating two resources: an EC2 instance and a security group attached to this instance. The CloudFormation stack can get four parameters:

- KeyName: The SSH key name that was already created inside the AWS EC2 service. This defaults to admin-key.

- InstanceType: The type of the instance we want to start with. It defaults to t4g.small.

- SSHCIDR: The **Classless Inter-Domain Routing (CIDR)** of IP address(es) that can connect to this machine via port 22. This defaults to 0.0.0.0/0. Here, we're validating the provided input against the regular expression and the length of the variable.

- LatestAmiId: The AMI ID to use as a base system to start the EC2 instance. This defaults to the latest AMI of Ubuntu Linux 22.04.

Next, there's a Resources section. Here, the EC2 instance is created using the AWS::EC2::Instance resource type and the security group is created with the AWS::EC2::SecurityGroup resource.

The final section is called `Outputs`; here, we can reveal IDs and other properties of the resources created. Here, we're exposing the instance ID, its public DNS name, and its public IP address.

It's possible to use these output values as input for another CloudFormation stack, which will make the YAML files of your CloudFormation code considerably smaller and easier to maintain.

AWS Cloud Development Kit

AWS **Cloud Development Kit (CDK)** is an open source software development framework that's used to define cloud infrastructure in code. With CDK, developers can write code in familiar programming languages such as TypeScript, Python, Java, C#, and JavaScript to create and manage cloud resources on AWS.

AWS CDK was first released in July 2018 as an open source project by AWS. It was designed to simplify the process of building and deploying cloud infrastructure by allowing developers to use their existing programming language skills and tools. With CDK, developers can define IaC and take advantage of the benefits of version control, automated testing, and **continuous integration/continuous deployment (CI/CD)** pipelines. Since its release, CDK has become a popular choice for building infrastructure on AWS, and it continues to receive updates and new features to this day.

Here's an example of some AWS CDK Python code that creates an EC2 instance:

```python
from aws_cdk import core
import aws_cdk.aws_ec2 as ec2

class MyStack(core.Stack):

    def __init__(self, scope: core.Construct, id: str, **kwargs) ->
None:
        super().__init__(scope, id, **kwargs)

        # VPC
        vpc = ec2.Vpc(self, "VPC",
            nat_gateways=0,
            subnet_configuration=[ec2.
SubnetConfiguration(name="public",subnet_type=ec2.SubnetType.PUBLIC)]
            )

        # Get AMI
        amzn_linux = ec2.MachineImage.latest_amazon_linux(
            generation=ec2.AmazonLinuxGeneration.AMAZON_LINUX_2,
            edition=ec2.AmazonLinuxEdition.STANDARD,
            virtualization=ec2.AmazonLinuxVirt.HVM,
            storage=ec2.AmazonLinuxStorage.GENERAL_PURPOSE
            )
```

```
# Create an EC2 instance
instance = ec2.Instance(self, "Instance",
    instance_type=ec2.InstanceType("t4g.small"),
    machine_image=amzn_linux,
    vpc = vpc
)
```

This code creates a new VPC and an EC2 instance with an instance type of `t4g.small` with Amazon Linux installed as the operating system. Note that you will need to have AWS CDK installed and configured to run this code.

Terraform

Terraform is a popular open source tool that's used for infrastructure automation, specifically for creating, managing, and provisioning cloud resources. It enables developers to define their IaC and automates the deployment of their infrastructure across multiple cloud platforms. With Terraform, users can write declarative configuration files in a simple and intuitive language that can be versioned, shared, and reused. This approach to infrastructure management ensures consistency and scalability and reduces the risk of manual errors. Terraform supports a wide range of cloud providers, including AWS, Azure, Google Cloud, and more, making it a popular choice for organizations with complex cloud infrastructure needs.

Terraform was created by HashiCorp, a company founded in 2012 by Mitchell Hashimoto and Armon Dadgar. The company is known for developing popular open source tools for infrastructure automation, including Vagrant, Consul, Nomad, and Vault. Terraform was first released in July 2014 and has since become one of the most widely adopted IaC tools in the industry. HashiCorp continues to maintain and develop Terraform, with regular releases and updates that address new cloud provider features, security vulnerabilities, and community feedback. The tool has a large and active community of contributors, which has helped to further enhance its functionality and support for new use cases.

Terraform is also a main topic for this chapter, so we will dig into the code and its internals in the *Terraform* section later.

Cloud Development Kit for Terraform

Cloud Development Kit for Terraform (CDKTF) is an open source software development framework for defining cloud infrastructure in code. It allows users to define infrastructure using familiar programming languages, such as TypeScript, JavaScript, Python, and C#. This provides developers with more flexibility and control over their infrastructure as they can leverage their existing programming skills and tools to define complex infrastructure. CDKTF was first released in 2019 as a collaboration between AWS and HashiCorp. Since then, it has gained popularity as a powerful tool for defining and deploying infrastructure using Terraform.

CDKTF supports a wide range of programming languages, making it easy for developers to use their preferred language. It uses constructs, which are reusable building blocks that represent AWS resources, to create infrastructure. Users can define constructs for each resource they want to create, which can be combined to form a more complex infrastructure. This allows users to define their infrastructure in a modular and reusable way, which simplifies the process of creating and maintaining infrastructure over time.

Here is the example code for creating an EC2 instance in AWS using CDKTF in Python:

```python
from constructs import Construct
from cdktf import App, TerraformStack
from imports.aws import AwsProvider, Instance, SecurityGroup


class MyStack(TerraformStack):
    def __init__(self, scope: Construct, ns: str):
        super().__init__(scope, ns)

        # Configure AWS provider
        aws_provider = AwsProvider(self, 'aws', region='us-east-1')

        # Create a security group
        security_group = SecurityGroup(self, 'web-server-sg',
            name='web-server-sg',
            ingress=[
                {
                    'from_port': 22,
                    'to_port': 22,
                    'protocol': 'tcp',
                    'cidr_blocks': ['0.0.0.0/0'],
                },
                {
                    'from_port': 80,
                    'to_port': 80,
                    'protocol': 'tcp',
                    'cidr_blocks': ['0.0.0.0/0'],
                },
            ],
        )

        # Create an EC2 instance
        Instance(self, 'web-server',
            ami='ami-0c55b159cbfafe1f0',
            instance_type='t4g.small',
            security_groups=[security_group.id],
```

```
            user_data="""
                #!/bin/bash
                echo "Hello, DevOps People!" > index.html
                nohup python -m SimpleHTTPServer 80 &
            """
        )

app = App()
MyStack(app, "my-stack")

app.synth()
```

The constructs, App, and `TerraformStack` classes are imported from the `cdktf` package, while the AWS resources are imported from the `imports.aws` module. The preceding code creates an EC2 instance with a security group and a basic user data script that starts a simple HTTP server. The resulting infrastructure can be deployed using the `cdktf deploy` command, which generates Terraform configuration files and executes the Terraform CLI.

You can read more about CDKTF at `https://developer.hashicorp.com/terraform/cdktf`.

Pulumi

Pulumi is an open source IaC tool that allows developers to build, deploy, and manage cloud infrastructure using familiar programming languages. Unlike traditional IaC tools that rely on declarative languages such as YAML or JSON, Pulumi uses real programming languages, such as Python, TypeScript, Go, and .NET, to define and manage infrastructure. This enables developers to leverage their existing skills and experience to create infrastructure using the same tools and processes they use for building applications. With Pulumi, developers can create, test, and deploy infrastructure changes just like they do with code changes – that is, by using version control and CI/CD tools.

The first release of Pulumi was in May 2018 and aimed to simplify the process of managing cloud infrastructure. Pulumi was founded by Joe Duffy, a former Microsoft engineer who previously worked on the development of the .NET runtime and compiler. Duffy saw an opportunity to use programming languages to manage infrastructure, providing a more flexible and powerful approach than traditional IaC tools. Since its release, Pulumi has gained popularity among developers, particularly those who work in cloud-native environments or who use multiple cloud providers.

Pulumi supports a variety of programming languages, including Python, TypeScript, Go, .NET, and Node.js. Pulumi also provides a rich set of libraries and tools for working with cloud resources, including support for popular cloud providers such as AWS, Azure, Google Cloud, and Kubernetes. Additionally, Pulumi provides integrations with popular CI/CD tools, such as Jenkins, CircleCI, and GitLab, enabling developers to easily incorporate infrastructure changes into their existing workflows.

Here's an example of Pulumi code in Python that creates an AWS EC2 instance:

```python
import pulumi
from pulumi_aws import ec2

# Create a new security group for the EC2 instance
web_server_sg = ec2.SecurityGroup('web-server-sg',
    ingress=[
        ec2.SecurityGroupIngressArgs(
            protocol='tcp',
            from_port=22,
            to_port=22,
            cidr_blocks=['0.0.0.0/0'],
        ),
    ],
)

# Create the EC2 instance
web_server = ec2.Instance('web-server',
    instance_type='t4g.small',
    ami='ami-06dd92ecc74fdfb36', # Ubuntu 22.04 LTS
    security_groups=[web_server_sg.name],
    tags={
        'Name': 'web-server',
        'Environment': 'production',
    },
)

# Export the instance public IP address
pulumi.export('public_ip', web_server.public_ip)
```

This code defines an AWS security group that allows inbound traffic on port 22 (SSH), and then creates an EC2 instance of the `t4g.small` instance type, using an Ubuntu 22.04 LTS AMI. The instance is associated with the security group we created earlier and is tagged with a name and environment. Finally, the public IP address of the instance is exported as a Pulumi stack output, which can be used by other resources in the stack or accessed by the user.

In this section, we introduced several IaC solutions: CDK, CDKTF, Terraform, and Pulumi. Some of them are targeted at one cloud provider, and some of them allow us to configure different cloud environments.

In the next section, we will get back to Terraform to explore what makes it tick and learn how to use IaC in practice. This will give us a foundation to quickly understand other solutions, including the CDKs we mentioned earlier.

Terraform

In this section, we are going to introduce Terraform, one of the most widely used IaC solutions in the wild.

Terraform is an IaC tool developed by HashiCorp. The rationale behind using it is similar to using Ansible to configure your systems: infrastructure configuration is kept in text files. They are not YAML, as with Ansible; instead, they are written in a special configuration language developed by HashiCorp: **HashiCorp Configuration Language** (**HCL**). Text files are easily versioned, which means that infrastructure changes can be stored in a version control system such as Git.

Actions performed by Terraform are more complicated than those you've seen in Ansible. A single HCL statement can mean setting up a whole bunch of virtual servers and routes between them. So, while Terraform is also declarative like Ansible, it is higher level than other tools. Also, contrary to Ansible, Terraform is state-aware. Ansible has a list of actions to perform and on each run, it checks which actions have already been applied. Terraform, on the other hand, records the last achieved state of the system and each time it is executed, it ensures the system will be as it is in the code.

To achieve this, Terraform creates and maintains a state file. It is a text file with the `.tfstate` extension that records the last known state of the infrastructure that the tool is aware of. The state file is versioned internally; Terraform maintains a special counter that allows it to know whether the file is the latest one. The state file is essential for Terraform to work properly. You should never corrupt or lose the state file. If you lose this file, Terraform will try to create things that already exist and maybe delete things it should not.

There are several ways to ensure the state file is safe. One of them is to store it in an object store (such as S3) that's been properly configured so that the state file cannot be deleted. For improved security, you can ensure that the file is versioned, which means that the storage will keep old copies of the file for eventual reuse.

There is one important thing to be aware of related to `.tfstate`: it will contain all information related to your infrastructure, as well as plaintext passwords, logins, access keys, and such. It is crucial to keep the file secret and exclude it from commits to version control systems (in Git, add it to the `.gitignore` file).

The code is developed in text files with the `.tf` extension. Contrary to Ansible, again, the order in which you place directives in the files is not important. Before execution, Terraform analyzes all `.tf` files in the current directory, creates a graph of dependencies between configuration elements, and orders them properly. It is a common occurrence that code is broken down into smaller `.tf` files that group related configuration directives. However, nothing is stopping you from keeping them all in one huge file, though it will quickly become too big for comfortable use.

Even though you are free to name your files as you wish, so long as their extension is .tf, there are best practices to be observed:

- main.tf: This is the main file where you develop your configuration code. It will contain the resources, modules, and other important information.

- variables.tf: This file will contain declarations of all the variables you wish to use in the main.tf file.

- outputs.tf: If the resources in your main.tf file produce any outputs, they are declared here.

- versions.tf: This file declares the required versions of the Terraform binary itself and providers. It is good practice to declare the lowest versions known to work properly with your code.

- providers.tf: If any of the providers need additional configuration, you would place them in this file.

- backend.tf: This file contains the configuration for where Terraform should store the state file. The state file is an integral part of IaC in Terraform. We will talk about this in more depth in the *Terraform state* subsection.

The heavy lifting in Ansible is done by Python programs called modules. In Terraform, it is done by providers. **Providers** are small Golang programs that consume the configuration plan prepared by Terraform, connect with services (the cloud, devices, and more) through said services' APIs, and execute the configuration. You can think of them as plugins. Providers deliver a set of resource types and, eventually, the data sources required to write configuration for the API that the provider connects to. The official explanation is that providers "*are a logical abstraction of an upstream API.*" Providers are usually published on Terraform Registry, which is a public repository of plugins maintained by HashiCorp. You can use separate registries, but providers published on Terraform Registry are known to be tested and trusted to work properly. Each provider published on this registry has extensive documentation with well-commented examples. Whenever you are using a new provider, you should visit the registry (https://registry.terraform.io/). An example of a provider is AWS. This provider exposes a very extensive number of resources that you would use to interact with AWS services to provision and configure them. Remember: the configuration is limited to infrastructure. You can couple Terraform (to provision VMs, for example) with Ansible (to install software and configure it within the VM) for a full experience.

Let's look at an example of an AWS provider from the Terraform Registry AWS provider documentation (https://registry.terraform.io/providers/hashicorp/aws/latest/docs):

```
terraform {
  required_providers {
    aws = {
      source = "hashicorp/aws" version = "~> 4.0"
    }
```

```
  }
}

# Configure the AWS Provider
provider "aws" {
  region = "us-east-1"
}

# Create a VPC
resource "aws_vpc" "example" {
  cidr_block = "10.0.0.0/16"
}
```

In the preceding snippet, we declare that we require the AWS provider that will be downloaded from Terraform Registry. Its version should be no lower than 4.0. Then, we configure the region that we are going to use (us-east-1). Finally, we create a virtual private network (**virtual private cloud (VPC)**) and declare an IP block for it.

The set of .tf files in a single directory is called a **module**. If you run Terraform commands in the directory where module files lie, it is called a **root module**.

Terraform modules

One of the key concepts in Terraform is the module. A Terraform module is a collection of resources and their dependencies that can be used to build a specific component of your infrastructure. Modules provide a way to organize your code and make it reusable across multiple projects. They can be shared with other users and teams, and even published to a public registry such as Terraform Registry.

When working with Terraform, it's important to ensure that you are using compatible versions of your infrastructure provider. Providers are responsible for managing the resources in your cloud environment, and different versions may have different features or behaviors. To avoid unexpected changes to your infrastructure, you can pin the version of a provider that you are using in your Terraform configuration. This can be done by specifying a version constraint in the provider block, using the Terraform version constraint syntax. When you run Terraform, it will download and use the specified version of the provider, ensuring that your infrastructure remains consistent and predictable.

Here's an example versions.tf file that pins the AWS provider to the latest version and requires at least version 1.0.0 of Terraform:

```
terraform {
  required_providers {
    aws = ">= 3.0.0"
  }
  required_version = ">= 1.0.0"
}
```

In this example, we are using the `required_providers` block to specify that we require at least version 3.0.0 of the AWS provider. By using the `>=` operator, we allow Terraform to use any version of the provider that is equal to or greater than 3.0.0, including the latest version.

Terraform will automatically download and use the latest version of the provider when we run `terraform init`. This command will also update or download other modules you might be using within the main one (or a root module). However, using a lot of modules that are using other modules is discouraged as it can cause dependency conflicts (for example, some old modules might require AWS provider version 1.23 when a root module requires version 3.0 or newer). We will get back to the CLI later in this chapter in the *Terraform CLI* subsection.

To refer to another module, you can use the `module` code block. Let's say we have a simple module in the `./module/aws_ec2` directory relative to the root module. The `aws_ec2` module expects to be fed with the `ami`, `subnet`, `vpc`, and `security_group` variables:

```
module "aws_ec2_instance" {
  source = "./modules/aws_ec2"

  ami               = "ami-06dd92ecc74fdfb36"
  subnet_id         = "subnet-12345678"
  vpc_id            = "vpc-12345678"
  security_group    = "sg-12345678"
}
```

If a module exposes some outputs (variables you can use as input for a resource or other module), you can refer to them with `module.NAME.OUTPUT_NAME`. In this case, we could expose the ID of the EC2 instance, so you could refer to it with the name `module.aws_ec2_instance.instance_id`.

There are several other ways to specify the source argument when referring to a module in Terraform than using a local path:

- The module can be stored and retrieved from a Git repository:

```
module "example" {
  source = "git::https://github.com/example-org/example-module.git"
}
```

  With a Git repository, you're also able to refer to a commit ID, branch, or tag:

```
module "example" {
  source = "git::https://github.com/example-org/example-module.git?ref=branch_name"
}
```

For the private repository, you will want to use SSH instead of HTTPS to clone it locally:

```
module "example" {
    source = "git::ssh://github.com/example-org/example-module.
    git?ref=branch_name"
}
```

- The module can be published and retrieved from Terraform Registry:

```
module "example" {
    source = "hashicorp/example-module/aws"
}
```

In this case, you can specify a module version with the version property, like so:

```
module "example" {
    source = "hashicorp/example-module/aws"
    version = "1.0.0"
}
```

- The module can be stored and retrieved from an S3 bucket:

```
module "example" {
    source = "s3::https://s3-eu-cental-1.amazonaws.com/example-
    bucket/example-module.zip"
}
```

You can find other possible sources in the official documentation: `https://developer.hashicorp.com/terraform/language/modules/sources`.

Terraform state

One of the essential concepts of Terraform is the state file. It is a JSON file that describes the current state of your infrastructure. This file is used to keep track of the resources that Terraform has created, updated, or deleted, and it also stores the configurations for each resource.

The purpose of the state file is to enable Terraform to manage your infrastructure consistently and reliably. By keeping track of the resources that Terraform has created or modified, the state file ensures that subsequent runs of Terraform will know the current state of your infrastructure and can make changes accordingly. Without the state file, Terraform would not know which resources are currently deployed and would be unable to make informed decisions about how to proceed with subsequent changes.

The state file is also used as a source of truth for Terraform's `plan` and `apply` operations. When you run `terraform plan` or `terraform apply`, Terraform will compare the current state of your infrastructure with the desired state, as defined in your Terraform code. The state file is used to determine what changes need to be made to bring your infrastructure into the desired state. Overall, the state file is a crucial component of Terraform's infrastructure management capabilities and enables Terraform to ensure the consistency and reliability of your infrastructure.

While the Terraform state file is a critical component of the tool, there are some downsides and challenges associated with using it.

The state file is a centralized file that is used to store information about your infrastructure. While this can be convenient, it can also create issues when working in a team setting, especially if multiple users are making changes to the same infrastructure concurrently. This can lead to conflicts and make it challenging to keep the state file up to date. It's mitigated by the use of the distributed locking mechanism. In AWS environments, it's simply a DynamoDB table with a lock entry with a state of 0 or 1.

Another downside of Terraform state is that the state file contains sensitive information about your infrastructure, such as passwords, keys, and IP addresses. Therefore, it's essential to protect the state file from unauthorized access. If the state file is compromised, an attacker could potentially gain access to your infrastructure or sensitive data. Inside AWS, the state file is usually kept in an S3 bucket, which is necessary with enabled encryption and blocked public access.

The state file can become large and unwieldy over time, especially if you're managing a large infrastructure with many resources. This can make it challenging to manage and maintain the state file, which can lead to errors and inconsistencies.

The next challenge we may encounter with state files is that Terraform state files are version-specific. This means that you must use the same version of Terraform to manage the state file as you did to create it. This can create issues when upgrading to a newer version of Terraform as you may need to migrate the state file to the new format.

Finally, Terraform's state file has some limitations, such as the inability to manage external resources or the difficulty of handling complex dependencies between resources. This can create challenges when working with certain types of infrastructure or when dealing with complex deployments.

Another feature that comes with the state file is enforcing the configuration of the resources Terraform manages. If someone makes a manual change, you will see it on the next `terraform plan` or `terraform apply` and those changes will be rolled back.

With all that in mind, Terraform is still one of the best solutions out there and most of these challenges can easily be mitigated when planning our infrastructure.

Here's an example `backend.tf` file that configures Terraform to use an S3 bucket named `state-files` for storing the state file and a DynamoDB table named `terraform` for state locking:

```
terraform {
  backend "s3" {
    bucket          = "state-files"
    key             = "terraform.tfstate"
    region          = "eu-central-1"
```

```
    dynamodb_table = "terraform"
  }
}
```

In this configuration, the backend block specifies that we want to use the s3 backend type, which is designed to store the state file in an S3 bucket. The bucket parameter specifies the name of the bucket where the state file should be stored, and the key parameter specifies the name of the state file within the bucket.

The region parameter specifies the AWS region where the bucket is located. You should set this to the region that makes the most sense for your use case.

Finally, the dynamodb_table parameter specifies the name of the DynamoDB table that will be used for state locking. This is an important feature of the S3 backend as it ensures that only one user at a time can make changes to the infrastructure.

Here's an example of the Terraform state file:

```
{
    "version": 3,
    "serial": 1,
    "lineage": "f763e45d-ba6f-9951-3498-cf5927bc35c7",
    "backend": {
        "type": "s3",
        "config": {
            "access_key": null,
            "acl": null,
            "assume_role_policy": null,
            "bucket": "terraform-states",
            "dynamodb_endpoint": null,
            "dynamodb_table": "terraform-state-lock",
            "encrypt": true,
            "endpoint": null,
            "external_id": null,
            "force_path_style": null,
            "iam_endpoint": null,
            "key": "staging/terraform.tfstate",
            "kms_key_id": null,
            "lock_table": null,
            "max_retries": null,
            "profile": null,
            "region": "eu-central-1",
            "role_arn": null,
            "secret_key": null,
            "session_name": null,
```

```
        "shared_credentials_file": null,
        "skip_credentials_validation": null,
        "skip_get_ec2_platforms": null,
        "skip_metadata_api_check": null,
        "skip_region_validation": null,
        "skip_requesting_account_id": null,
        "sse_customer_key": null,
        "sts_endpoint": null,
        "token": null,
        "workspace_key_prefix": null
      },
      "hash": 1619020936
    },
    "modules": [
      {
        "path": [
          "root"
        ],
        "outputs": {},
        "resources": {},
        "depends_on": []
      }
    ]
}
```

By using the S3 backend with DynamoDB state locking, you can ensure that your Terraform deployments are safe and consistent, even in a team environment where multiple users may be making changes to the same infrastructure concurrently.

In the next subsection, we will discuss how to use the Terraform CLI and interact with our infrastructure and state files.

Terraform CLI

The core of Terraform is its command-line tool called, aptly, `terraform`. We linked the guide for installation when introducing Terraform to you. While there are tools to automate the workflow that remove the necessity of using the CLI, the use of the tool is simple and there is a lot of useful knowledge to be gained from working with it. In this section, we are going to introduce the most common options and workflows for the `terraform` command.

Initializing the work environment

The very first subcommand of `terraform` you will use is `terraform init`. After you write your first portion of the `main.tf` file (if you follow the proposed module structure), you will run `terraform init` to download the required plugins and create some important directories and helper files.

Let's consider a portion of the first piece of code we used previously:

```
terraform {
  required_providers {
    aws = {
      source = "hashicorp/aws" version = "~> 4.0"
    }
  }
}
```

This code informs Terraform about the required plugin to be downloaded and its minimum version. Now, let's run the `terraform init` command:

```
admin@myhome:~$ terraform init

Initializing the backend...

Initializing provider plugins...
- Finding latest version of hashicorp/aws...
- Installing hashicorp/aws v4.58.0...
- Installed hashicorp/aws v4.58.0 (signed by HashiCorp)
Terraform has been successfully initialized!
```

We have shortened the output for brevity, but the most important part is there. You will see Terraform telling you what actions have been performed. The backend is the storage area for keeping the `.tfstate` file. If you don't specify a storage area, the `.tfstate` file will be saved into `terraform.tfstate` in the local directory. There's also a new subdirectory called `.terraform` where required plugins were installed. Finally, a `.terraform.lock.hcl` file exists, where Terraform records the versions of the providers that have been used so that you can keep them for compatibility reasons.

The `terraform init` command is a safe command. You can run it as many times as you wish; it won't break anything.

Planning changes

The next command to run would be `terraform fmt`. This command formats your `.tf` files according to the best practices in place. Using it increases the readability and maintainability of your code, making all source files adhere to the same formatting strategy in all Terraform projects you will see. Running `terraform fmt` on our example will yield the following output:

```
admin@myhome:~$ terraform fmt

| Error: Missing attribute separator
|
|   on main.tf line 4, in terraform:
|    3:      aws = {
|    4:          source = "hashicorp/aws" version = "~> 4.0"
|
| Expected a newline or comma to mark the beginning of the next
attribute.
```

You will notice that `fmt` caught a glaring error in my `main.tf` file. It is not only a readability issue; it may also introduce errors in code interpretation with some providers. I have put two attributes on one line. Editing it so that it looks like this fixes the issue:

```
terraform {
  required_providers {
    aws = {
      source = "hashicorp/aws"
      version = "~> 4.0"
    }
  }
}
```

With this change, `fmt` is satisfied and we can proceed to the next step.

The plan of action is built by using the `terraform plan` command. It takes the last known recorded state of your infrastructure (`terraform.tfstate`), compares it to the code in the directory, and prepares steps to make them match. Running `terraform plan` with our preceding example code yields the following output:

```
admin@myhome:~$ terraform plan

No changes. Your infrastructure matches the configuration.

Terraform has compared your real infrastructure against your
configuration and found no differences, so no changes are needed.
```

Since we have not introduced any code that would create any resource, Terraform informed us that no changes are planned.

This, however, is not very interesting. Therefore, we are going to show you something that will create some resources in AWS.

> **Note**
>
> Before you follow this example, please consult your liability for the Free Tier services in AWS. Running these examples can incur a cost and neither the authors of this guide nor the publisher can claim any responsibility for them if they do occur.

If you want to follow these examples, you will need to have an AWS account (which is free at the time of writing this book). Then, you will need to create a role and generate AWS_ACCESS_KEY_ID and AWS_SECRET_ACCESS_KEY. Doing so is outside the scope of this chapter.

We have edited the preceding example slightly. The required_providers block has been moved to providers.tf. We have also added another provider block there. The file looks like this:

```
terraform {
  required_providers {
    aws = {
      source  = "hashicorp/aws"
      version = "~> 4.0"
    }
  }
}

provider "aws" {
  region = "us-west-2"
}
```

Notice that the new block is configuring a new provider resource called aws. The name (aws) is actually up to us and could be anything. Remember to give meaningful names, ones that will help you understand the code later on. We have provided a bare minimum of configuration for this provider, which is specifying a region where our resource will be started.

We do the actual work in the newly created empty main.tf file:

```
resource "aws_instance" "vm_example" {
  ami           = "ami-830c94e3"
  instance_type = "t2.micro"

  tags = {
    Name = "DevOpsGuideTerraformExample"
  }
}
```

Here, we inform Terraform that we want to create a new resource of the aws_instance type. We are calling it vm_example. Next, we tell the tool to use a VM image (AMI) called ami-830c94e3. The instance's type (how much RAM it is going to have, how many CPU cores, how big the system drive is, and more) is t2.micro. Finally, we add a tag that can help us identify and search for this instance.

Let's call `terraform plan` and apply it:

```
admin@myhome:~$ terraform plan

Terraform used the selected providers to generate the following
execution plan. Resource actions are indicated with the following
symbols:
  + create

Terraform will perform the following actions:

  # aws_instance.vm_example will be created
  + resource "aws_instance" "vm_example" {
      + ami                                  = "ami-830c94e3"
[...]
      + tags                                 = {
          + "Name" = "DevOpsGuideTerraformExample"
        }
      + tags_all                             = {
          + "Name" = "DevOpsGuideTerraformExample"
        }
      + tenancy                              = (known after apply)
[...]
      + vpc_security_group_ids               = (known after apply)
    }

Plan: 1 to add, 0 to change, 0 to destroy.
```

Note

You didn't use the `-out` option to save this plan, so Terraform can't guarantee to take exactly these actions if you run `terraform apply` now.

We've cut out a lot of the output from the plan. However, you can see that it differs from the previous example. Terraform noticed that we don't have a VM with the specified parameters (remember, it compares to the `.tfstate` file). So, it is going to create one. We can always see the summary on the line starting with `Plan`. In the plan, all attributes starting with + (plus sign) will be created. Everything starting with - (minus sign) will be destroyed and everything with ~ (tilde) in front of it will be modified.

Be cautious with Terraform when changing attributes of already created resources. More often than not, it will treat it as something new, especially if you change names. This will result in destroying the VM with the old name and creating a new one with a new name. This is probably not what you want.

Applying changes

The plan is being implemented by calling `terraform apply`. This command will introduce changes to your environment if your `.tf` files differ from your `.tfstate` file. Also, if your actual running infrastructure differs from your `.tfstate` file, `terraform apply` will do its best to re-align the live infrastructure with the Terraform state file:

```
admin@myhome:~$ terraform apply


Terraform used the selected providers to generate the following
execution plan. Resource actions are indicated with the following
symbols:
  + create

Terraform will perform the following actions:

  # aws_instance.vm_example will be created
  + resource "aws_instance" "vm_example" {
      + ami                          = "ami-830c94e3"
[...]
      + subnet_id                    = (known after apply)
      + tags                         = {
          + "Name" = "DevOpsGuideTerraformExample"
        }
      + tags_all                     = {
          + "Name" = "DevOpsGuideTerraformExample"
        }
      + tenancy                      = (known after apply)
[...]
      + vpc_security_group_ids       = (known after apply)
    }

Plan: 1 to add, 0 to change, 0 to destroy.

Do you want to perform these actions?
  Terraform will perform the actions described above.
  Only 'yes' will be accepted to approve.

  Enter a value: yes
```

```
aws_instance.vm_example: Creating...
aws_instance.vm_example: Still creating... [10s elapsed]
aws_instance.vm_example: Still creating... [20s elapsed]
aws_instance.vm_example: Still creating... [30s elapsed]
aws_instance.vm_example: Still creating... [40s elapsed]
[...]
aws_instance.vm_example: Still creating... [1m20s elapsed]
aws_instance.vm_example: Creation complete after 1m29s [id=i-
0a8bee7070b7129e5]


Apply complete! Resources: 1 added, 0 changed, 0 destroyed.
```

Again, a lot of the output has been cut out for brevity.

The `terraform apply` command created a plan again. We can avoid this by recording `terraform plan` into a file and later feeding the file to the `apply` step.

The interesting part is the confirmation step, where Terraform asks you to type in `yes` before it proceeds. Then, it is going to print out a summary of performed actions every 10 seconds. After you've spent some time working with Terraform, you will usually guess whether the action was successful by the time it takes it to finish.

In the AWS console, in the **Instances** menu, we can observe that the VM has been created:

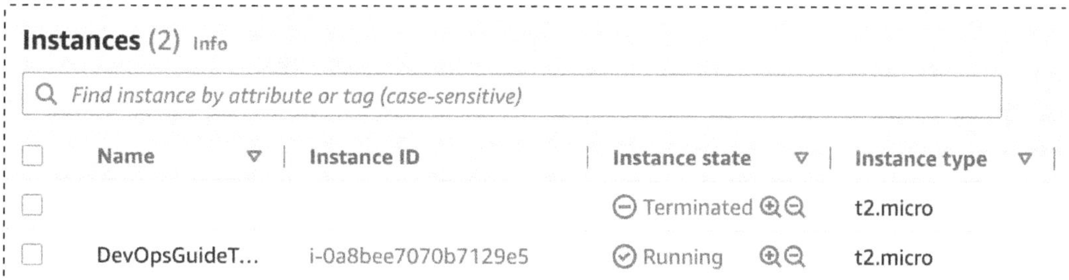

	Name			Instance ID		Instance state			Instance type	
☐		▽				⊖ Terminated ⊕⊖	▽		t2.micro	▽
☐	DevOpsGuideT...			i-0a8bee7070b7129e5		⊘ Running ⊕⊖			t2.micro	

Instances (2) Info

Q *Find instance by attribute or tag (case-sensitive)*

Figure 12.1 – New VM instances created via Terraform

We can delete all the infrastructure we have just created by running `terraform destroy`.

An interesting thing to note is that at no point in our workflow have we told Terraform what files it should interpret. This is because, as mentioned previously, it will read all `.tf` files in the current directory and create a proper plan of execution.

If you are interested in seeing the hierarchy of steps, Terraform provides the `terraform graph` command, which will print it out for you:

```
admin@myhome:~$ terraform graph
digraph {
        compound = "true"
        newrank = "true"
        subgraph "root" {
                "[root] aws_instance.vm_example (expand)" [label = "aws_
instance.vm_example", shape = "box"]
                "[root] provider[\"registry.terraform.io/hashicorp/aws\"]"
[label = "provider[\"registry.terraform.io/hashicorp/aws\"]", shape =
"diamond"]
                "[root] aws_instance.vm_example (expand)" -> "[root]
provider[\"registry.terraform.io/hashicorp/aws\"]"
                "[root] provider[\"registry.terraform.io/hashicorp/aws\"]
(close)" -> "[root] aws_instance.vm_example (expand)"
                "[root] root" -> "[root] provider[\"registry.terraform.io/
hashicorp/aws\"] (close)"
        }
}
```

There are tools out there that can create a nice visual representation of the graph produced by Terraform.

Modifying Terraform state

Occasionally, there's a need to modify the resources in the state file. In the older versions of Terraform, this had to be done manually and was error-prone. Thankfully, Terraform developers added some CLI commands that could help us with that.

The most useful commands are as follows:

- `Terraform state rm`: This command removes the resource from the state. This is useful when we've removed the resource manually and from the Terraform code but it persists in the state.

- `terraform state mv`: This command changes the name of the resource. This is useful when we change the resource name to prevent the removal and creation of a new resource, which often isn't the desired behavior

- `terraform taint`: This command forces the resource to be recreated.

Importing existing resources

Importing existing resources into Terraform allows you to include those resources in your Terraform state, which is a snapshot of the resources managed by Terraform.

The `terraform import` command is used to add an existing resource to your Terraform state. This command maps the existing resource to a configuration block in your Terraform code, which allows you to manage the resource using Terraform.

The syntax for the `terraform import` command is as follows:

```
terraform import [options] resource_in_code resource_identifier
```

The two important arguments for the `terraform import` command are as follows:

- `resource_in_code`: The address of the resource in your Terraform code
- `resource_identifier`: The unique identifier of the resource you want to import

For example, let's say you have an existing AWS S3 bucket with an **Amazon Resource Names (ARN)** of `arn:aws:s3:::devopsy-bucket`. To import this resource into your Terraform state, you can run the following command:

```
terraform import aws_s3_bucket.devopsy_bucket arn:aws:s3:::devopsy-bucket
```

Importing resources is useful when you have pre-existing infrastructure that you want to manage using Terraform. This is often the case when you are starting to use Terraform on an existing project or when you have resources that were created outside of Terraform. Importing resources allows you to bring those resources under Terraform management, so you can use Terraform to make changes to them in the future.

Not all resources can be imported into Terraform. The resource you intend to import must have a unique identifier that Terraform can use to locate it in the remote service. Additionally, the resource must be supported by the provider you are using in Terraform.

Workspaces

Terraform has a notion of a workspace. A **workspace** is similar to the version of the state. Workspaces allow you to store different states for the same code. To be able to use workspaces, you have to store your state file in a backend that supports them. The list of backends that support workspaces is quite long and covers the most popular cloud providers.

These workspaces are available in `.tf` files via the `${terraform.workspace}` sequence. Along with conditional expressions, this allows you to create varying environments. For example, you can use different IP addresses, depending on the workspace, allowing you to specify testing and production environments.

There is always one workspace present: `default`. It can't be removed. Workspace manipulation can be done using the `terraform workspace` command.

We can easily inspect what workspaces we have and which one is active using the `terraform` `list` command:

```
admin@myhome:~$ terraform workspace list
* default
```

The one with an asterisk in front of it is the current one. If we're only interested in seeing the current workspace and not the whole list, we can run the `terraform show` command:

```
admin@myhome:~$ terraform workspace show
default
```

Each workspace will have a state file. Let's experiment: we will create a new workspace called `testing` and apply Terraform to the testing workspace.

First, we must call `terraform workspace new` to create the workspace:

```
admin@myhome:~$ terraform workspace new testing
Created and switched to workspace "testing"!

You're now on a new, empty workspace. Workspaces isolate their state,
so if you run "terraform plan" Terraform will not see any existing
state
for this configuration.
```

Now, we must confirm that we are actually in the new workspace and run `terraform apply` in it using our previous example:

```
admin@myhome:~$ terraform apply

Terraform used the selected providers to generate the following
execution plan. Resource actions are indicated with the following
symbols:
  + create

Terraform will perform the following actions:

  # aws_instance.vm_example will be created
  + resource "aws_instance" "vm_example" {
      + ami                                 = "ami-830c94e3"
[...]
Plan: 1 to add, 0 to change, 0 to destroy.

Do you want to perform these actions?
  Terraform will perform the actions described above.
  Only 'yes' will be accepted to approve.
```

```
    Enter a value: yes

aws_instance.vm_example: Creating...
aws_instance.vm_example: Still creating... [10s elapsed]
aws_instance.vm_example: Still creating... [20s elapsed]
aws_instance.vm_example: Still creating... [30s elapsed]
aws_instance.vm_example: Still creating... [40s elapsed]
aws_instance.vm_example: Still creating... [50s elapsed]
aws_instance.vm_example: Creation complete after 57s [id=i-
06cf29fde369218e2]


Apply complete! Resources: 1 added, 0 changed, 0 destroyed.
```

As you can see, we successfully created the VM. However, when we change the workspace back to the default one, suddenly, Terraform will want to create it again:

```
admin@myhome:~$ terraform workspace switch default
Switched to workspace "default".
admin@myhome:~$ terraform apply


Terraform used the selected providers to generate the following
execution plan. Resource actions are indicated with the following
symbols:
  + create


Terraform will perform the following actions:

  # aws_instance.vm_example will be created
  + resource "aws_instance" "vm_example" {
[...]
```

Even though the creation of the resource was successful, Terraform will want to create it anew. When we inspect the directory where our .tf files live, we will spot the .tfstate file that is connected to the default workspace and a new directory called terraform.tfstate.d/ where .tfstate files are stored, each in its own subdirectory named aptly after the workspace. So, for the testing workspace, the state file will be stored in terraform.tfstate.d/testing:

```
admin@myhome:~$ ll
total 40
-rw-r--r--  1 trochej  staff    159B Mar 21 13:11 main.tf
-rw-r--r--  1 trochej  staff    158B Mar 21 12:27 providers.tf
-rw-r--r--  1 trochej  staff    4.4K Mar 21 21:17 terraform.tfstate
-rw-r--r--  1 trochej  staff    180B Mar 21 21:15 terraform.tfstate.
backup
```

```
drwxr-xr-x  3 trochej  staff    96B Mar 21 21:07 terraform.tfstate.d
admin@myhome:~$ ll terraform.tfstate.d
total 0
drwxr-xr-x  3 trochej  staff    96B Mar 21 21:20 testing
admin@myhome:~$ ll terraform.tfstate.d/testing
total 8
-rw-r--r--  1 trochej  staff   180B Mar 21 21:18 terraform.tfstate
```

How can we use this to our advantage in Terraform code? As we have mentioned, there's a special sequence (let's call it a variable) that will be expanded to the name of the current workspace in our code:

```
resource "aws_instance" "vm_example" {
  ami           = "ami-830c94e3"
  instance_type = terraform.workspace == "default" ? "t2.micro" : "t2.
nano"

  tags = {
    Name = "DevOpsGuideTerraformExample"
  }
}
```

With this small change (if `terraform.workspace` is the default, then the instance will be `t2.micro`; otherwise, it will be `t2.nano`) we have introduced a conditional change related to the workspace we start the VM in.

Let's quickly confirm this with `terraform plan`:

```
admin@myhome:~$ terraform workspace show
default
admin@myhome:~$ terraform plan | grep instance_type
    + instance_type                = "t2.micro"
admin@myhome:~$ terraform workspace select testing
Switched to workspace "testing".
admin@myhome:~$ terraform plan | grep instance_type
    + instance_type                = "t2.nano"
```

As shown in the preceding output, depending on the workspace we select, different types of an instance will be created.

In this section, we dived deep into the Terraform IaC tool. We explained what providers and modules are, as well as the role of the state file. We also demonstrated simple Terraform configuration and its interaction with the AWS cloud.

In the next section, we are going to cover the HashiCorp Configuration Language (HCL) in more detail, which was created specifically to write those configurations.

HCL in depth

HCL is a configuration language that's used by several HashiCorp tools, including Terraform, to define and manage IaC.

HCL is designed to be easy to read and write for both humans and machines. It uses a simple syntax that is similar to JSON but with a more relaxed structure and support for comments. HCL files typically have an `.hcl` or `.tf` file extension.

HCL uses curly braces to define blocks of code, and each block has a label that identifies its type. Within each block, we define attributes using a `key-value` syntax, where the key is the attribute name and the value is the attribute value. We can also define objects using curly braces, as shown in the example with the `tags` object.

Variables

In HCL, variables are defined using the `variable` block. Here's an example of how to define a variable in HCL:

```
variable "region" {
  type = string
  default = "eu-central-1"
}
```

In this example, we define a variable named `region` of the `string` type and a default value of `us-west-2`. We can reference this variable later in our code using the `${var.region}` syntax.

HCL supports several data types for variables, including `string`, `number`, `boolean`, `list`, `map`, and `object`. We can also specify a description for the variable using the `description` argument in the `variable` block.

Variables can be assigned values in a variety of ways, including through default values, command-line arguments, or environment variables. When using Terraform, we can also define variables in a separate file and pass them in during execution using a `.tfvars` file extension (for example, `variables.tfvars`) or through command-line arguments.

Once variables have been defined, they can't be changed, but HCL also allows local variables that can be defined within the `locals` block. Local variables are useful for simplifying complex expressions or calculations within a module or resource block as they allow us to break down the logic into smaller, more manageable pieces. They can also make it easier to maintain our code as we can define local variables for values that may change frequently or need to be updated across multiple resources or modules.

Here's an example `locals` block that defines the **availability zones (AZ)** in the `eu-central-1` region and generates subnets in every AZ:

```
locals {
  azs          = ["eu-central-1a", "eu-central-1b", "eu-central-1c"]
  cidr_block   = "10.0.0.0/16"
  subnet_bits  = 8
  subnets      = {
    for idx, az in local.azs : az => {
      name          = "${var.environment}-subnet-${idx}"
      cidr_block = cidrsubnet(local.cidr_block, local.subnet_bits,
idx)
      availability_zone = az
    }
  }
}
```

In this example, we define a `locals` block that includes the following variables:

- `azs`: A list of the availability zones in the `eu-central-1` region
- `cidr_block`: The CIDR block for the VPC
- `subnet_bits`: The number of bits to allocate for subnets within the CIDR block
- `subnets`: A map that generates subnets for each availability zone in the `azs` list using a `for` expression

The `for` expression in the `subnets` map generates a subnet for each availability zone in the `azs` list. The subnet's name includes the environment variable (which can be passed in as a variable) and the index of the availability zone in the list. The `cidrsubnet` function is used to calculate the CIDR block for each subnet based on the `cidr_block` variable and the `subnet_bits` variable.

The resulting `subnets` map will contain a key-value pair for each availability zone in the `azs` list, where the key is the availability zone name and the value is a map that includes the subnet name, CIDR block, and availability zone.

Comments

Comments in HCL can be written in two ways: single-line and multi-line comments. Single-line comments begin with the # symbol and continue until the end of the line. Multi-line comments, on the other hand, start with /* and end with */. Multi-line comments can span multiple lines and are often used to provide longer explanations or temporarily disable sections of code.

The following is an example of a single-line comment:

```
# This is a single-line comment in HCL
```

The following is an example of a multi-line comment:

```
/*
This is a multi-line comment in HCL
It can span multiple lines and is often used
to provide longer explanations or to temporarily disable sections of
code.
*/
```

Terraform meta-arguments

In Terraform, **meta-arguments** are special arguments that can be used to modify the behavior of resource blocks. They are called meta-arguments because they apply to the resource block as a whole, rather than to specific attributes within the block.

Meta-arguments are used to configure things such as the number of instances of a resource to create (count), the names of resources (name), the dependencies between resources (depends_on), and more.

count

count allows you to create multiple instances of a resource based on a numeric value. This can be useful for creating multiple instances of a resource, such as EC2 instances in AWS, without having to repeat the entire block of code.

For example, let's say you want to create three EC2 instances in your AWS account. Instead of creating three separate aws_ec2_instance resource blocks, you can use the count meta-argument to create multiple instances of the same block. Here's an example:

```
resource "aws_ec2_instance" "example" {
  ami           = "ami-0c55b159cbfafe1f0"
  instance_type = "t2.micro"
  count         = 3
}
```

In this example, we're creating three EC2 instances using the same ami and instance_type. The count meta-argument is set to 3, which means that Terraform will create 3 instances of the aws_ec2_instance resource block. Each instance will be given a unique identifier, such as aws_ec2_instance.example[0], aws_ec2_instance.example[1], and aws_ec2_instance.example[2].

for_each

The for_each meta-argument is similar to the count meta-argument in that it allows you to create multiple instances of a resource. However, for_each is more flexible than count because it allows you to create instances based on a map or set of values, rather than just a numeric value.

For example, let's say you have a map of AWS security groups that you want to create in your Terraform code. Instead of creating multiple aws_security_group resource blocks, you can use for_each to create them all in a single block. Here's an example:

```
variable "security_groups" {
  type = map(object({
    name        = string
    description = string
    ingress     = list(object({
      from_port   = number
      to_port     = number
      protocol    = string
      cidr_blocks = list(string)
    }))
  }))
}

resource "aws_security_group" "example" {
  for_each = var.security_groups

  name_prefix = each.value.name
  description = each.value.description

  ingress {
    from_port   = each.value.ingress[0].from_port
    to_port     = each.value.ingress[0].to_port
    protocol    = each.value.ingress[0].protocol
    cidr_blocks = each.value.ingress[0].cidr_blocks
  }
}
```

In this example, we're using the for_each meta-argument to create multiple instances of the aws_security_group resource block based on the security_groups variable, which is a map of objects. Each instance will have a unique identifier based on the key of the map. We're also using the name_prefix attribute to set the name of each security group, and the description attribute to set the description. Finally, we're using the ingress block to define the inbound traffic rules for each security group.

Using for_each can simplify your Terraform code and make it more reusable, especially when you're dealing with maps or sets of values. However, it's important to be aware of any potential dependencies between instances and to make sure that your code is properly structured to handle multiple instances.

lifecycle

The lifecycle meta-argument is used to define custom behavior for creating, updating, and deleting resources. It allows you to control the life cycle of a resource and its dependencies in a more fine-grained manner than the default behavior.

The lifecycle meta-argument can be used to define the following attributes:

- create_before_destroy: If set to true, Terraform will create the new resource before destroying the old one, which can prevent downtime in some cases.

- prevent_destroy: If set to true, Terraform will prevent the resource from being destroyed. This can be useful for protecting critical resources from accidental deletion.

- ignore_changes: A list of attribute names that Terraform should ignore when determining whether a resource needs to be updated.

- replace_triggered_by: A list of dependencies that will cause the resource to be recreated.

Here's an example of using the lifecycle meta-argument to prevent the destruction of an S3 bucket:

```
resource "aws_s3_bucket" "example" {
  bucket = "example-bucket"
  acl    = "private"

  lifecycle {
    prevent_destroy = true
  }
}
```

In this example, the lifecycle block is used to set the prevent_destroy attribute to true, which means that Terraform will prevent the aws_s3_bucket resource from being destroyed. This can be useful for protecting critical resources from being accidentally deleted.

depends_on

The depends_on meta-argument is used to define dependencies between resources. It allows you to specify that one resource depends on another resource, which means that Terraform will create the dependent resource after the resource it depends on has been created.

However, it's important to note that in most cases, Terraform can create a dependency tree automatically by analyzing your resource configurations. This means that using depends_on should be avoided unless absolutely necessary as it can lead to dependency cycles that can cause errors and make your Terraform code harder to manage.

If you do need to use depends_on, it's important to be aware of the potential for dependency cycles and to structure your code in a way that avoids them. This might involve splitting your resources into smaller modules or using other techniques to reduce complexity and avoid circular dependencies.

Here's an example of using depends_on to specify a dependency between an EC2 instance and a security group:

```
resource "aws_security_group" "example" {
  name_prefix = "example"
  ingress {
    from_port = 22
    to_port   = 22
    protocol  = "tcp"
    cidr_blocks = ["0.0.0.0/0"]
  }
}

resource "aws_instance" "example" {
  ami             = "ami-0c55b159cbfafe1f0"
  instance_type = "t2.micro"
  depends_on = [aws_security_group.example]
}
```

In this example, we're using depends_on to specify that the aws_instance resource depends on the aws_security_group resource. This means that Terraform will create the security group before creating the instance.

You can find out more about the HCL language by reading the official documentation: https://developer.hashicorp.com/terraform/language.

Terraform examples with AWS

In this section, we will create two sample modules to demonstrate how you would go about creating one and what you will need to consider when choosing the way it is supposed to create resources. The module we are going to create will be able to create one or more EC2 instances, a security group attached to it, and other needed resources, such as an instance profile. It will do almost everything we went through in *Chapter 10*, but with the use of the AWS CLI.

EC2 instance module

Let's create a module that will be able to create EC2 instances. Consider the following directory structure:

```
├── aws
│       └── eu-central-1
└── modules
```

The `modules` directory is where we will put all our modules, `aws` is where we will keep our AWS infrastructure, and `eu-central-1` is the code of the infrastructure for the Frankfurt AWS region. So, let's go ahead and start with creating the EC2 module. Let's create a directory to hold it and the basic files we will need, as we described earlier:

```
admin@myhome:~$ cd modules
admin@myhome:~/modules$ mkdir aws_ec2
admin@myhome:~/modules$ cd aws_ec2
admin@myhome:~/modules/aws_ec2$ touch versions.tf main.tf variables.tf
outputs.tf providers.tf
admin@myhome:~/modules/aws_ec2$ ls -l
total 0
-rw-r--r--  1 admin   admin   0 Mar 16 13:02 main.tf
-rw-r--r--  1 admin   admin   0 Mar 16 13:02 outputs.tf
-rw-r--r--  1 admin   admin   0 Mar 16 13:02 providers.tf
-rw-r--r--  1 admin   admin   0 Mar 16 13:02 variables.tf
-rw-r--r--  1 admin   admin   0 Mar 16 13:02 versions.tf
admin@myhome:~/modules/aws_ec2$
```

Notice that we didn't create the backend configuration file. This is because the backend will be configured in the root module instead. Modules don't have state files as the resources created by the module will be using the state file from the root (or in other words, main) module. Let's start configuring providers. In this case, we will only need the AWS provider at this moment. In our example, we will use the `eu-central-1` region:

```
provider "aws" {
   region = "eu-central-1"
}
```

Next, let's configure the versions of Terraform and the AWS provider we will use in the `versions.tf` file:

```
terraform {
  required_version = ">= 1.0.0"
  required_providers {
    aws = {
      source  = "hashicorp/aws"
```

```
        version = ">= 3.0.0"
    }
  }
}
```

In this example, the `required_version` attribute is set to `>= 1.0.0` to require Terraform version 1.0.0 or greater. The `required_providers` attribute is used to specify the AWS provider, with the `source` attribute set to `hashicorp/aws` and the `version` attribute set to `>= 3.0.0` to require the latest version of the AWS provider.

Now, we can do more interesting things, such as adding an actual `aws_instance` resource. For that, we will start filling the variables this resource needs:

```
resource "aws_instance" "test_instance" {
  ami           = "ami-1234567890"
  instance_type = "t3.micro"

  tags = {
    Name = "TestInstance"
  }
}
```

After saving all changes to the module files, we can go back to the `aws/eu-central-1` directory and create a similar set of files as in the module:

```
admin@myhome:~/modules/aws_ec2$ cd ../../aws/eu-central-1
admin@myhome:~/aws/eu-central-1$ touch versions.tf main.tf variables.
tf providers.tf
admin@myhome:~/aws/eu-central-1$ ls -l
total 0
-rw-r--r--  1 admin  admin  0 Mar 16 13:02 main.tf
-rw-r--r--  1 admin  admin  0 Mar 16 13:02 providers.tf
-rw-r--r--  1 admin  admin  0 Mar 16 13:02 variables.tf
-rw-r--r--  1 admin  admin  0 Mar 16 13:02 versions.tf
admin@myhome:~/aws/eu-central-1$
```

This time, we will only need `main.ft`, `providers.tf`, `variables.tf`, and `versions.tf`. To simplify things, we can just copy the contents of the providers and the versions files:

```
admin@myhome:~/aws/eu-central-1$ cp ../../modules/aws_ec2/providers.
tf .
admin@myhome:~/aws/eu-central-1$ cp ../../modules/aws_ec2/versions.tf
.
```

Now, we can focus on the main.tf file, where we're going to try and use our first version of the module. The main.tf file will look like this:

```
module "test_instance" {
  source = "../../modules/aws_ec2"
}
```

The module we've created doesn't require any variables, so this is all we need in this file.

Since this is our root module, we will also need to configure the location of the Terraform state file. For simplicity, we will use a local state file, but in real life, we recommend using an S3 bucket with a distributed lock configured. If there is no backend block, Terraform will create a local file. We're ready to test our module (the output has been shortened for brevity):

```
admin@myhome:~/aws/eu-central-1$ terraform init
Initializing modules...
- test_instance in ../../modules/aws_ec2

Initializing the backend...

Terraform has been successfully initialized!

You may now begin working with Terraform. Try running "terraform plan"
to see
any changes that are required for your infrastructure. All Terraform
commands
should now work.

If you ever set or change modules or backend configuration for
Terraform,
rerun this command to reinitialize your working directory. If you
forget, other
commands will detect it and remind you to do so if necessary.
```

Once you've run terraform init (you only have to rerun it if you update modules or backend configuration), you can execute terraform plan to see what changes have to be applied:

```
admin@myhome:~/aws/eu-central-1$ terraform plan
Terraform used the selected providers to generate the following
execution plan. Resource actions are indicated with the following
symbols:
  + create

Terraform will perform the following actions:

  # module.test_instance.aws_instance.test_instance will be created
```

```
+ resource "aws_instance" "test_instance" {
    + ami                                    = "ami-1234567890"
# Some of the output removed for readability
Plan: 1 to add, 0 to change, 0 to destroy.
```

> **Note**
>
> You didn't use the `-out` option to save this plan, so Terraform can't guarantee it will take exactly these actions if you run `terraform apply` now.

With this plan, Terraform confirmed that our module will create an EC2 instance for us. Unfortunately, this plan is not ideal as it doesn't check whether the AMI actually exists, or whether the subnet is there. Those errors will come up when we run `terraform apply`. The AMI we've provided, for instance, is bogus, so Terraform will fail in creating the instance. Let's get back to the module and improve it by automatically getting a correct Ubuntu Linux AMI. For this, the Terraform AWS provider provides a data resource. This special resource enables us to *ask* AWS for various resources through their API. Let's add an AMI data resource to the `modules` directory's `main.tf` file:

```
data "aws_ami" "ubuntu" {
  most_recent = true
  owners = ["099720109477"] # Canonical
  filter {
    name    = "name"
    values = ["ubuntu/images/hvm-ssd/ubuntu-jammy-22.04-amd64-
server-*"]
  }
}


resource "aws_instance" "test_instance" {
  ami             = data.aws_ami.ubuntu.id
  instance_type = "t3.micro"

  tags = {
    Name = "TestInstance"
  }
}
```

The (`aws_ami`) block of code uses the `aws_ami` data source to fetch the latest Ubuntu AMI owned by Canonical from the AWS Marketplace. It does this by setting the `most_recent` parameter to `true` and filtering the results using the name attribute of the AMI. It looks for an AMI with a specific name pattern: `ubuntu/images/hvm-ssd/ubuntu-jammy-22.04-amd64-server-*`.

The second block of code creates an AWS EC2 instance using the AMI that was fetched in the first block of code. It sets the instance type to `t3.micro`, which is a small instance type that's suitable for testing purposes.

It also adds a tag to the EC2 instance with a key called `Name` and a value called `TestInstance` so that it can easily be identified in the AWS Management Console.

You can read more about the `aws_ami` data resource in the documentation: `https://registry. terraform.io/providers/hashicorp/aws/latest/docs/data-sources/ami`.

After making this modification, we can run `terraform plan` and see whether something changes:

```
admin@myhome:~/aws/eu-central-1$ terraform plan
Terraform used the selected providers to generate the following
execution plan. Resource actions are indicated with the following
symbols:
  + create


Terraform will perform the following actions:

  # module.test_instance.aws_instance.test_instance will be created
  + resource "aws_instance" "test_instance" {
      + ami                              = "ami-050096f31d010b533"
  # Rest of the output removed for readability
```

The plan succeeded and it seems it found a recent AMI for Ubuntu Linux 22.04. There are a few other issues we'll need to consider if we want to make sure we'll be able to connect to this new EC2 instance. Of course, it will be created if we apply the changes, but we don't have a way to connect to it. First, let's connect the EC2 instance to the correct network: we will use a default VPC and a public subnet, which will allow us to connect to this instance directly.

To figure out the ID of the default VPC and a public subnet, once again, we will use a data resource:

- **VPC documentation**: `https://registry.terraform.io/providers/hashicorp/ aws/latest/docs/data-sources/vpc`
- **Subnet documentation**: `https://registry.terraform.io/providers/hashicorp/ aws/latest/docs/data-sources/subnets`

The question is whether we want to automatically put all created instances in a default VPC (and a public subnet) or not. Usually, the answer to this question is *no*. In that case, we will need to add some variables to be passed to this module.

Let's add another file to the root module, in which we will put all the data resources, called `data.tf`:

```
data "aws_vpc" "default" {
  filter {
    name   = "isDefault"
    values = ["true"]
  }
}

data "aws_subnets" "public" {
  filter {
    name   = "vpc-id"
    values = [data.aws_vpc.default.id]
  }

  filter {
    name   = "map-public-ip-on-launch"
    values = ["true"]
  }
}
```

Now, we can create an input variable in our module. Go back to the `modules/aws_ec2` directory and edit the `variables.tf` file:

```
variable "public_subnet_id" {
  description = "Subnet ID we will run our EC2 instance"
  type        = string
}
```

Now, when you run `terraform plan`, you will see the following error:

```
| Error: Missing required argument
|
|   on main.tf line 1, in module "test_instance":
|    1: module "test_instance" {
|
| The argument "public_subnet_id" is required, but no definition was
found.
```

Here, we've created a mandatory variable we will need to provide to the module. Let's do so by editing the `main.tf` file in our root module (the `aws/eu-central-1` directory):

```
module "test_instance" {
  source = "../../modules/aws_ec2"
```

```
    public_subnet_id = data.aws_subnets.public.ids[0]
}
```

Notice `data.aws_subnets.public.ids[0]`. We've used a list notation, where we're choosing the first element of the list (which is a string since the module expects it to be). This is because there are multiple subnets and `aws_subnets` returned a list of those subnets for us.

Running a plan again should give us one resource to be added. Great! Now, our instance will get a public IP address we can connect to. But we're still missing a firewall rule that will allow us to connect to port `22` (SSH). Let's create a **security group (SG)**.

Again, we could decide to create an SG in the root module, which would allow us to modify it without changing the EC2 module. Alternatively, we could add an SG inside the EC2 module, which would mean that the module would have full control over it, but it would lack some flexibility. It's also possible to create a module that would do both: attach an SG injected from the root module and, at the same time, use an SG predefined in the module, but this is outside the scope of this chapter. In this case, we're going to create an SG inside the module for simplicity's sake.

To create an SG, we will use the `aws_security_group` resource, which requires a VPC ID. There are two possibilities: we will need to introduce another variable to our EC2 module, or we will use another data resource to automatically get the VPC ID from the provided subnet. This time, a more elegant solution would be to use data resources. Let's add it to `main.tf` in our module:

```
data "aws_subnet" "current" {
  id = var.public_subnet_id
}
```

With that in place, we will be able to add an SG now:

```
resource "aws_security_group" "allow_ssh" {
  name        = "TestInstanceSG"
  description = "Allow SSH traffic"
  vpc_id      = data.aws_subnet.current.vpc_id

  ingress {
    description      = "SSH from the Internet"
    from_port        = 22
    to_port          = 22
    protocol         = "tcp"
    cidr_blocks      = ["0.0.0.0/0"]
  }

  egress {
    from_port        = 0
    to_port          = 0
```

```
    protocol        = "-1"
    cidr_blocks     = ["0.0.0.0/0"]
    ipv6_cidr_blocks = ["::/0"]
  }

  tags = {
    Name = "TestInstanceSG"
  }
}
```

The preceding code creates an AWS SG that allows SSH traffic to the EC2 instance we created earlier.

The aws_security_group resource is used to create an SG for the EC2 instance. It sets the name of the SG to TestInstanceSG and provides a brief description.

The vpc_id attribute is set to the VPC ID of the current subnet. It uses the aws_subnet data source called current to fetch the current subnet's VPC ID.

The ingress block defines the inbound rules for the security group. In this case, it allows SSH traffic from any IP address (0.0.0.0/0) by specifying from_port, to_port, protocol, and cidr_blocks.

The egress block defines the outbound rules for the security group. In this case, it allows all outbound traffic by specifying from_port, to_port, protocol, and cidr_blocks. It also allows all IPv6 traffic by specifying ipv6_cidr_blocks.

The tags attribute sets a tag for the SG with a key called Name and a value called TestInstanceSG so that it can be easily identified in the AWS Management Console.

Now, we're ready to attach this SG to our instance. We will need to use the security_groups option for the aws_instance resource:

```
resource "aws_instance" "test_instance" {
  ami           = data.aws_ami.ubuntu.id
  instance_type = "t3.micro"

  security_groups = [aws_security_group.allow_ssh.id]

  tags = {
    Name = "TestInstance"
  }
}
```

Now, after running `terraform plan`, you will see two resources to be added:

```
Plan: 2 to add, 0 to change, 0 to destroy.
```

At this point, we need to add our public SSH key to AWS and configure EC2 to use it for a default Ubuntu Linux user (`ubuntu`). Assuming you have already generated your SSH key, we will create a variable with said key, create a resource that will make the key available for the EC2 instance, and finally, add it to the instance configuration.

> **Important note**
> In some AWS regions, it's required to use an old RSA key format. Whenever available, we recommend the newest format according to up-to-date recommendations. At the time of writing this book, it's recommended to use the `ED25519` key.

Let's add a variable to the root module:

```
variable "ssh_key" {
  description = "SSH key attached to the instance"
  type = string
  default = "ssh-rsa AAASomeRSAKEY""
}
```

Let's add a similar one for the EC2 module:

```
variable "ssh_key" {
  description = "SSH key attached to the instance"
  type = string
}
```

This one is without the default value to make this variable required for every module in use. Now, let's add the key to AWS inside the EC2 module (`main.tf`):

```
resource "aws_key_pair" "deployer" {
  key_name   = "ssh_deployer_key"
  public_key = var.ssh_key
}
```

Then, we can use it in the `aws_instance` resource:

```
resource "aws_instance" "test_instance" {
  ami           = data.aws_ami.ubuntu.id
  instance_type = "t3.micro"

  security_groups = [aws_security_group.allow_ssh.id]
```

```
  key_name = aws_key_pair.deployer.key_name

  tags = {
    Name = "TestInstance"
  }
}
```

We will need to use the new variable inside the root module in `main.tf`:

```
module "test_instance" {
  source = "../../modules/aws_ec2"

  public_subnet_id = data.aws_subnets.public.ids[0]
  ssh_key          = var.ssh_key
}
```

Finally, running `terraform plan` will give us three resources:

```
Plan: 3 to add, 0 to change, 0 to destroy.
```

Great! Running `terraform apply` and accepting any changes will deploy the EC2 instance in a public subnet with our key. However, we still won't know the IP address for the instance unless we go to the AWS console and check it manually.

To get this information, we will need to export these variables from the EC2 module and then once again in the root module. For this, we have another code block called `output`. Its syntax is very similar to the `variable` syntax, but additionally, you can mark the `output` variable as sensitive so that it's not shown by default when running `terraform plan` or `terraform apply` commands.

Let's define outputs showing us the public IP address of the EC2 instance. In the EC2 module in the `outputs.tf` file, place the following code:

```
output "instance_public_ip" {
  value       = aws_instance.test_instance.public_ip
  description = "Public IP address of the EC2 instance"
}
```

In the root module, create the `outputs.tf` file and put the following code there:

```
output "instance_public_ip" {
  value       = module.test_instance.instance_public_ip
  description = "Public IP address of the instance"
}
```

Now, when you run `terraform plan`, you will see a change in the output:

```
Plan: 3 to add, 0 to change, 0 to destroy.

Changes to Outputs:
  + instance_public_ip = (known after apply)
```

This way, we've created a simple module by creating a single EC2 instance. If we run `terraform apply`, the instance will be created and the output will show us the IP address.

From here, the next steps would involve adding more functionality to the module, adding the ability to create more than one instance with the `count` meta-argument, or creating a set of different EC2 instances by using the `for_each` meta-argument.

Summary

In this chapter, we introduced the concept of IaC. We explained why it is an important method of managing and developing your infrastructure. We also introduced some tools that are quite popular in this way of working. As a tool of choice, we explained Terraform – probably the most widely used one.

In the next chapter, we are going to show you how you can leverage some online tools and automation to build pipelines for **CI** and **CD**.

Exercises

Try out the following exercises to test what you've learned in this chapter:

1. Create a module that will create an S3 bucket with enabled server-side encryption.
2. Add an instance profile to the module we've created using the same IAM policy that we used in *Chapter 10*.
3. Use the `count` meta-argument to create two instances.

13

CI/CD with Terraform, GitHub, and Atlantis

In this chapter, we are going to build on the previous chapters in this book by introducing pipelines for **continuous integration** (**CI**) and **continuous deployment** (**CD**). There are many CI and CD tools available for you, both open source and closed source, as well as self-hosted and **Software-as-a-Service** (**SaaS**). We are going to demonstrate an example pipeline, starting from committing source to the repository where we store **Terraform** code to applying changes in your infrastructure. We will do this automatically but with reviews from your team.

In this chapter, we are going to cover the following topics:

- What is CI/CD?
- Continuously integrating and deploying your infrastructure
- CI/CD with Atlantis

Technical requirements

For this chapter, you will need the following:

- A Linux box
- A free account on GitHub or similar platform (GitLab or Bitbucket)
- The latest version of Terraform
- The **AWS CLI**
- Git

What is CI/CD?

CI/CD is a set of practices, tools, and processes that allow software development teams to automate the building, testing, and deployment of their applications, enabling them to release software more frequently and with greater confidence in its quality.

Continuous integration (**CI**) is a practice where developers regularly integrate their code changes into a repository, and each integration triggers an automated build and test process. This helps catch errors early and ensures that the application can be built and tested reliably.

For example, using **Docker**, a developer can set up a **CI pipeline** that automatically builds and tests their application whenever changes are pushed to the code repository. The pipeline can include steps to build a Docker image, run automated tests, and publish the image to a Docker registry.

Continuous delivery is the practice of getting software to be available for deployment after the successful integration process. For example, with a Docker image, delivery would be pushing the image to the Docker registry from where it could be picked up by the deployment pipeline.

Finally, **continuous deployment** (**CD**) is the practice of automatically deploying continuous delivery process code artifacts (Docker images, Java JAR files, ZIP archives, and so on) to production as soon as they are tested and validated. This eliminates the need for manual intervention in the deployment process.

Let's look at some common deployment strategies:

- **Rolling deployment**: This strategy involves deploying changes to a subset of servers at a time, gradually rolling out the changes to the entire infrastructure. This allows teams to monitor the changes and quickly roll back if any issues arise.

- **Blue-green deployment**: In this strategy, two identical production environments are set up, one active (blue) and the other inactive (green). Code changes are deployed to the inactive environment and tested before switching the traffic to the new environment. This allows for zero downtime deployment.

- **Canary deployment**: This strategy involves deploying changes to a small subset of users while keeping the majority of the users on the current version. This allows teams to monitor the changes and gather feedback before rolling out the changes to the entire user base.

- **Feature toggles/feature switches**: With this strategy, changes are deployed to production but hidden behind a feature toggle. This toggle is then gradually turned on for select users or environments, allowing teams to control the rollout of new features and gather feedback before making them available to everyone.

All of these strategies can be automated using **CD tools**, such as **Jenkins**, **CircleCI**, **GitHub** and **GitLab Actions**, **Travis CI**, and many others.

While we're talking about the deployment of applications, we need to at least mention **GitOps**. GitOps is a new approach to infrastructure and application deployment that uses **Git** as the single source of truth for declarative infrastructure and application specifications. The idea is to define the desired state of the infrastructure and applications in Git repositories and use a GitOps tool to automatically apply those changes to the target environment.

In GitOps, every change to the infrastructure or application is made via a Git commit, which triggers a pipeline that applies the changes to the target environment. This provides a complete audit trail of all changes and ensures that the infrastructure is always in the desired state.

Some of the tools that help with enabling GitOps include the following:

- **FluxCD**: This is a popular GitOps tool that can automate the deployment and scaling of applications and infrastructure, using Git as the single source of truth. It integrates with **Kubernetes**, **Helm**, and other tools to provide a complete GitOps workflow.

- **ArgoCD**: This is another popular GitOps tool that can deploy and manage applications and infrastructure, using Git as the source of truth. It provides a web-based UI and **CLI** to manage the GitOps pipeline and integrates with Kubernetes, Helm, and other tools.

- **Jenkins X**: This is a CI/CD platform that includes GitOps workflows for building, testing, and deploying applications to Kubernetes clusters. It uses GitOps to manage the entire pipeline, from source code to production deployment.

Now that we know what CI/CD is, we can look into various tools that could be used to build such pipelines. In the next section, we will provide you with some examples of a pipeline for cloning the latest version of your repository, building a Docker image, and running some tests.

An example of CI/CD pipelines

Let's look at a few examples of automation pipelines that will apply our Terraform changes to the different CD tools out there.

Jenkins

Jenkins is the most popular open source CI/CD tools out there. It transformed from the clicked-through configuration to the **CaaC** approach. Its go-to language for defining pipelines is **Groovy** and **Jenkinsfile**. The following is an example of Jenkinsfile that will run a plan, ask the user for the input, and run `apply` if the user approves the change:

```
pipeline {
  agent any
  environment {
    AWS_ACCESS_KEY_ID = credentials('aws-key-id')
    AWS_SECRET_ACCESS_KEY = credentials('aws-secret-key')
  }
```

The preceding code opens a new pipeline definition. It's defined to run in any available Jenkins agent. Next, we're setting up environment variables to be used within this pipeline. The environment variables text is being retrieved from the `aws-key-id` and `aws-secret-key` Jenkins credentials. These need to be defined before we can run this pipeline.

Next, we'll define that each step pipeline will run inside the `stages` block:

```
stages {
    stage('Checkout') {
        steps {
            checkout scm
        }
    }
```

First, we'll clone our Git repository; the step that does this is `checkout scm`. The URL will be configured in the Jenkins UI directly:

```
stage('TF Plan') {
    steps {
        dir('terraform') {
            sh 'terraform init'
            sh 'terraform plan -out terraform.tfplan'
        }
    }
}
```

Next, we'll run `terraform init` to initialize the Terraform environment. Here, we're running the plan and saving the output to the `terraform.tfplan` file, which we will run `apply` with in the final step:

```
stage('Approval') {
    steps {
        script {
            def userInput = input(id: 'confirm', message: 'Run
apply?', parameters: [ [$class: 'BooleanParameterDefinition',
defaultValue: false, description: 'Running apply', name: 'confirm'] ])
        }
    }
}
```

This step defines user input. We'll need to confirm this by running `apply` after we review the plan's output. We're defining the default as `false`. The pipeline will wait for your input on this step:

```
stage('TF Apply') {
    steps {
        dir('terraform') {
```

```
                    sh 'terraform apply -auto-approve -input=false
    terraform.tfplan'
                    sh 'rm -f terraform.tfplan'
                }
            }
        }
    }
}
```

Finally, if you've confirmed that the pipeline will run `apply` without asking for further user input (`-input=false option`) and if `apply` ran without any errors, it will remove the `terraform.tfplan` file that was created in the plan step.

GitHub Actions basics

It's possible to create a similar pipeline with **GitHub Actions** that corresponds to the preceding Jenkinsfile, but unfortunately, GitHub Actions does not support user input while an action is running. We could use the `workflow_dispatch` option, but it will ask the user for input before the Action runs (see the official documentation for a reference: `https://github.blog/changelog/2021-11-10-github-actions-input-types-for-manual-workflows/`). So, instead, let's create an action that will run `plan` and `apply`:

```
name: Terraform Apply

on:
  push:
    branches: [ main ]
```

The preceding code defines that the GitHub Action will only be triggered on changes to the main branch:

```
env:
  AWS_ACCESS_KEY_ID: ${{ secrets.AWS_ACCESS_KEY_ID }}
  AWS_SECRET_ACCESS_KEY: ${{ secrets.AWS_SECRET_ACCESS_KEY }}
```

Here, we're defining environment variables similarly to the Jenkins pipeline. AWS access and secret keys are coming from secrets stored in GitHub that we will need to add beforehand. This won't be required if our GitHub runner is running inside the AWS environment or we are using GitHub OpenID Connect. You can read about the latter by checking out the GitHub documentation: `https://docs.github.com/en/actions/deployment/security-hardening-your-deployments/configuring-openid-connect-in-amazon-web-services`.

Next, we can define the GitHub Action steps within the `jobs` block:

```
jobs:
  terraform_apply:
    runs-on: ubuntu-latest
```

Here, we're defining a job named `terraform_apply` that will run on the latest version of the Ubuntu runner that's available in Github Actions:

```
    steps:
    - name: Checkout code
      uses: actions/checkout@v2
```

This step checks out the code. We're using a predefined Action available in GitHub instead of creating a script that will run the Git command line. The exact code it will run is available at `https://github.com/actions/checkout`:

```
    - name: Setup Terraform
      uses: hashicorp/setup-terraform@v1
```

The `Setup Terraform` step will download Terraform for us. By default, it'll download the latest available version of Terraform, but we can pin an exact version if we need to. The code for the step is available at `https://github.com/hashicorp/setup-terraform`:

```
    - name: Terraform Plan
      working-directory: terraform/
      run: |
        terraform init
        terraform plan -out terraform.tfplan
```

In the `Terraform Plan` step, we're initializing Terraform and running the plan in the same way we did for the Jenkins pipeline:

```
    - name: Terraform Apply
      working-directory: terraform/
      run: |
        terraform apply -auto-approve -input=false terraform.tfplan
        rm -f terraform.tfplan
```

Finally, the `Terraform Apply` step is applying infrastructure changes from the previously saved Terraform plan file, `terraform.tfplan`, and removing the plan file.

If you want to create a more robust piece of code that will work in every CI/CD tool out there, it's possible to create a **Bash script** to do the heavy lifting. With Bash scripting, it's also much easier to embed some testing before even running the plan. Here's a sample Bash script that will run the Terraform plan and apply it for you:

```
#!/usr/bin/env bash
set -u
set -e
```

Here, we're setting Bash as a default shell for running this script. In the next few lines, we're modifying the default settings for the script so that it stops executing when it encounters any unbound variables or errors:

```
# Check if Terraform binary is in PATH
if command -v terraform &> /dev/null; then
  TERRAFORM_BIN="$(command -v terraform)"
else
  echo "Terraform not installed?"
  exit 1
fi
```

This code block checks whether Terraform is available in the system and saves the full path to it inside the TERRAFORM_BIN variable, which we will be using later:

```
# Init terraform backend
$TERRAFORM_BIN init -input=false
```

Initialize the Terraform environment before running the plan:

```
# Plan changes
echo "Running plan..."
$TERRAFORM_BIN plan -input=false -out=./terraform.tfplan
```

Run plan and save it to a file for later use:

```
echo "Running Terraform now"
if $TERRAFORM_BIN apply -input=false ./terraform.tfplan; then
    echo "Terraform finished successfully"
    RETCODE=0
else
    echo "Failed!"    fi
fi
```

The preceding code block executes Terraform apply and checks the return code of the command. It also displays appropriate information.

Other major CI/CD solutions use a very similar approach. The biggest difference is between Jenkins, corporate tools, and open source solutions where **YAML** configuration is the most common. In the next section, we will dig a bit deeper into each stage of the pipeline, focusing on the integration testing for Terraform and deploying changes to the infrastructure.

Continuously integrating and deploying your infrastructure

Testing application code is now a de facto standard, especially since the adoption of **test-driven development** (**TDD**). TDD is a software development process in which developers write automated tests before writing code.

These tests are designed to fail initially, and developers then write code to make them pass. The code is continuously refactored to ensure it is efficient and maintainable while passing all tests. This approach helps reduce the number of bugs and increase the reliability of the software.

Testing infrastructure is not as easy as that as it's hard to check whether Amazon **Elastic Compute Cloud** (**EC2**) will be successfully started without actually starting the instance. It's possible to mock **API** calls to AWS, but it won't guarantee that the actual API will return the same results as your testing code. With AWS, it would also mean that testing will be slow (we will need to wait for this EC2 instance to come up) and probably generate additional cloud costs.

There are multiple infrastructure testing tools, both integrated with Terraform and third-party software (which is also an open source software).

Integration testing

There are multiple basic tests we could run in our CI pipeline. We can detect drift between our actual code and what's running in the cloud, we can lint our code according to the recommended format, and we can test whether the code is to the letter of our compliance policies. We can also estimate AWS costs from the Terraform code. Harder and more time-consuming processes include unit testing and end-to-end testing of our code. Let's take a look at the available tools we could use in our pipeline.

Most basic tests we could be running from the start involve simply running `terraform validate` and `terraform fmt`. The former will check whether the syntax of the Terraform code is valid (meaning there are no typos in resources and/or variables, all required variables are present, and so on). The `fmt` check will update the formatting of the code according to the Terraform standards, which means that all excessive white spaces will be removed, or some white spaces may be added to align the = sign for readability. This may be sufficient for simpler infrastructure and we recommend adding those tests from the start as it's pretty straightforward to do. You can reuse parts of the code we provided earlier to bootstrap the process for your existing code.

Infrastructure costs

Infrastructure costs are not a functional, static, or unit test you could be running in your testing pipeline. Although it's very useful to monitor that aspect of your infrastructure, your manager will also be happy to know when the AWS budget is checking out with predictions.

Infracost.io is a cloud cost estimation tool that allows you to monitor and manage infrastructure costs by providing real-time cost analysis for cloud resources. With Infracost.io, you can estimate the cost of infrastructure changes and avoid any unexpected costs by providing cost feedback at every stage of the development cycle.

Integrating Infracost.io into GitHub Actions is a straightforward process that involves creating an Infracost.io account and generating an API key that will allow you to access the cost estimation data.

Next, you need to install the Infracost.io CLI on your local machine. The CLI is a command-line tool that allows you to calculate and compare infrastructure costs.

After installing the CLI, you can add an Infracost.io action to your GitHub workflow by creating a new action file – for example, `.github/workflows/infracost.yml`.

In the Infracost.io action file, you need to specify the Infracost.io API key and the path to your Terraform configuration files:

```
name: Infracost

on:
  push:
    branches:
      - main

jobs:
  infracost:
    runs-on: ubuntu-latest
    steps:
      - name: Checkout code
        uses: actions/checkout@v2
      - name: Run Infracost
        uses: infracost/infracost-action@v1
        with:
          api_key: ${{ secrets.INFRACOST_API_KEY }}
          terraform_dir: ./terraform
```

Finally, commit and push the changes to your GitHub repository. Whenever a new Terraform configuration file is pushed to the repository, the Infracost.io action will automatically calculate the cost estimation and provide feedback on the GitHub Actions page.

Infracost is free for open source projects, but it's also possible to create your own service for monitoring cloud costs. A GitHub repository for Infracost can be found at `https://github.com/infracost/infracost`.

By integrating it into your CI pipeline, you can proactively monitor and manage your infrastructure costs and make informed decisions on infrastructure changes before deploying it to your cloud account.

Drift testing

One of the challenges of managing infrastructure with Terraform is ensuring that the actual state of the infrastructure matches the desired state defined in the Terraform configuration files. This is where the concept of *drift* comes in.

Drift occurs when there is a difference between the desired state of the infrastructure and its actual state. For example, if a resource that was created using Terraform is manually modified outside of Terraform, the actual state of the infrastructure will differ from the desired state defined in Terraform configuration files. This can cause inconsistencies in the infrastructure and may lead to operational issues.

To detect drift in the infrastructure, Terraform provides a command called `terraform plan`. When this command is run, Terraform compares the desired state defined in the Terraform configuration files with the actual state of the infrastructure and generates a plan of the changes that need to be made to bring the infrastructure back into the desired state. If there are any differences between the desired and actual states, Terraform will show them in the plan output.

It's also possible to use third-party tools that will extend this feature of Terraform. One of these tools is `Driftctl`.

`Driftctl` is an open source tool that helps detect drift in cloud infrastructure managed by Terraform. It scans the actual state of the infrastructure resources and compares them against the desired state defined in Terraform configuration files to identify any discrepancies or differences. `Driftctl` supports a wide range of cloud providers, including AWS, **Google Cloud**, **Microsoft Azure**, and Kubernetes.

`Driftctl` can be used in various ways to detect drift in infrastructure. It can be integrated with a CI/CD pipeline to automatically detect drift and trigger corrective actions. It can also be run manually to check for drift on demand.

Here's an example GitHub pipeline that utilizes `Driftctl` to detect infrastructure drift:

```
name: Detect Infrastructure Drift

on:
  push:
    branches:
      - main
```

The preceding code indicates that this pipeline will run only on the `main` branch.

Here, we're defining a job called `detect-drift` that will run on the latest Ubuntu Linux runner available:

```
jobs:
  detect-drift:
    runs-on: ubuntu-latest
    steps:
    - name: Checkout code
      uses: actions/checkout@v2
```

Next, we're starting to define what we're going to do in each step of the pipeline – first, we will use a predefined action that will run `git clone` on the runner:

```
    - name: Install Terraform
      run: |
        sudo apt-get update
        sudo apt-get install -y unzip
        curl -fsSL https://apt.releases.hashicorp.com/gpg | sudo apt-
key add -
        sudo apt-add-repository "deb [arch=amd64] https://apt.
releases.hashicorp.com $(lsb_release -cs) main"
        sudo apt-get update && sudo apt-get install terraform
```

The next step we're defining is a shell script that will install, unzip, and download Terraform from the public repository published by HashiCorp:

```
    - name: Install Driftctl
      run: |
        curl https://github.com/cloudskiff/driftctl/releases/download/
v0.8.2/driftctl_0.8.2_linux_amd64.tar.gz -sSLo driftctl.tar.gz
        tar -xzf driftctl.tar.gz
        sudo mv driftctl /usr/local/bin/
```

In this step, we will install the `Driftctl` tool by downloading an archive from the public release on GitHub. We will extract it and move the binary to the `/usr/local/bin` directory:

```
    - name: Initialize Terraform
      run: terraform init

    - name: Check Terraform Configuration
      run: terraform validate
```

The preceding steps simply involve running `terraform init` and `terraform validate` to verify whether we can access the Terraform backend and whether the code we intend to check in the next few steps is valid from a syntax perspective:

```
    - name: Detect Drift with Driftctl
      run: |
          driftctl scan -from tfstate://./terraform.tfstate -output json
  > drift.json

      - name: Upload Drift Report to GitHub
        uses: actions/upload-artifact@v2
        with:
          name: drift-report
          path: drift.json
```

The final two steps are running the `Driftctl` tool and saving their findings inside the `driftctl.json` file, which is uploaded to GitHub artifacts with the name `drift-report`.

To summarize, this pipeline runs on the `main` branch and performs the following steps:

- Checks out the code from the GitHub repository.

- Installs `terraform` and `driftctl`.

- Initializes Terraform and validates the Terraform configuration files.

- Uses `driftctl` to scan the actual state of the infrastructure resources and compares them against the desired state defined in the Terraform configuration files to detect any drift. The output of this scan is saved to a JSON file called `drift.json`.

- Uploads the `drift.json` file as an artifact to GitHub, making it available for further analysis.

Additionally, this pipeline can be customized to meet specific requirements, such as integrating with a CI/CD pipeline or running on a schedule to regularly check for drift in the infrastructure.

The need for installing `driftctl` and `terraform` on every pipeline run is not desired, so we recommend that you prepare your own Docker image with preinstalled proper versions of those tools and use that instead. It will also increase your security along the way.

You can find more information about the project on the website: `https://driftctl.com/`.

Security testing

Testing infrastructure security is an essential aspect of maintaining secure and stable systems. As modern infrastructure is often defined and managed as code, it is necessary to test it just like any other code. **Infrastructure as Code** (IaC) testing helps identify security vulnerabilities, configuration issues, and other flaws that may lead to system compromise.

There are several automated tools available that can aid in infrastructure security testing. These tools can help identify potential security issues, such as misconfigured security groups, unused security rules, and unsecured sensitive data.

There are several tools we can use to test security as a separate process from the CI pipeline:

- **Prowler**: This is an open source tool that scans AWS infrastructure for security vulnerabilities. It can check for issues such as AWS **Identity and Access Management (IAM)** misconfigurations, open security groups, and **S3 bucket** permission issues.

- **CloudFormation Guard**: This is a tool that validates **AWS CloudFormation** templates against a set of predefined security rules. It can help identify issues such as open security groups, unused IAM policies, and non-encrypted S3 buckets.

- **OpenSCAP**: This is a tool that provides automated security compliance testing for Linux-based infrastructure. It can scan the system for compliance with various security standards, such as the **Payment Card Industry Data Security Standard (PCI DSS)** or **National Institute of Standards and Technology Special Publication 800-53 (NIST SP 800-53)**.

- **InSpec**: This is another open source testing framework that can be used for testing infrastructure compliance and security. It has built-in support for various platforms and can be used to test against different security standards, such as the **Health Insurance Portability and Accountability Act (HIPAA)** and **Center for Internet Security (CIS)**.

Here, we're focusing on integrating some security testing within the CI. The tools we could integrate here are as follows:

- **tfsec** is an open source static analysis tool for Terraform code. It scans Terraform configurations to detect potential security issues and provides suggestions for remediation. It has built-in support for various cloud providers and can help identify issues such as weak authentication, insecure network configurations, and unencrypted data storage. Its GitHub repository can be found at `https://github.com/aquasecurity/tfsec`.

- **Terrascan** is an open source tool for static code analysis of IaC files. It supports various IaC file formats, including Terraform, Kubernetes YAML, and Helm charts, and scans them for security vulnerabilities, compliance violations, and other issues. Terrascan can be integrated into a CI/CD pipeline and helps ensure that infrastructure deployments are secure and compliant with industry standards. Its GitHub repository can be found at `https://github.com/tenable/terrascan`.

- **CloudQuery** is an open source tool that enables users to test security policies and compliance across different cloud platforms, including AWS, **Google Cloud Platform (GCP)**, and Microsoft Azure. It provides a unified query language and interface to access cloud resources, allowing users to analyze configurations and detect potential security vulnerabilities. CloudQuery integrates with various CI/CD pipelines, making it easy to automate testing for security policies and compliance. Users can also customize queries and rules based on their specific needs and standards. You can read more about this topic in their blog post: https://www.cloudquery.io/how-to-guides/open-source-cspm.

Let's look at the example GitHub pipeline that's integrating the terrascan tool:

```
name: Terrascan Scan

on: [push]

jobs:
  scan:
    runs-on: ubuntu-latest

    steps:
    - name: Checkout code
      uses: actions/checkout@v2

    - name: Install Terrascan
      run: |
        wget https://github.com/accurics/terrascan/releases/latest/
download/terrascan_linux_amd64.zip
        unzip terrascan_linux_amd64.zip
        rm -f terrascan_linux_amd64.zip
        sudo mv terrascan /usr/local/bin/

    - name: Run Terrascan
      run: |
        terrascan scan -f ./path/to/infrastructure/code
```

In this workflow, the on: [push] line specifies that the workflow should be triggered whenever changes are pushed to the repository.

The jobs section contains a single job called scan. The runs-on key specifies that the job should run on an Ubuntu machine.

The steps section contains three steps:

1. It checks out the code from the repository using the actions/checkout action.

2. It downloads and installs Terrascan on the machine using the wget and unzip commands. Note that this step assumes that you're running the workflow on a Linux machine.

3. It runs Terrascan to scan the infrastructure code. You'll need to replace ./path/to/infrastructure/code with the actual path to your infrastructure code.

Once you've created this workflow and pushed it to your GitHub repository, GitHub Actions will automatically run the workflow whenever changes are pushed to the repository. You can view the results of the Terrascan scan in the workflow logs.

Let's move on to **infrastructure unit testing**. There are generally two options at the moment: **Terratest** when testing **HCL** code directly or **CDKTF/Pulumi** if you want to use more advanced programming languages to maintain your IaC.

Testing with Terratest

Terratest is an open source testing framework for infrastructure code, including HCL code for Terraform. It was first released in 2017 by **Gruntwork**, a company that specializes in infrastructure automation and offers a set of pre-built infrastructure modules called the Gruntwork IaC Library.

Terratest is designed to simplify testing infrastructure code by providing a suite of helper functions and libraries that allow users to write automated tests written in **Go**, a popular programming language for infrastructure automation. Terratest can be used to test not only Terraform code but also infrastructure built with other tools such as **Ansible**, **Packer**, and **Docker**.

One of the key benefits of Terratest is that it allows developers to test their infrastructure code in a production-like environment, without the need for a dedicated test environment. This can be achieved by using tools such as Docker and Terraform to spin up temporary infrastructure resources for testing purposes.

Terratest also provides a range of test types, including **unit tests**, **integration tests**, and **end-to-end tests**, allowing users to test their infrastructure code at different levels of abstraction. This helps ensure that code changes are thoroughly tested before they are deployed to production, reducing the risk of downtime or other issues.

An example of testing an `aws_instance` resource we created in a module in the previous chapter would look like this:

```
package test

import (
  "testing"

  "github.com/gruntwork-io/terratest/modules/aws"
  "github.com/gruntwork-io/terratest/modules/terraform"
  "github.com/stretchr/testify/assert"
)

func TestAwsInstance(t *testing.T) {
  terraformOptions := &terraform.Options{
    TerraformDir: "../path/to/terraform/module",
```

```
  }

  defer terraform.Destroy(t, terraformOptions)
  terraform.InitAndApply(t, terraformOptions)

  instanceID := terraform.Output(t, terraformOptions, "aws_instance_
id")
  instance := aws.GetEc2Instance(t, "us-west-2", instanceID)

  assert.Equal(t, "t3.micro", instance.InstanceType)
  assert.Equal(t, "TestInstance", instance.Tags["Name"])
}
```

In this example, we first define a test function called `TestAwsInstance` using the standard **Go** testing library. Inside the function, we create a `terraformOptions` object that specifies the directory of our Terraform module.

Then, we use the `terraform.InitAndApply` function to initialize and apply the Terraform configuration, creating the **AWS EC2** instance resource.

Next, we use the `aws.GetEc2Instance` function from the **Terratest AWS module** to retrieve information about the instance that was created, using its ID as input.

Finally, we use the `assert` library from the `testify` package to write assertions that validate the properties of the instance, such as its instance type and tags. If any of the assertions fail, the test will fail.

To run this example, you will need to ensure that you have installed the Terratest and AWS Go modules, and have valid AWS credentials set up in your environment.

Unit testing with CDKTF

AWS **Cloud Development Kit for Terraform** (**CDKTF**) is an open source framework for Terraform that allows developers to define the infrastructure and services inside any cloud solution Terraform supports using programming languages such as **TypeScript**, **Python**, and **C#**. It enables the creation of IaC using high-level object-oriented abstractions, reducing the complexity of writing and maintaining infrastructure code.

CDKTF was initially released in March 2020, and it is a collaboration between AWS and HashiCorp, the company behind Terraform. CDKTF leverages the best of both worlds: the familiarity and expressiveness of modern programming languages, and the declarative, multi-cloud capabilities of Terraform.

TypeScript is the most popular language used with CDKTF, and it provides a type-safe development experience with features such as static type-checking, code completion, and refactoring.

As an example, let's reuse the Terraform code from *Chapter 12*:

```
resource "aws_instance" "vm_example" {
  ami           = "ami-830c94e3"
  instance_type = "t2.micro"

  tags = {
    Name = "DevOpsGuideTerraformExample"
  }
}
```

The equivalent code in CDKTF in Python would look like this:

```
#!/usr/bin/env python
from constructs import Construct
from cdktf import App, TerraformStack
from cdktf_cdktf_provider_aws.provider import AwsProvider
from cdktf_cdktf_provider_aws.instance import Instance
from cdktf_cdktf_provider_aws.data_aws_ami import DataAwsAmi,
DataAwsAmiFilter
```

Here, we're importing all required modules. The first line of the script says that the default interpreter for it should be a Python interpreter that's available in the system:

```
class MyStack(TerraformStack):
    def __init__(self, scope: Construct, id: str):
        super().__init__(scope, id)

        AwsProvider(self, "AWS", region="eu-central-1")
```

In the preceding line, we're configuring a default provider to use the `eu-central-1` region. Let's see what's next:

```
        ec2_instance = Instance(
            self,
            id_="ec2instanceName",
            instance_type="t2.micro"
            ami="ami-830c94e3",
        )


app = App()
MyStack(app, "ec2instance")

app.synth()
```

The following is an example unit test for the preceding code that uses the `unittest` Python module and the usual syntax for TDD in Python:

```python
import unittest
from your_cdk_module import MyStack  # Replace 'your_cdk_module' with
the actual module name containing the MyStack class

class TestMyStack(unittest.TestCase):
    def test_ec2_instance(self):
        app = App()
        stack = MyStack(app, "test-stack")

        # Get the EC2 instance resource created in the stack
        ec2_instance = stack.node.try_find_child("ec2instance")

        # Assert EC2 instance properties
        self.assertEqual(ec2_instance.ami, "ubuntuAMI")
        self.assertEqual(ec2_instance.instance_type, "t3.micro")
        self.assertEqual(ec2_instance.key_name, "admin_key")
        self.assertEqual(ec2_instance.subnet_id, "subnet-1234567890")

if __name__ == '__main__':
    unittest.main()
```

As stated previously, unfortunately, the instance must be running in AWS for us to test whether we have the desired tags and other properties available through the CDKTF. The instance is being created by the `setUp()` function and terminated with the `tearDown()` function. Here, we're using a small instance that is **Free Tier eligible**, but with bigger instances, it will generate some costs.

Experimental Terraform testing module

The final but very interesting option is using the testing Terraform module. This module allows you to write Terraform (HCL) code tests, also in the same language. This will potentially make it much easier to write tests as the current options we've already gone through cover writing tests in Golang or by using CTKTF.

At the time of writing, the module is considered highly experimental, but it's worth keeping an eye on how it will develop in the future.

The module's website can be found at `https://developer.hashicorp.com/terraform/language/modules/testing-experiment`.

Other integration tools worth mentioning

There is a galaxy of other testing tools out there, and more are being developed as you read this. The following is a short list of tools worth mentioning that we didn't have enough space to describe properly:

- **Checkov** (`https://www.checkov.io/`): This is an open source IaC static analysis tool that helps developers and **DevOps** teams identify and fix security and compliance issues early in the development life cycle.

- **Super-linter** (`https://github.com/github/super-linter`): This is an open source code linting tool that can automatically detect and flag issues in various programming languages and help maintain consistent code quality.

- **Trivy** (`https://github.com/aquasecurity/trivy`): This is a vulnerability scanner for container images that helps developers and DevOps teams identify and fix vulnerabilities in their containerized applications.

- **Kitchen-Terraform** (`https://github.com/newcontext-oss/kitchen-terraform`): This tool is part of the Test Kitchen plugin collection and allows systems to utilize **Test Kitchen** to apply and validate Terraform configurations with **InSpec** controls.

- **RSpec-Terraform** (`https://github.com/bsnape/rspec-terraform`): This tool provides RSpec tests for Terraform modules. RSpec is a **behavior-driven development** (**BDD**) testing framework for **Ruby** that allows developers to write expressive and readable tests in a **domain-specific language** (**DSL**).

- **Terraform-Compliance** (`https://github.com/terraform-compliance/cli`): This is a BDD testing tool designed for Terraform files.

- **Clarity** (`https://github.com/xchapter7x/clarity`): This is a declarative Terraform test framework that specializes in unit testing.

In conclusion, we have a lot of testing options available. This could be overwhelming for some. Testing the Terraform code is still in development, similar to the other IaC solutions we've mentioned. When implementing CI pipelines, it's best to focus on the low-hanging fruits at the beginning (format checking, running `terraform plan`, and so on) and add more testing during development. We realize it's a hard thing to accomplish, but we believe that investing in this will enable us to make infrastructure changes with more confidence and protect us from unintentional outages.

In the next section, we will focus on various CD solutions, both SaaS and self-hosted.

Deployment

Continuous deployment in terms of Terraform comes down to running `terraform apply`. This can be done automatically with a simple Bash script, but the hard part is to integrate this into the CD tool to ensure we won't remove any data by mistake with confidence. Assuming we have done a great job with our integration testing and we're confident enough to do it without any further user interaction, we can enable it to run automatically.

Previously, we've shown examples of using Jenkins, GitHub Actions, and even a Bash script that we could embed into the process to automate it. We can successfully use these solutions to deploy our changes to the infrastructure; however, there are already dedicated solutions for that. Let's take a look at the most popular ones, starting with SaaS offerings from the company behind Terraform – HashiCorp.

HashiCorp Cloud and Terraform Cloud

HashiCorp is a software company that provides various tools for infrastructure automation and management. Two of the most popular products offered by HashiCorp are **HashiCorp Cloud** and **Terraform Cloud**.

HashiCorp Cloud is a cloud-based service that provides a suite of tools for infrastructure automation and management. It includes HashiCorp's popular tools such as Terraform, **Vault**, **Consul**, and **Nomad**. With HashiCorp Cloud, users can create and manage their infrastructure using the same tools that they use locally.

On the other hand, Terraform Cloud is a specific offering from HashiCorp that focuses solely on the IaC tool Terraform. Terraform Cloud provides a central place for teams to collaborate on infrastructure code, store configuration state, and automate infrastructure workflows. It offers several features, such as workspace management, version control, and collaboration tools, that make it easier for teams to work together on large-scale infrastructure projects.

One of the key differences between HashiCorp Cloud and Terraform Cloud is that the former offers a suite of tools for infrastructure automation and management, while the latter is specifically focused on Terraform.

Scalr

Scalr is a cloud management platform that provides enterprise-level solutions for cloud infrastructure automation and management. It was founded in 2007 by Sebastian Stadil to simplify the process of managing multiple cloud environments.

Scalr is a multi-cloud management platform. It comes in two flavors: commercial, which is hosted by the parent company as a SaaS solution, and open source, which you can deploy yourself. It can run your Terraform code, but it has more features, such as cost analysis, which will present the estimated bill from the cloud provider for the infrastructure you are deploying. It comes with a web UI, abstracting a lot of work that needs to be done when working with IaC. As we mentioned previously, it is a multi-cloud solution and it comes with a centralized **single sign-on** (**SSO**) that lets you view and manage all your cloud environments from one place. It comes with roles, a modules registry, and so on. It is a good solution if you are looking for more than just a centralized IaC tool.

Spacelift

Spacelift is a cloud-native IaC platform that helps development teams automate and manage their infrastructure with Terraform, Pulimi, or **CloudFormation**. It also supports automation of Kubernetes using **kubectl** and **Ansible CaaC**.

The platform offers a range of features, such as version control, automated testing, and continuous delivery, allowing teams to accelerate their infrastructure deployment cycles and reduce the risk of errors. Spacelift also provides real-time monitoring and alerts, making it easy to identify and resolve issues before they cause downtime or affect user experience.

Spacelift was founded in 2020 by a team of experienced DevOps and infrastructure engineers who recognized the need for a better way to manage IaC. The company has since grown rapidly, attracting customers from a wide range of industries, including healthcare, finance, and e-commerce.

The official website can be found at `https://spacelift.com`.

Env0

Env0 is a SaaS platform that enables teams to automate their infrastructure and application delivery workflows with Terraform. It was founded in 2018 by a team of experienced DevOps engineers who recognized the need for a streamlined and easy-to-use solution for managing IaC.

Env0 offers a variety of features and integrations to help teams manage their Terraform environments, including automated environment provisioning, integration with popular CI/CD tools such as Jenkins and **CircleCI**, and support for multiple cloud providers, including AWS, Azure, and Google Cloud.

As a private company, Env0 does not publicly disclose its financial information. However, they have received significant funding from venture capital firms, including a $3.5 million seed round in 2020 led by Boldstart Ventures and Grove Ventures.

Env0 has quickly established itself as a leading SaaS provider for managing Terraform environments and streamlining DevOps workflows and looks like a very interesting option to us in your environment.

The official website can be found at `https://www.env0.com/`.

Atlantis

Atlantis is an open source project that aims to simplify the management of Terraform infrastructure code by providing a streamlined workflow for creating, reviewing, and merging pull requests. The first release of Atlantis was made in 2018 and it has since gained popularity among developers and DevOps teams who use Terraform as their IaC tool.

Atlantis works by integrating with your existing version control system, such as GitHub or GitLab, and continuously monitors for pull requests that contain Terraform code changes. When a new pull request is opened, Atlantis automatically creates a new environment for the changes and posts a comment in the pull request with a link to the environment. This allows reviewers to quickly and easily see the changes in a live environment and provide feedback. Once the changes have been reviewed and approved, Atlantis can automatically merge the pull request and apply the changes to the target infrastructure.

This open source tool is the one we will dive deeper into. Since its source code is available for free, you will be able to download it yourself and deploy it on your local environment or inside the public cloud. Let's deploy Atlantis in AWS and configure a simple infrastructure to be managed by it.

CI/CD with Atlantis

Armed with the knowledge about tooling and principles around CI/CD (both delivery and deployment), we will create a CI/CD pipeline with the use of Git and the open source tool Atlantis. We will automatically test and deploy changes to our AWS infrastructure with it and do basic testing along the way.

Deploying Atlantis to AWS

We will use the Terraform module created by Anton Bobenko from the `terraform-aws-modules` project on GitHub. Here is the Terraform Registry link to the module: `https://registry.terraform.io/modules/terraform-aws-modules/atlantis/aws/latest`.

You can use this module in two ways. First, which is natural, is using it in your existing Terraform code to deploy it in AWS. The second, which we will use for this demonstration, is using the module as a standalone project. The module will also create a new **Virtual Private Cloud** (**VPC**) for you in the **eu-west** AWS zone and Atlantis will be running inside the AWS ECS service. This will generate some infrastructure costs.

To do so, we need to clone the GitHub repository:

```
git clone git@github.com:terraform-aws-modules/terraform-aws-atlantis.
git
Cloning into 'terraform-aws-atlantis'...
Host key fingerprint is SHA256:+Aze234876JhhddE
remote: Enumerating objects: 1401, done.
remote: Counting objects: 100% (110/110), done.
remote: Compressing objects: 100% (101/101), done.
remote: Total 1400 (delta 71), reused 81 (delta 52), pack-reused 1282
Receiving objects: 100% (1401/1401), 433.19 KiB | 1.12 MiB/s, done.
Resolving deltas: 100% (899/899), done.
```

Next, we will have to create a Terraform variables file. We have some boilerplate in the repository in the `terraform.tfvars.sample` file. Let's copy it:

```
cp terraform.tfvars.sample terraform.tfvars
```

Before proceeding, make sure you've created a GitHub repository that will hold all your Terraform code. We will be creating a Webhook for this repository when Atlantis is deployed, but you will need to add it to the `terraform.tfvars` file before applying it.

Let's take a look at the variables in the `terraform.tfvars` file we will be able to change:

```
cidr = "10.10.0.0/16"
azs = ["eu-west-1a", "eu-west-1b"]
private_subnets = ["10.10.1.0/24", "10.10.2.0/24"]
public_subnets = ["10.10.11.0/24", "10.10.12.0/24"]
route53_zone_name = "example.com"
ecs_service_assign_public_ip = true
atlantis_repo_allowlist = ["github.com/terraform-aws-modules/*"]
atlantis_github_user = ""
atlantis_github_user_token = ""

tags = {
  Name = "atlantis"
}
```

`atlantis_repo_allowlist` is the first one you would need to update to match the repositories you'd like Atlantis to be able to use. Make sure it's pointing to your repository. `route53_zone_name` should be changed as well to something similar, such as `automation.yourorganisation.tld`. Note that it needs to be a public domain – GitHub will use it to send webhooks over to Atlantis to trigger builds. You will need to create the **Route53** hosted DNS zone in your Terraform code or use the web console.

Two more variables you will need to update are `atlantis_github_user` and `atlantis_github_user_token`. The first one is self-explanatory, but for the second, you will need to visit the GitHub website and generate your **personal access token** (**PAT**). This will allow Atlantis to access the repository you want to use. To do that, you will need to follow the guidelines on the GitHub documentation pages: `https://docs.github.com/en/authentication/keeping-your-account-and-data-secure/creating-a-personal-access-token`.

After updating the `terraform.tfvars` file, we're ready to run `terraform init` and `terraform plan`:

```
admin@myhome~/aws$ terraform init
Initializing modules...
# output truncated for readability- Installed hashicorp/random v3.4.3
(signed by HashiCorp)
```

```
Terraform has been successfully initialized!
```

Terraform has created a lock file called .terraform.lock.hcl to record the provider selections it made. Include this file in your version control repository so that Terraform will make the same selections by default when you run terraform init in the future.

Now, we can run the following terraform plan command:

```
admin@myhome~/aws$ terraform plan
Terraform used the selected providers to generate the following
execution plan. Resource actions are indicated with the following
symbols:
  + create
 <= read (data resources)

Terraform will perform the following actions:

# Output truncated for readability

  # aws_cloudwatch_log_group.atlantis will be created
  + resource "aws_cloudwatch_log_group" "atlantis" {
      + arn               = (known after apply)
      + id                = (known after apply)
      + name              = "atlantis"
      + name_prefix       = (known after apply)
      + retention_in_days = 7
      + skip_destroy      = false
      + tags              = {
          + "Name" = "atlantis"
        }
      + tags_all          = {
          + "Name" = "atlantis"
        }
    }
# Removed some output for readability
Plan: 49 to add, 0 to change, 0 to destroy.
```

If you see a similar output, you can apply it. The module will also return a lot of information about the created resources. It's worth paying attention to them.

After running terraform apply (it will take a couple of minutes), you will see an output similar to the following:

```
Apply complete! Resources: 49 added, 0 changed, 0 destroyed.
```

```
Outputs:

alb_arn = "arn:aws:elasticloadbalancing:eu-central-
1:673522028003:loadbalancer/app/atlantis/8e6a5c314c2936bb"
# Output truncated for readabilityatlantis_url = "https://atlantis.
atlantis.devopsfury.com"
atlantis_url_events = "https://atlantis.atlantis.devopsfury.com/
events"
public_subnet_ids = [
  "subnet-08a96bf6a15a65a20",
  "subnet-0bb98459f42567bdb",
]
webhook_secret = <sensitive>
```

If you've done everything correctly, you should be able to access the Atlantis website under the `atlantis.automation.yourorganisation.tld` domain we created previously. The module added all the necessary records to the Route53 zone for us.

If everything has gone well up to this point, when you visit `https://atlantis.automation.yourorganisation.tld`, you will see the following Atlantis panel:

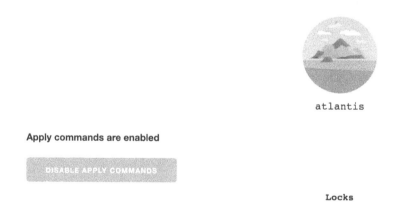

Figure 13.1 – Atlantis website after successfully deploying it using the Terraform module

The `webhook_secret` output that's marked as sensitive in the preceding output will be used to set up a webhook on the GitHub repository side. To view it, you will need to run the following command:

```
admin@myhome:~/aws$ terraform output webhook_secret
"bf3b20b285c91c741eeff34621215ce241cb62594298a4cec44a19ac3c70ad3333c-
c97d9e8b24c06909003e5a879683e4d07d29efa750c47cdbeef3779b3eaef"
```

We can also automate it with Terraform by using the module available in the same repository as the Atlantis one.

Here's the full URL to the module: `https://github.com/terraform-aws-modules/terraform-aws-atlantis/tree/master/examples/github-repository-webhook`.

Alternatively, you can create a webhook manually for testing by going to the GitHub website and following the documentation: `https://docs.github.com/en/webhooks-and-events/webhooks/creating-webhooks`.

Remember to use the secret generated automatically by Terraform in the output variable – that is, `webhook_secret`.

Creating webhook documentation is also well described in the Atlantis documentation: `https://www.runatlantis.io/docs/configuring-webhooks.html#github-github-enterprise`.

You may encounter that Atlantis won't come up as expected and you will see an `HTTP 500 error` issue when accessing the web panel. To track down any issues with this service, such as Atlantis is still unavailable or responding with errors to the GitHub webhook, you can go to the AWS console and find the ECS service. From there, you should see a cluster named **atlantis**. If you click on it, you'll see the configuration and status of the cluster, as shown in the following figure:

Figure 13.2 – Amazon Elastic Container Service (ECS) Atlantis cluster information

If you go to the **Tasks** tab (visible in the preceding screenshot) and click on the task ID (for example, **8ecf5f9ced3246e5b2bf16d7485e981c**), you will see the following information:

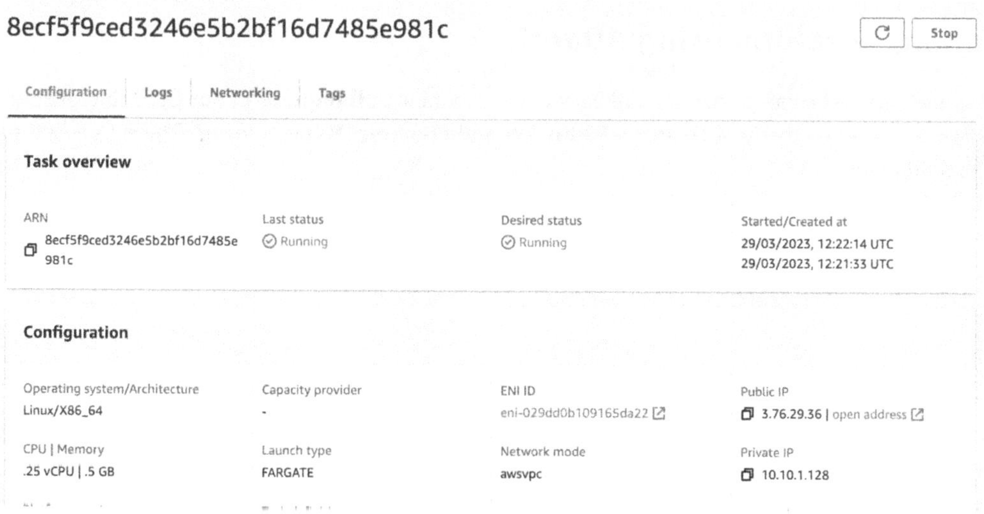

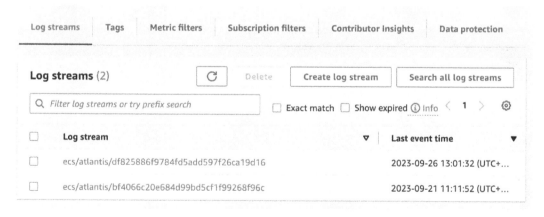

Figure 13.3 – Details of the task inside the Atlantis ECS cluster

The **Logs** tab will show you all recent events.

You can view more detailed log information in the **CloudWatch** service when you go to the **Logs | Log groups** section and find the **atlantis** log group. Inside, you will be able to see log streams containing all the logs from your task. If you already have multiple streams, it's possible to quickly track down the correct stream by its **Task ID**:

Figure 13.4 – CloudWatch log streams containing logs from ECS tasks running Atlantis

If everything has worked fine so far, we're ready to test whether Atlantis can run `terraform plan` and `terraform apply`. Let's get back to our code.

Running Terraform using Atlantis

To execute `terraform plan`, we will have to create a new **pull request** on our GitHub repository. Let's create a module that will create an **S3 bucket** and **DynamoDB** table for us. The `main.tf` file will look like this:

```
# Configure the AWS Provider
provider "aws" {
  region = var.region
}
```

The preceding code configures the AWS provider to use the region specified inside the `region` variable.

This Terraform code block configures a required Terraform version and where the Terraform state file is located. In this example, we're using local storage, but in a production environment, we should use a remote state location.

```
terraform {
  required_version = ">=1.0"

  backend "local" {
    path = "tfstate/terraform.local-tfstate"
  }
}

resource "aws_s3_bucket" "terraform_state" {
  bucket        = "terraform-states"
  acl           = "private"
  force_destroy = false

  versioning {
    enabled = true
  }
```

The preceding block defines an S3 bucket where we intend to store a Terraform state. It is a private S3 bucket with enabled versioning. This is the recommended setup for storing state files.

```
server_side_encryption_configuration {
  rule {
    apply_server_side_encryption_by_default {
```

```
    sse_algorithm = "AES256"
  }
 }
 }
}
```

The preceding code also configures **server-side encryption** (SSE) for the S3 bucket.

```
resource "aws_dynamodb_table" "dynamodb-terraform-state-lock" {
  name           = "terraform-state-lock"
  hash_key       = "LockID"
  read_capacity  = 1
  write_capacity = 1

  attribute {
    name = "LockID"
    type = "S"
  }

  tags = {
    Name = "DynamoDB Terraform State Lock Table"
  }
}
```

The preceding code block defines a DynamoDB table that's used for Terraform state locking. The `variables.tf` file will only contain one variable:

```
variable "region" {
  type    = string
  default = "eu-central-1"
}
```

After adding these files to Git and committing the changes and creating a new pull request on GitHub, you will be able to ask Atlantis to run a plan and apply for you. If you run a plan, Atlantis will lock the module you've modified and no one else will be able to apply any changes to it unless you unlock or apply your changes.

To trigger a plan for your new pull request, just add a comment to your pull request stating `atlantis plan`. After a while, depending on how big the module is, you will get a plan output similar to what's shown in the following screenshot:

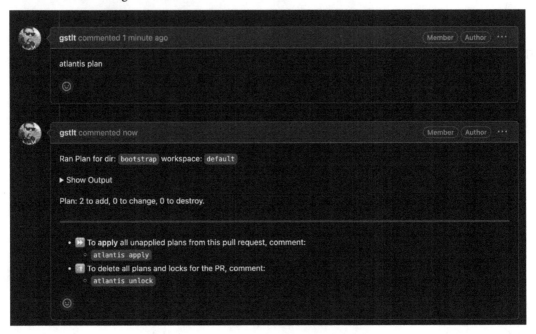

Figure 13.5 – Interaction with Atlantis on GitHub

At the time of writing, Atlantis doesn't support automatically applying the changes. However, it's possible to automate at the CI level. For example, when using GitHub, you could create a GitHub Action that would, after a successful test, add the `atlantis apply` comment, which would trigger Atlantis to apply changes and report back with the status.

At this point, we can give developers the power to modify our infrastructure without directly allowing them to run Terraform on their local machines. At the same time, we're removing the possibility of applying changes by many users at the same time, which, without a distributed locking mechanism, can be very destructive. Furthermore, working with Terraform will be easier as no one will have to install it on local machines, no one will have to have direct access to our AWS account, and we will gain more visibility of changes to our infrastructure.

Building CI/CD with Terraform still has a long way to go. IaC is still behind testing features available in other programming languages, but many developers are working on it. We're looking forward to making testing infrastructure easier for everybody.

Summary

In this chapter, we explored the benefits of using Terraform for IaC and discussed the importance of incorporating CI/CD processes in Terraform workflows. We covered testing infrastructure and various tools for automating deployment.

In the final section, we explained how to deploy Atlantis, an open source tool for automated Terraform pull request previews, to AWS and configure GitHub to trigger `terraform plan` and `terraform apply`. With Atlantis, Terraform users can collaborate on infrastructure changes through GitHub pull requests, allowing for infrastructure changes to be reviewed and approved before they are applied to production. By incorporating Atlantis into your Terraform workflow, you can improve collaboration, reduce errors, and achieve faster and more secure infrastructure changes.

In the final chapter, we will slow down a little and talk about DevOps misconceptions and antipatterns, and how to avoid them.

Exercises

Try out the following exercises to test your knowledge of this chapter:

1. Try to deploy Atlantis locally by following the documentation found at `https://www.runatlantis.io/guide/testing-locally.html`.

2. Create a repository and configure the webhook and PAT for yourself. Run a plan for your new repository (hint: instead of AWS resources, you can use a **null resource** for testing).

3. Create an account on one of the CD solution websites and try to run a plan using this SaaS. There's usually a free plan for public repositories.

Avoiding Pitfalls in DevOps

This final chapter focuses on DevOps pitfalls and antipatterns that can hinder the successful implementation of DevOps practices. We will highlight the importance of adopting a collaborative culture and prioritizing continuous improvement and discuss various common pitfalls, such as neglecting testing and **quality assurance** (**QA**), over-reliance on automation, poor monitoring and feedback loops, inadequate security and compliance measures, and lack of scalability and flexibility.

We will also emphasize the importance of proper documentation and knowledge sharing and discuss strategies to overcome resistance to change. By highlighting these common pitfalls and antipatterns, the chapter aims to provide guidance to organizations on how to successfully implement DevOps practices and avoid common mistakes. These are also probably things organizations struggle the most with, apart from the technical parts.

We will cover the following topics in this chapter:

- Too much or not enough automation
- Not understanding the technology
- Failure to adopt a collaborative culture
- Neglecting testing and QA
- Poor monitoring and feedback loops
- Inadequate security and compliance measures
- Lack of scalability and flexibility
- Lack of proper documentation and knowledge sharing
- Overcoming resistance to change

Technical requirements

There are no technical requirements for this chapter. It is more of a discussion and does not provide any instructions to be followed on a system.

Too much or not enough automation

Automation is a core tenet of DevOps. Let's face it—automation is the greatest way of making our work easier, more efficient, and fun.

But organizations can sometimes become too reliant on it, leading to a lack of human oversight and accountability. If you automate too many things, you will fail to catch any errors that could have been spotted by a human if you'd embedded them into the process. That's why we have peer review processes in place to ensure we don't miss something that tests or any integration testing tool didn't catch. That's also why many organizations prefer to manually sign off the `terraform apply` process before it actually gets deployed.

On the other hand, if you don't automate anything, you expose yourself to accidental errors, as we humans are not very good with boring repeatable tasks. And that's the point here: *identify repeatable tasks for automation.*

To identify what can be automated, we'd suggest focusing on low-hanging fruits first. With that in mind, let's identify which tasks are easiest to automate to ensure a successful automation strategy.

Consider the following checklist ordered from the easiest to most difficult strategies for how to get around automation:

- Repetitive tasks
- Time-consuming tasks
- Manual and error-prone tasks
- Tasks with version control integration
- Tasks with repeatable patterns
- Tasks with well-defined APIs or interfaces
- Tasks with clear and well-defined requirements

Let us now look at the strategies.

Repetitive tasks

Look for tasks that are repetitive and can be executed with a consistent set of steps. These tasks are prime candidates for automation, as they can save time and reduce the risk of human error. These tasks include provisioning and managing AWS resources using Terraform, backup jobs, creating and managing GitHub repositories, or setting up build and deployment pipelines with **GitHub Actions** that can be automated to streamline the DevOps workflow.

Time-consuming tasks

Look for tasks that are time-consuming and can benefit from automation. These include long-running tasks such as data synchronization, compilation, building Docker images, and security audits that can be done by one of many tools, both commercial in a SaaS model or open source (**Prowler** is one example).

Manual and error-prone tasks

Identify tasks that are error-prone when performed manually. These tasks often involve multiple steps or configurations, which can be tedious and prone to mistakes. Automating these tasks using Python scripts or **Infrastructure as Code** (**IaC**) tools such as Terraform can help minimize human errors and ensure consistency across environments. Automating those tasks with clear code will also have the benefit of documenting them. You've heard so many times that your code is your documentation, and especially in this case it is true.

Tasks with version control integration

Identify tasks that can be integrated with version control systems such as GitHub. GitHub Actions, for example, provides a powerful automation framework that can be used to trigger workflows automatically based on events in GitHub repositories, such as code pushes, pull requests, or issue updates. This allows you to automate tasks such as building and deploying applications, running tests, or creating documentation as part of your DevOps workflow.

Tasks with repeatable patterns

Identify tasks that follow a repeatable pattern or can be templated. These tasks can be automated using Python scripts, Terraform modules, or GitHub Actions templates. For example, tasks such as creating similar AWS resources for multiple environments (for example, dev, test, prod), managing multiple GitHub repositories with similar settings, or deploying the same application to multiple AWS accounts can be automated using templates or scripts to reduce duplication and increase efficiency.

Tasks with well-defined APIs or interfaces

Identify tasks that have well-documented APIs or interfaces. These tasks can be easily automated using Python libraries, Terraform providers, AWS SDKs, or GitHub APIs. For example, AWS provides comprehensive SDKs for various programming languages, including Python, which makes it easy to automate tasks such as managing AWS resources, configuring AWS services, or monitoring AWS resources.

Tasks with clear and well-defined requirements

Look for tasks that have clear and well-defined requirements, inputs, and expected outputs. These tasks are easier to automate as they can be precisely defined in scripts, templates, or configurations. For example, tasks such as provisioning AWS resources using Terraform, setting up AWS CloudFormation stacks, or configuring GitHub repository settings can be automated using declarative IaC templates or configuration files.

By considering these criteria, you can identify which tasks are easiest to automate in your organization. Keep in mind that this is usually a process and can take months or even years. As a rule of thumb, try to create solutions that involve a lot of small (*atomic*) steps to build a more complex system, instead of building complex solutions from the start. For example, the deployment process could be broken down into smaller steps, such as setting up the environment, building, testing, uploading artifacts, and creating a manifest that will update the version of the application, ultimately leading to deployment on the server side.

Automating common tasks is something that will make your work easier and more satisfying, yet not understanding what's behind the scenes and how the technology you're working with is functioning can lead to some unforeseen problems.

Not understanding the technology

You don't have to know how your TV works to use it. But it's necessary to understand the principles behind it, what's the input interface, what are the outputs, and so on. Similarly, with any other technology, you don't need to 100% understand what makes it tick, but you need to know what the core use case is and what its purpose is.

As a DevOps professional, it's crucial to have a deep enough understanding of the underlying technologies that power modern software development and operations. From databases to message queues, notifications to block and object storage, each of these technologies plays a critical role in building and maintaining reliable and scalable software systems. Yet one of the most common pitfalls in DevOps is not fully grasping the technology behind these common tasks.

Why is it so important to understand technology? The answer is simple—DevOps is not just about using tools and automation to streamline the software development and deployment process. It's about understanding how these tools work, what they do, and how they interact with each other. Without this understanding, DevOps practices can become superficial and ineffective, resulting in suboptimal outcomes and increased risks of failures and downtime. Remember: *DevOps is a way of working*. Tools are just that—tools to help you in your everyday work. Understanding your tools will let you use them efficiently and effectively, shaping them up to your needs.

For example, consider a scenario where you are tasked with setting up a highly available and performant database for a new application. Without a solid understanding of database technologies, the team might rely solely on default configurations or follow outdated practices, leading to poor performance, data loss, or even system failures. On the other hand, a team with a strong grasp of database principles would be able to make informed decisions about data modeling, indexing, caching, replication, and other critical aspects, resulting in a robust and scalable database architecture.

Similarly, understanding messaging queues, notifications, block and object storage, and other technologies is essential for designing and implementing reliable and efficient communication patterns, data processing pipelines, and storage strategies. It allows you to optimize system performance, ensure data integrity, and plan for future growth.

On the other hand, you don't necessarily need to know the details of the implementation. For the database, you should not worry about its source code.

So, how can you quickly learn new and unknown concepts? Here are some tips you could follow:

- Stay curious and proactive
- Start with the basics
- Learn hands-on
- Collaborate and share knowledge
- Stay updated

Let us look at these tips in detail.

Stay curious and proactive

Embrace a growth mindset and actively seek out opportunities to learn. Don't wait for a problem to arise before diving into a new technology. Stay curious, explore documentation, tutorials, and online resources, and proactively experiment with different tools and technologies in a safe and controlled environment. We cannot stress it enough - experimentation is very important. *"It has always been done that way"* is one of the worst things you can say as a DevOps. Without a sandbox and testing new set ups, new workflows, new tools, you won't be able to improve your infrastructure and pipelines.

Start with the basics

Don't be intimidated by complex concepts. Start with the fundamentals and build your understanding from there. Familiarize yourself with the basic principles, terminologies, and concepts of the technology you want to learn. Once you have a solid foundation, you can gradually delve into more advanced topics.

Hands-on learning

Theory is important, but hands-on experience is invaluable. Set up a sandbox environment, experiment with different configurations, and try building small projects or prototypes to apply what you've learned. Learning by doing will help you gain practical skills and reinforce your understanding of the technology.

Collaborate and share knowledge

DevOps is a collaborative field, and learning from your peers can be incredibly valuable. Engage with your team, participate in online communities, attend meetups or conferences, and share your knowledge with others. Teaching others is a powerful way to reinforce your own understanding and learn from different perspectives.

Stay updated

Technology is constantly evolving, and it's essential to stay updated with the latest trends, best practices, and updates in the field. Follow industry blogs, subscribe to newsletters, and participate in relevant forums or communities to stay informed about the latest developments. Core technologies usually stay the same, but use cases, the way of using them, and interacting with the technology are always changing. Did you know that you can make a PostgreSQL database to be available using a RESTful API (`https://postgrest.org/en/stable/`)? Or that with the use of the Multicorn extension (`https://multicorn.org/`), you could query (and join tables) from multiple sources (even Twitter) using one endpoint?

We like to think about technology as the ultimate playground for innovation and increasing the productivity of teams. If it ain't fun, why do it? On that note, consider working with a siloed team that is only dealing with its own *piece of the puzzle*; it's definitely not fun. Collaboration in DevOps is one of the fundamental principles, and collaboration is a part of the culture in your organization.

Failure to adopt a collaborative culture

DevOps is all about collaboration and breaking down silos between teams. However, many organizations struggle to adopt a culture of collaboration, leading to miscommunications, delays, and, ultimately, project failures. The importance of collaboration in DevOps with examples of how a lack of collaboration can derail projects would certainly be a good idea for a separate chapter.

Silos refer to isolated teams or departments that do not communicate or collaborate effectively with each other, leading to poor coordination and hindering overall productivity and efficiency. Several possible reasons why organizations fail to adopt a collaborative culture in DevOps include the following: lack of leadership, siloed structure from the get-go, lack of trust, and lack of communication. Let's take a look at each of these and try to find a good way out.

Lack of leadership

A common challenge is when leadership does not prioritize or actively promote a collaborative culture. This can result in teams focusing solely on their individual tasks without considering the broader organizational goals. To address this, it is crucial to have leadership buy-in and support for DevOps practices, including fostering a collaborative culture. Leaders should set the tone by actively promoting collaboration, establishing shared goals, and providing the necessary resources and support for cross-team collaboration.

Good leaders are scarce, so it's also a good idea to identify potential leaders within your organization and promote them by supporting their growth and development as a person and a leader. Simply promoting a person who was only a contributor up to now will set them up for failure as they don't yet have the needed tools for the job.

Siloed organizational structure

Organizations with a hierarchical and siloed structure can impede collaboration. Teams may operate in isolated silos with their own goals, processes, and tools, leading to a lack of visibility and coordination across teams. To overcome this, organizations should restructure to foster cross-functional teams that have end-to-end ownership of their applications or services. Creating multidisciplinary teams with representatives from different departments, such as development, operations, and QA, can promote collaboration and enable better communication and coordination.

Another very effective strategy is building a culture of collaboration. This involves promoting a mindset that values open communication, transparency, and teamwork. Encourage cross-functional teams to work together, share information, and collaborate on projects. Recognize and reward collaboration, and create forums for knowledge sharing and learning across different teams and departments.

You can also create shared goals and metrics. Let the team define shared goals and metrics that align with the overall business objectives and require collaboration across different teams. This will encourage them to work together toward common outcomes and helps them see the bigger picture beyond their individual silos. Regularly review progress toward these shared goals and metrics in cross-functional meetings to foster accountability and alignment.

Also, leadership plays a crucial role in breaking down silos. Leaders should model the behavior they expect from their teams by actively promoting collaboration, communication, and alignment across different teams and departments. This includes setting clear expectations for collaboration, recognizing and rewarding collaborative behaviors, and providing support and resources to enable cross-functional collaboration.

Lack of trust and communication

Without trust and effective communication, collaboration can suffer. Teams may be hesitant to share information or ideas due to fear of criticism or competition, leading to silos. Building a culture of trust and open communication is essential. This can be achieved through regular team meetings, cross-team workshops, and fostering an environment where team members feel comfortable sharing their perspectives and ideas without fear of judgment. Encouraging open and transparent communication channels, such as chat platforms or collaborative documentation tools, can also facilitate communication and collaboration.

Building trust between teams and team members is critical for effective collaboration and communication. Trust is the foundation upon which healthy relationships and successful teamwork are built. Among the different strategies you can learn in other books, the most effective are: establishing clear expectations, fostering open communication, promoting transparency, knowledge sharing, and building personal connections.

Let's break this down:

- **Clearly define roles, responsibilities, and expectations for each team and team member**: This helps to avoid misunderstandings and promotes accountability. Ensure that expectations are realistic, achievable, and aligned with the overall goals of the organization. Regularly review and update expectations as needed.

- **Open and inclusive communication**: To create a safe and inclusive environment where team members feel comfortable expressing their thoughts, opinions, and concerns openly without fear of judgment or reprisal. Encourage active listening and respect for diverse perspectives. Avoid blame or finger-pointing when issues arise and focus on collaborative problem-solving.

- **Fostering a culture of transparency**: Information is shared openly and consistently across teams and team members, including sharing relevant updates, progress, and challenges related to projects, processes, and goals. Transparent communication builds trust by ensuring that everyone has access to the same information and is on the same page.

- **Collaborative mindset**: To encourage collaboration and knowledge sharing, you will need to foster a collaborative mindset where teams and team members actively collaborate and share knowledge with each other. Encourage cross-functional collaboration, pair programming, and cross-training opportunities. Create a space outside of work, sponsor events, and encourage team members to share their knowledge and experiences. Recognize and reward collaborative behaviors to reinforce their importance.

- **Address conflicts proactively**: Conflicts are inevitable in any team or organization, but they can erode trust if not addressed promptly. Encourage team members to address conflicts in a constructive and timely manner. Provide tools and resources for conflict resolution, such as mediation or facilitated discussions, to help teams resolve conflicts and rebuild trust.

- **Role of leaders**: Leaders play a critical role in building trust. Leaders should model open and unbiased communication, actively listen to team members, and demonstrate trustworthiness through their actions and decisions. Leaders should also promote trust-building behaviors and hold themselves accountable for building trust within their teams and across teams.

- **Provide feedback and recognition regularly**: Acknowledge others' efforts, celebrate their successes in public, and provide constructive feedback to help them improve privately. This fosters a positive feedback loop, promotes trust, and encourages open communication.

There are many more strategies you can use to develop trust between team members and that in turn will result in better communication.

There's also another pitfall that ultimately will create silos in your organization. It happens when you neglect the cultural aspects of the organization and focus solely on tools.

Tool-centric approach

Organizations may focus too heavily on implementing DevOps tools without addressing the underlying cultural aspects. While tools are important, they are not a substitute for a collaborative culture. Adopting a tool-centric approach can lead to teams working in isolation and relying solely on automated processes, which can hinder effective collaboration. To overcome this, organizations should prioritize building a collaborative culture first and then select tools that align with their culture and facilitate collaboration. It is essential to provide training and support to ensure teams are proficient in using the selected tools to foster collaboration.

To remediate that, you can use strategies you've already talked about and additionally provide shared communication channels, encourage—and even organize yourself—cross-team meetings and events, and provide training and resources (especially time).

A final great remediation method is to foster cross-team roles and responsibilities. Define and encourage cross-team roles and responsibilities that promote collaboration. This can include roles such as liaisons or ambassadors who facilitate communication and coordination between teams. These roles can help bridge the gap between teams and promote collaboration by acting as a conduit for information sharing.

Software QA is also quite often neglected. As with the culture, quality needs to be consciously developed and encouraged at the organizational level and at the team level.

It's best to explain the preceding method with an example, and there's a perfect one in the Linux world: the **Linux kernel project**.

It started famously on August 25, 1991, with the following post to a newsgroup:

> *"I'm doing a (free) operating system (just a hobby, won't be big and professional like gnu) for 386(486) AT clones. This has been brewing since April, and is starting to get ready. I'd like any feedback on things people like/dislike in minix, as my OS resembles it somewhat (same physical layout of the file-system (due to practical reasons) among other things). I've currently ported bash(1.08) and gcc(1.40), and things seem to work. This implies that I'll get something practical within a few months [...] Yes – it's free of any minix code, and it has a multi-threaded fs. It is NOT protable [sic] (uses 386 task switching etc), and it probably never will support anything other than AT-harddisks, as that's all I have :-(."*

As you can see, from the very beginning, Linus Torvalds, the creator of Linux, asked other enthusiasts to join his small hobby project and help develop it. This spirit of collaboration was visible from the very first day. Everyone can join the project, and their contributions are evaluated on a technical basis. The medium of communication is an open and public mailing list named—aptly—the **Linux kernel mailing list** (**LKML**), where roadmaps, patches, new ideas, and all things related to the project are discussed. Everyone can read the list's archives and join the list and the discussion.

While the discussion is pretty much open to everyone and patches (or pull requests) can be submitted by anyone (although the acceptance is another thing, as the project has to adhere to code quality and legal and licensing requirements, which we will not be covering here), there is a hierarchy of leadership, although it's pretty flat. Kernel subsystems have maintainers, which are people who decide if the new code is going to be accepted or rejected. The final code merge into the kernel is done by Linus Torvalds, who has the final word in the code acceptance process, although he most often relies on the subsystem maintainers to make the decision.

The preceding structure inherently saves the Linux kernel project from a siloed structure, as there are not many levels of management. All knowledge is open and freely accessible; every person in the project management chain can be easily contacted.

The source code of the Linux kernel is publicly available in a Git repository. Everyone can clone and modify the kernel, as long as they don't break the letter of the license under which the Linux kernel is being published.

Communication and trust are direct results of the open communication model and the open source code repository adopted by the project. There is no *"behind closed door"* communication; decisions are made on a technical basis, and thus one can trust the developers and leaders.

Neglecting testing and QA

Testing and QA are critical components of any DevOps workflow, yet many organizations fail to prioritize them, leading to buggy software, user dissatisfaction, and lost revenue. In the world of DevOps, where software development and operations are closely integrated, testing and QA are critical components of the development process. Neglecting these aspects can lead to various issues that can have serious consequences for software development projects. Let's explore some potential pitfalls of neglecting testing and QA in DevOps, and propose solutions to address them:

- Increased software defects

- Deployment failures

- Security vulnerabilities

- Lack of documentation

- Inadequate test coverage

- Lack of continuous improvement

Let us check these pitfalls in detail.

Increased software defects

Without proper testing and QA, software defects may—and usually do—go unnoticed, leading to the release of poor-quality software into production. This can result in increased customer complaints, decreased user satisfaction, and loss of revenue.

It's essential to implement comprehensive testing processes, including unit testing, integration testing, and **end-to-end** (**E2E**) testing, to identify and fix defects at different stages of the development life cycle. Start with basics (**linting**, **static code analysis**) and gradually add more testing into your workflow. It's essential to collaborate with developers on that unless you're willing to write all tests for your application yourself.

Deployment failures

Without thorough testing and QA, software deployments can fail, causing system downtime and disrupting business operations. This can result in financial losses, reputational damage, and increased customer churn. To avoid deployment failures, it's crucial to have automated testing and deployment pipelines in place that include rigorous testing and QA checks before releasing code to production. This helps catch issues early and ensures that only stable and reliable software is deployed.

Security vulnerabilities

Neglecting QA and testing can leave software vulnerable to security threats, such as code injections, **cross-site scripting** (**XSS**), and other types of attacks. This can result in data breaches, compliance violations, and legal liabilities. To address this, security testing should be an integral part of the testing and QA processes. This includes vulnerability assessments, penetration testing, and other security testing techniques to identify and fix security flaws in the software.

Lack of documentation

Proper documentation is essential for maintaining software quality and facilitating troubleshooting, maintenance, and future development. Neglecting QA and testing can lead to incomplete or outdated documentation, making it challenging to understand and maintain the software. To mitigate this, documentation should be considered an essential deliverable of the testing and QA process.

Documentation should be regularly updated to reflect the changes made during development and testing and should be easily accessible to the development and operations teams. To achieve that, documentation will need to be as close to the code as possible to make it easy to update while updating the code. Technical documentation (for example, classes, code interfaces, and so on) should be automated and available for everyone interested.

Inadequate test coverage

Without proper testing and QA, there may be gaps in the test coverage, resulting in untested or poorly tested code. This can lead to unanticipated issues and defects slipping through to production. To address this, it's essential to establish clear testing objectives, define test coverage criteria, and use tools for code coverage analysis to ensure that all critical code paths are thoroughly tested.

Lack of continuous improvement

Neglecting testing and QA can lead to a complacent culture where quality is not prioritized. This can result in a lack of **continuous improvement** in the software development process, leading to a decline in software quality over time. To overcome this, it's essential to establish a culture of continuous improvement where feedback from testing and QA is used to identify and address process gaps, improve testing practices, and enhance overall software quality.

To mitigate these pitfalls, it's crucial to implement comprehensive and automated testing processes, prioritize security testing, maintain up-to-date documentation, ensure adequate test coverage, and foster a culture of continuous improvement. By addressing these challenges, organizations can ensure the delivery of high-quality software that meets customer expectations, drives business success, and results in a happy team—let's not forget that.

While QA will do its best to catch any errors and defects going into the production system, nothing beats good monitoring and alerting.

Poor monitoring and feedback loops

Monitoring and feedback loops are essential for identifying issues and making improvements in DevOps workflows, yet many organizations fail to implement effective monitoring and feedback mechanisms.

In the context of DevOps, a **feedback loop** refers to the continuous exchange of information between different stages of the software development and operations life cycle. It involves collecting data, analyzing it, and providing insights that drive improvements in the development and operations processes.

Feedback loops play a critical role in enabling teams to identify and rectify issues early in the software delivery life cycle, leading to faster development cycles, improved quality, and increased on-call staff satisfaction as they won't be woken up at night.

A characteristic of good monitoring is its ability to provide timely, accurate, and relevant information about the health, performance, and behavior of the system. The key characteristics of good monitoring include many different aspects, such as the following:

- Real-time
- Comprehensive
- Scalable
- Actionable
- Continuously improving

Let's break this down.

Real-time

Good monitoring provides real-time visibility into the system's state, allowing teams to quickly detect and address issues before they escalate into critical problems. Real-time monitoring helps to identify anomalies, trends, and patterns that may indicate potential issues or bottlenecks, enabling proactive troubleshooting and resolution.

Comprehensive

Good monitoring covers all critical components of the system, including infrastructure, applications, services, and dependencies. It provides a holistic view of the entire system, helping teams to understand the relationships and dependencies between different components and their impact on system performance and availability.

Additionally, monitoring needs to be able to deliver only relevant information, not only dry alerting data. For instance, if a server is down, good monitoring will not deliver alerts about missing data on CPU or RAM usage, a service being down, and so on. The core issue is that the server is not responsive—flooding your on-call team with irrelevant information will lead to slower response times.

Scalable

Good monitoring can handle large volumes of data and can scale horizontally to accommodate the growing needs of the system. It can collect and process data from multiple sources and integrate it with different tools and technologies, providing a unified and consolidated view of the system's health.

Actionable

Good monitoring provides actionable insights that enable teams to make informed decisions and take timely actions. It includes rich visualization, reporting, and analytics capabilities that help teams to identify trends, correlations, and anomalies, and take appropriate actions to optimize system performance and availability.

Continuously improving

On top of what we've discussed in this section, we need to also add that going from good to great monitoring will involve continuous reviews of the current state and implementation of improvements.

This review should be done every month if your organization is changing rapidly, or quarterly if you already have well-established monitoring that is doing its job properly. This review involves the following:

- Review of the most frequently triggered alerts in the last period and comparing alerts with previous periods
- Review of newly added monitoring metrics to check if they're relevant
- Review of alerts that didn't trigger over a long period of time (3-4 review periods)

With this data, you can decide which alerts are adding more noise, ensure that new metrics are actually what you expect to be monitored, and review alerts that didn't trigger, which will make you aware of metrics you probably don't need to monitor at all.

Even though when talking about monitoring we're thinking about things such as response times or memory consumption, it goes even further than that. You can trace the interaction between classes inside software code, latency between your system and a database system, or you can measure security. What happens if you don't monitor your security posture? You get vulnerabilities in your systems without even knowing about them. Let's look into security and compliance measures.

Inadequate security and compliance measures

Security and compliance are critical concerns for any team, yet many organizations fail to adequately address these issues. In the world of DevOps, security is a critical aspect that must be integrated into every step of the software development life cycle. However, many organizations still struggle with inadequate security and compliance measures, which can lead to serious consequences such as data breaches, regulatory fines, and reputational damage. In this chapter, we will explore common misconceptions and pitfalls related to security measures in DevOps, and discuss the characteristics of good security measures that organizations should strive for.

We will talk here about the following aspects:

- What are security measures?
- Characteristics of good security measures

What are security measures?

Security measures refer to the practices, processes, and tools used to protect software systems, applications, and data from unauthorized access, breaches, or other security threats. In the context of DevOps, security measures are implemented throughout the entire software development pipeline, from code creation and testing to deployment and operations.

Common security measures in DevOps include the following:

- Authentication and authorization to ensure that only authorized users have access to the system and that they have appropriate permissions to perform their tasks
- Encrypting sensitive data to protect it from being intercepted or accessed by unauthorized users
- Regularly scanning for vulnerabilities in software components and applying patches or updates to fix them
- Monitoring and recording activities within the system to detect and investigate security incidents
- Network security that includes implementing firewalls, **intrusion detection systems** (**IDS**) and/or **intrusion prevention systems** (**IPS**), and **virtual private networks** (**VPNs**) to protect the network from external threats
- Reviewing code for security vulnerabilities and using static analysis tools to identify potential weaknesses and prevent leaking secrets such as passwords and access tokens

Having implemented some measures of security, we will need to make sure the quality of those measures is as high as we can make them and, at the same time, hit the compliance our organization needs to meet legally. Let's look into the characteristics of security measures.

Characteristics of good security measures

Effective security measures in DevOps should possess certain characteristics to ensure that they provide adequate protection to the software systems and data.

Good security measures are *proactive* rather than *reactive*, meaning they are designed to prevent security breaches rather than just detecting and mitigating them after the fact. Proactive security measures may include regular vulnerability assessments, code reviews, and automated testing to identify and fix security issues before they become critical.

Security measures should also cover all aspects of the software development life cycle, from code creation and testing to deployment and operations. This includes securing the development environment, the code repository, the build and deployment process, and the production environment where the software is deployed and operated.

Scalability is another characteristic of good security measures, meaning they can be applied to different types of software applications, environments, and technologies. They should be flexible enough to adapt to the changing needs of the organization and the evolving threat landscape.

Leveraging automation to enable quick and consistent security checks and responses is a theme that always repeats itself when talking about DevOps practices. Automation can help identify security vulnerabilities, apply patches, and enforce security policies in a timely and efficient manner, reducing the risk of human error.

Many organizations need to comply with industry regulations, standards, and best practices. Compliance with relevant regulations, such as the **General Data Protection Regulation (GDPR)**, the **Health Insurance Portability and Accountability Act (HIPAA)**, or the **Payment Card Industry Data Security Standard (PCI DSS)**, is crucial to avoid legal and financial risks associated with non-compliance. Even if your organization is not regulated by these standards, it's still a good idea to choose the most appropriate and follow it whenever it makes sense. It will bring your security level higher than it would be if you didn't follow anything in particular.

Finally, all security measures need to be continuously updated and improved to address emerging security threats and technologies. The threat landscape is constantly changing, and security measures must be agile and adaptive to effectively protect against new vulnerabilities and attacks.

As always, choose the lowest hanging fruit and start there to establish good security measures in your team. Revisit it early and often and improve over time.

Lack of scalability and flexibility

Many organizations fail to design their DevOps workflows to be scalable and flexible, leading to issues when projects grow in size or complexity, even though those two things are crucial aspects of DevOps as they enable organizations to respond to changing business requirements and handle increasing workloads efficiently. However, you can fall into the trap of overlooking these factors, which can lead to serious misconceptions and pitfalls. Let's delve into the importance of scalability and flexibility in DevOps and explore some common misconceptions and pitfalls, such as the following:

- DevOps is only for small teams or projects
- Inability to scale infrastructure
- Flexibility compromises stability
- Lack of flexibility in release management

Let us look at these misconceptions and pitfalls in detail.

DevOps is only for small teams or projects

One common misconception is that DevOps is only relevant for small teams or projects. Some organizations believe that DevOps practices are not necessary for larger teams or projects, as they assume that traditional development and operations practices can handle the scale.

In reality, DevOps is not limited to the size of the team or project. It is a set of principles and practices that can be applied to organizations of all sizes. In fact, as teams and projects grow, the need for DevOps becomes even more critical to ensure smooth collaboration, faster delivery, and efficient operations.

Inability to scale infrastructure

It's easy to overlook the scalability of the infrastructure, resulting in system failures, performance issues, and unplanned downtime.

Underestimating the infrastructure requirements for the DevOps process will lead to many problems in the future, from poor user experience to losing opportunities for organizations to earn money. For example, deploying applications in an environment with limited resources or inadequate capacity can lead to performance issues and system failures when the workload increases. Similarly, not planning for future growth or changes in business requirements can result in the need for costly and time-consuming infrastructure upgrades or migrations. This is true for both an on-premise setup and cloud infrastructure. Furthermore, a shortage of electronic devices due to the global pandemic has shown that AWS and other cloud providers may sometimes be missing hardware resources. This may impact your organization when you try to add more resources to your infrastructure. For an on-prem setup, you can control hardware resources more closely.

To avoid this pitfall, teams should carefully assess the scalability needs of their applications and infrastructure, plan for future growth, and ensure that the infrastructure is designed and provisioned to handle increased workloads efficiently. This may involve adopting practices such as IaC, automated provisioning, and horizontal scaling, which can enable teams to scale their infrastructure quickly and easily to meet changing demands. For a cloud setup, you may want to pay upfront for some capacity to reserve resources.

Flexibility compromises stability

Another misconception about DevOps is that flexibility compromises stability. Some organizations fear that introducing flexibility in the development and operations process may result in an unstable environment, leading to increased risks and vulnerabilities. As a result, they may adopt a rigid approach to DevOps, emphasizing stability over flexibility.

However, this misconception can hinder the agility and innovation that DevOps aims to achieve. Flexibility is essential in DevOps, as it enables teams to respond quickly to changing business requirements, experiment with new ideas, and iterate on applications and infrastructure. In fact, DevOps practices such as **continuous integration and continuous deployment (CI/CD)** and automated testing are designed to ensure that changes are thoroughly tested and validated before being deployed to production, thereby maintaining stability while enabling flexibility.

Lack of flexibility in release management

A common pitfall is also the lack of flexibility in release management. **Release management** involves the process of deploying changes to production, and you may adopt a rigid approach to this, resulting in delays, complexity, and increased risks.

For example, following a fixed release schedule or inflexible change management process may hinder the ability to respond quickly to business requirements or customer feedback, resulting in missed opportunities or increased customer dissatisfaction. Similarly, not allowing for experimentation or fast rollback options can limit the ability to iterate on changes and quickly address issues that may arise in production.

To avoid this, you should focus on establishing a flexible and agile release management process. This may involve implementing practices such as feature toggles, dark launches, canary deployments, and blue-green deployments, which allow for gradual and controlled rollouts of changes and provide options for rollback or rollback options in case of issues. Additionally, adopting automated release pipelines, version control, and monitoring can help teams gain visibility and control over the release process.

Building flexible and scalable systems is not a trivial thing to do. On top of that, you need to count on changes from the business perspective that will affect your current processes. If your process is hard to change or you can't make it scale your systems in response to high traffic, you will experience delays and unstable systems respectively.

To understand and identify the current weak points of your processes, you will need proper documentation and visualizations, such as network or workflow graphs. In the next section, we're going to talk about this part of the processes you've established.

Lack of proper documentation and knowledge sharing

Documentation and knowledge sharing are critical for maintaining consistency and avoiding errors in DevOps workflows, yet many organizations fail to prioritize these activities.

In any software development project, documentation plays a crucial role in ensuring its success. It serves as a reference guide, provides insights into the project's architecture, design, and implementation details, and aids in maintaining and troubleshooting the software. One of the common pitfalls in DevOps is the lack of proper and up-to-date documentation, which can lead to confusion, delays, and mistakes. To address this issue, it's essential to understand the different types of documentation in software projects and their intended audiences. These are set out here:

- Technical documentation
- API documentation
- User documentation
- Process documentation
- Operational documentation
- Release notes and changelogs

Let us delve into them in detail.

Technical documentation

Technical documentation is targeted toward developers, operations teams, and other technical stakeholders involved in the software development and deployment process. It includes documentation related to system architecture, the code base, APIs, database schemas, deployment scripts, configuration files, and other technical details. Technical documentation helps in understanding the software's internal workings, making it easier to maintain, troubleshoot, and enhance the system.

Some of this documentation, such as for the code base and APIs, can be automated with specialized software. You can ensure your development teams can write self-documenting code and, additionally, automate documentation of the code with software such as **Doxygen** (https://www.doxygen.nl/), **Swimm** (https://swimm.io/), or **Redoc** (https://github.com/Redocly/redoc). To document your API, you can use a project based on OpenAPI, such as **Swagger** (https://github.com/swagger-api/swagger-ui).

API documentation

API documentation is focused on documenting the APIs exposed by the software, which are used for integration with other systems or for building extensions or plugins. It includes documentation related to API endpoints, request and response formats, authentication and authorization mechanisms, error handling, and other API-related details. API documentation helps developers understand how to interact with the software programmatically, enabling seamless integration with other systems.

User documentation

User documentation is aimed at end users of the software, including customers, clients, and other stakeholders who interact with the software. It includes user manuals, guides, tutorials, and other resources that explain how to install, configure, and use the software effectively. User documentation should be written in a clear and concise manner using non-technical language and should cover all the necessary functionality and features of the software.

Ensure that end users have the ability to reach out easily to your support teams right from the documentation.

Process documentation

Process documentation focuses on documenting the workflows, processes, and procedures followed in the software development and deployment life cycle. It includes documentation related to coding standards, version control, build and deployment processes, testing methodologies, release management, and other development practices. Process documentation helps maintain consistency, repeatability, and efficiency in the software development process, ensuring that best practices are followed consistently across the team.

Operational documentation

Operational documentation is intended for operations teams responsible for deploying, configuring, and managing the software in production environments. It includes documentation related to installation instructions, configuration guides, monitoring and troubleshooting procedures, **disaster recovery (DR)** plans, and other operational tasks. Operational documentation helps operations teams in effectively managing and maintaining the software in production, ensuring its availability, performance, and reliability.

Release notes and changelogs

Release notes and changelogs document the changes and updates made to the software in each release. They provide a summary of new features, bug fixes, and other changes, along with instructions on how to upgrade or migrate to the latest version. Release notes and changelogs help keep stakeholders informed about the software's progress, and they serve as a historical record of changes made to the software over time.

As you can see, documentation can be tricky to organize as you first need to know your intended audience and purpose. With the information we've been talking about in this section, you should be able to identify that very quickly. Also, it's worth adding that documentation is never set in stone and needs to be updated regularly to reflect what's the current situation of your application.

In the next section, we'll be talking about resistance to change. The bigger the organization, the bigger it gets because of inertia. Let's look into it from the DevOps point of view.

Overcoming resistance to change

DevOps requires a significant cultural shift in many organizations, and resistance to change can be a significant obstacle to successful implementation. Resistance to change is a known element playing a major role in the implementation of new processes, new tools, and other cultural changes in any organization. This is why just a few pages ago we said that *"It has always been done that way"* is one of the worst things to say. Improvement requires changes, and changes require an open mind and a readiness to destroy the status quo.

There are several sources of change resistance. One of them is the fear of change. Changing the organization will introduce difficulties: a new process increases the probability of failure. It also requires learning new things and dropping already well-known and tested solutions. For most people, this is an area beyond their comfort zone.

The other change resistance factor is organizational inertia. Quite often, the way that any change is introduced in an organization requires a lot of paperwork and acknowledgment from upper management. Read the previous paragraph about the fear of change. In companies, one of the metrics of performance is the amount of work pushed out. Anyone who accepts a change that leads to delays will be in a hot seat.

There are some strategies to overcome this resistance. The basis for all of them is transparent communication that goes both ways. Anyone introducing the change must communicate it in a clear way with heads-up time provided. People whose work will be influenced by the change must have time to consider what is going to happen and how they fit into it. They must be able to give their opinion and feel that they are being heard.

The biggest chance of a change implementation failing comes if it feels forced and makes people feel like they don't matter at all throughout the whole process.

Summary

The final chapter of this book delved into potential pitfalls and misconceptions that can impede the successful implementation of DevOps practices. We've emphasized the significance of fostering a collaborative culture and prioritizing continuous improvement to achieve desired outcomes.

We've gone through various common pitfalls, including neglecting testing and QA, relying excessively on automation, overlooking proper monitoring and feedback loops, inadequately addressing security and compliance measures, failing to achieve scalability and flexibility, and not aligning with business objectives.

One of the key areas of focus in this chapter was the importance of documentation and knowledge sharing, as well as strategies to overcome resistance to change. Many organizations struggle with these non-technical aspects of DevOps implementation, and the chapter provided practical guidance on how to address them effectively.

Another crucial aspect highlighted in this chapter was the need for robust monitoring and feedback loops to provide timely insights into the performance and stability of the DevOps pipeline. Without proper monitoring, it can be challenging to identify and rectify issues promptly, leading to prolonged downtimes and decreased productivity.

We hope that you will be able to have an influence on some aspects of issues you might encounter during your DevOps journey and will successfully straighten the path for others to join the organization after you. With the knowledge embedded in this publication, you're well equipped to take the challenge and have well-grounded knowledge to build upon.

Good luck!

Index

A

Packtpub.com

Subscribe to our online digital library for full access to over 7,000 books and videos, as well as industry leading tools to help you plan your personal development and advance your career. For more information, please visit our website.

Why subscribe?

- Spend less time learning and more time coding with practical eBooks and Videos from over 4,000 industry professionals

- Improve your learning with Skill Plans built especially for you

- Get a free eBook or video every month

- Fully searchable for easy access to vital information

- Copy and paste, print, and bookmark content

Did you know that Packt offers eBook versions of every book published, with PDF and ePub files available? You can upgrade to the eBook version at packtpub.com and as a print book customer, you are entitled to a discount on the eBook copy. Get in touch with us at customercare@packtpub.com for more details.

At www.packtpub.com, you can also read a collection of free technical articles, sign up for a range of free newsletters, and receive exclusive discounts and offers on Packt books and eBooks.

Other Books You May Enjoy

If you enjoyed this book, you may be interested in these other books by Packt:

Go for DevOps

John Doak, David Justice

ISBN: 978-1-80181-889-6

- Understand the basic structure of the Go language to begin your DevOps journey
- Interact with filesystems to read or stream data
- Communicate with remote services via REST and gRPC
- Explore writing tools that can be used in the DevOps environment
- Develop command-line operational software in Go
- Work with popular frameworks to deploy production software
- Create GitHub actions that streamline your CI/CD process
- Write a ChatOps application with Slack to simplify production visibility

Automating DevOps with GitLab CI/CD Pipelines

Christopher Cowell, Nicholas Lotz, Chris Timberlake

ISBN: 978-1-80323-300-0

- Gain insights into the essentials of Git, GitLab, and DevOps
- Understand how to create, view, and run GitLab CI/CD pipelines
- Explore how to verify, secure, and deploy code with GitLab CI/CD pipelines
- Configure and use GitLab Runners to execute CI/CD pipelines
- Explore advanced GitLab CI/CD pipeline features like DAGs and conditional logic
- Follow best practices and troubleshooting methods of GitLab CI/CD pipelines
- Implement end-to-end software development lifecycle workflows using examples

Packt is searching for authors like you

If you're interested in becoming an author for Packt, please visit `authors.packtpub.com` and apply today. We have worked with thousands of developers and tech professionals, just like you, to help them share their insight with the global tech community. You can make a general application, apply for a specific hot topic that we are recruiting an author for, or submit your own idea.

Share Your Thoughts

Now you've finished *The Linux DevOps Handbook*, we'd love to hear your thoughts! Scan the QR code below to go straight to the Amazon review page for this book and share your feedback or leave a review on the site that you purchased it from.

`https://packt.link/r/1803245662`

Your review is important to us and the tech community and will help us make sure we're delivering excellent quality content.

Download a free PDF copy of this book

Thanks for purchasing this book!

Do you like to read on the go but are unable to carry your print books everywhere?

Is your eBook purchase not compatible with the device of your choice?

Don't worry, now with every Packt book you get a DRM-free PDF version of that book at no cost.

Read anywhere, any place, on any device. Search, copy, and paste code from your favorite technical books directly into your application.

The perks don't stop there, you can get exclusive access to discounts, newsletters, and great free content in your inbox daily

Follow these simple steps to get the benefits:

1. Scan the QR code or visit the link below

https://packt.link/free-ebook/9781803245669

2. Submit your proof of purchase
3. That's it! We'll send your free PDF and other benefits to your email directly

www.ingramcontent.com/pod-product-compliance
Lightning Source LLC
Chambersburg PA
CBHW060648060326
40690CB00020B/4564